DEC 1 3

D1217844

Harvesting the Biosphere

Also by Vaclav Smil

China's Energy
Energy in the Developing World (edited with W. Knowland)
Energy Analysis in Agriculture (with P. Nachman and T. V. Long II)
Biomass Energies
The Bad Earth
Carbon Nitrogen Sulfur
Energy Food Environment
Energy in China's Modernization
General Energetics
China's Environmental Crisis
Global Ecology
Energy in World History
Cycles of Life
Energies
Feeding the World
Enriching the Earth
The Earth's Biosphere
Energy at the Crossroads
China's Past, China's Future
Creating the 20th Century
Transforming the 20th Century
Energy: A Beginner's Guide
Oil: A Beginner's Guide
Energy in Nature and Society
Global Catastrophes and Trends
Why America Is Not a New Rome
Energy Transitions
Energy Myths and Realities
Prime Movers of Globalization
Japan's Dietary Transition and Its Impacts (with K. Kobayashi)

Harvesting the Biosphere

What We Have Taken from Nature

Vaclav Smil

The MIT Press
Cambridge, Massachusetts
London, England

MIT Press books may be purchased at special quantity discounts for business or sales promotional use. For information, please email special_sales@mitpress.mit.edu or write to Special Sales Department, The MIT Press, 55 Hayward Street, Cambridge, MA 02142.

This book was set in Sabon by Toppan Best-set Premedia Limited. Printed on recycled paper and bound in the United States of America.

Library of Congress Cataloging-in-Publication Data

Smil, Vaclav.
Harvesting the biosphere : what we have taken from nature / Vaclav Smil.
  p. cm.
Includes bibliographical references and index.
ISBN 978-0-262-01856-2 (hardcover : alk. paper)
1. Biomass. 2. Biosphere. 3. Natural resources—Accounting. 4. Environmental auditing. 5. Earth—Surface. I. Title.
TP360.S55 2013
333.95—dc23
2012021381

10 9 8 7 6 5 4 3 2 1

Contents

Preface

The Earth's biosphere—that thin envelope of life permeating the planet's hydro-sphere, the lowermost part of its atmosphere, and a small uppermost volume of its lithosphere—is of surprisingly ancient origin: the first simple organisms appeared nearly four billion years ago (the planet itself was formed about 4.6 billion years ago), metazoan life (the first multicellular organisms belonging to the kingdom of animals) is more than half a billion years old, and complex terrestrial ecosystems have been around for more than 300 million years. Many species have exerted enormous influence on the biosphere's character and productivity, none more so than (on the opposite ends of the size spectrum) oceanic cyanobacteria and the large trees of the tropical, temperate, and boreal forests. But no species has been able to transform the Earth in such a multitude of ways and on such a scale as *Homo sapiens*—and most of these transformations can be traced to purposeful harvesting or destruction of the planet's mass of living organisms and reduction, as well as improvement, of their productivity.

These transformations long predate the historical period that began about five millennia ago and was preceded by millennia of gradual domestication of wild plant and animal species and by the evolution of sedentary agriculture. Humans eventually created entirely new landscapes of densely populated areas through intensive agri-culture, industrialization, and urbanization. These processes reached an unprece-dented intensity and extent beginning in the latter half of the nineteenth century as industrialization was accompanied by improved food supplies, greater personal consumption, expanded trade, and a doubling of the global population in 100 years (from nearly 1.3 billion in 1850 to 2.5 billion in 1950), followed by a 2.4-fold increase (to six billion) by the year 2000. Fossil fuels have energized this latest, industrial and postindustrial stage of human evolution, whose accomplishments would have been impossible without tapping an expanding array of other mineral resources or deploying many remarkable technical innovations.

But the metabolic imperatives of human existence remain unchanged, and harvesting phytomass for food is still the quintessential activity of modern civilization. What has changed is the overall supply and the variety and quality of typical diets: increasing populations and improved standards of living have meant greater harvests of the Earth's primary production, digestible photosynthates suitable for consuming directly as food crops or indirectly (after feed crops and natural vegetation are consumed by domesticated or wild vertebrates) as the milk, eggs, and meat of terrestrial animals or as the highly nutritious tissues of aquatic invertebrates, fishes, and mammals. Harvests of woody phytomass were initially undertaken to feed the hominin fires and make simple weapons. Sedentary cultures had a much greater demand for firewood (they burned crop residues, too), as well as for wood as a principal construction material. Industrialization increased such demands, and acquiring wood for pulp has been the third major motivation for tree harvests since the latter half of the nineteenth century.

The demand for food could not be met just by increasing yields but required the substantial conversion of forests, grasslands, and wetlands to new cropland. This led to a net loss of phytomass stores as well as to losses of overall primary production; in turn, some of the best agricultural lands were lost to expanding cities and industrial and transportation infrastructures. Similar losses of potential productivity have followed as substantial areas of natural ecosystems have been converted to pastures or affected by the grazing of domesticated herbivores, and as secondary tree growth or inferior woodlands replaced original forests.

Food, feed, fiber, and wood are the key phytomass categories that must be harvested to meet basic human needs. Harvests of furs, ornamental and medicinal plants, and companion animals may be important for their impact on particular ecosystems and species, but their overall removal has been (in mass terms) much smaller than the aggregate of many uncertainties that complicate the quantification of phytomass belonging to the four principal categories. The steeply ascending phase of phytomass harvests has yet to reach its peak, but I do not forecast when it may do so. Instead, I will review the entire spectrum of harvests and present the best possible quantifications of past and current global removals and losses as a way to assess the evolution and extent of human claims on the biosphere. Although some of the claims can be appraised with satisfactory accuracy, in many other cases phytomass accounting can do no better than suggest the correct orders of magnitude. But even that is useful, as our actions should be guided by the best available quantifications rather than by strong but unfounded qualitative preferences or wishes.

I

The Earth's Biomass: Stores, Productivity, Harvests

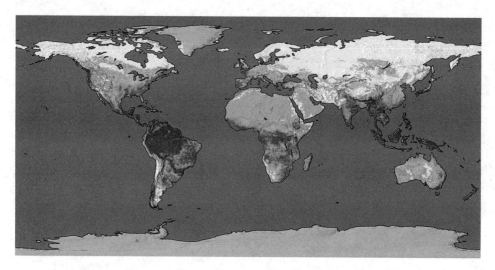

NASA's global map shows the intensity of the Earth's primary (photosynthetic) productivity. The darkest shading in the Amazon and in Southeast Asia indicates net annual primary productivity approaching 1 kg C/m². A high-resolution image in color can be downloaded at http://earthobservatory.nasa.gov/Features/GlobalGarden/Images/npp_199912_bump_lrg.jpg.

In December 1990, the *Galileo* spacecraft came as close as 960 km to the Earth's surface in order to get a gravitational assist from the planet on its way to Jupiter. This flyby was used by Sagan et al. (1993) as an experiment in the remote detection of life on Earth: just imagine that the spacecraft, equipped with assorted detection devices, belonged to another civilization, and registers what its beings could sense. The three phenomena indicating that this planet was very different from all others in its star system were a widespread distribution of a pigment, with a sharp absorption edge in the red part of the visible spectrum; an abundance of molecular oxygen in the Earth's atmosphere; and the broadcast of narrow-band, pulsed, amplitude-modulated signals.

The light-harvesting pigment absorbing in red light is chlorophyll, a key enabling agent of photosynthesis, the planet's defining energy conversion mechanism, which releases oxygen and produces an astonishing variety of biomass, directly as plants and indirectly as the energizer of all nonphotosynthesizing organisms, from the simplest heterotrophic bacteria to humans, the species that began, at the turn of the twentieth century, sending modulated signals into space as yet another demonstration of its mental prowess. These realities circumscribe and define the subject of this book: there would be no story of human evolution and advancing civilization without a parallel story of harvesting the biomass, and the increasing intensity and extent of biomass harvests are in turn changing the very foundations of civilization's well-being.

Human harvesting of the biosphere has transformed landscapes on vast scales, altered the radiative properties of the planet, impoverished as well as improved soils, reduced biodiversity as it exterminated many species and drove others to a marginal existence, affected water supply and nutrient cycling, released trace gases and particulates into the atmosphere, and played an important role in climate change. These harvests started with our hominin ancestors hundreds of thousands of years ago, intensified during the era of Pleistocene hunters, assumed entirely new forms with the adoption of sedentary lifeways, and during the past two centuries transformed into global endeavors of unprecedented scale and intensity.

The story of biomass harvests spans all of human evolution and all of our history, and its continuation will be a key determinant of our future success or failure as a civilization. And yet, comprehensive inquiries into this complex harvesting process are of surprisingly recent origin (only a few decades old), and interest still remains heavily tilted toward particulars rather than toward a global examination of this now planetwide transformation. But an important chore must be taken care of before I proceed. The careless or vague use of undefined (or poorly defined) terms is all too common in modern scientific discourse. While putting in place a fairly comprehensive foundation of accurate definitions of many relevant variables is not a captivating way to start a book, for the sake of clarity I will briefly do just that.

1

Biomass: Definitions and Compositions

The classic life sciences—botany, zoology, plant and animal anatomy and physiology—were for centuries preoccupied with classification. This concern was later extended to life's assemblages, that is (in ascending order), communities, ecosystems, and biomes. Such a focus is now seen as antiquated: the preoccupation has shifted to the intricacies of genetic makeup and metabolism and to the dynamic processes of evolution and adaptation. An undesirable side effect of this shift has been a lack of attention to the precise meanings of many variables used to describe organisms and their evolution. This declining rigor is not a trivial matter, as even apparently straightforward definitions hide some complications; settling these matters (or explaining clearly how unsettled they must remain) is an essential precondition for using the variables with the least possible ambiguity. This clarification must proceed along two tracks because biomass stores and productivities are not expressed in mass terms only but are also quantified as reservoirs and fluxes of carbon, a choice that requires reliable data on the element's content of major biopolymers.

Key Variables

Biomass is the mass of any living organism. The term can be applied on scales ranging from that of tissues and organs (the biomass of leaves, of muscles) through that of individual species (the peak biomass of corn to be harvested for silage, the biomass of migrating wildebeests) and narrowly circumscribed ecosystems (such as the total plankton biomass of a small pond) all the way to that of the aggregates of a biome (for example, the biomass of boreal forests), and on to the planetary level (the biomass of all photosynthesizing organisms). Complications and uncertainties arise as soon as we start analyzing the term, be it according to the two major classes of life forms that it embraces or according to the physical or chemical composition of biomass.

Most of the world's biomass is in forests, and most of the forest biomass is in the dead wood of large tree trunks. This photograph shows a massive trunk of a giant sequoia (*Sequoiadendron giganteum*) in the Tuolumne Grove in Yosemite National Park. Photograph by V. Smil.

Life's great dichotomy is between *autotrophs,* organisms that can nourish themselves, and *heterotrophs,* or life forms that must feed on other organisms. And for comparisons among different organisms, their biomasses must be adjusted for widely differing water content and expressed in absolutely dry terms or, alternatively, their carbon content may be calculated using specific conversion rates. Autotrophs (whose sizes span nearly ten orders of magnitude, from the smallest bacteria at 0.01 μm to the tallest trees at more than 100 m) can transform inorganic carbon (drawn overwhelmingly from atmospheric CO_2), hydrogen (mostly from H_2O), and smaller amounts of nitrogen and mineral nutrients into complex organic compounds.

Most autotrophs are *phototrophs,* organisms that rely on solar radiation to power complex photosynthetic processes, but some bacteria are *chemoautotrophs,* which derive the needed energy by oxidation of sulfur or H_2S, and an even smaller group of facultative autotrophs (most notably methanogenic bacteria) can switch to a heterotrophic existence in the presence of suitable organic substrates. *Autotrophic biomass* is the correct term for all tissues produced by photosynthesizers and chemotrophs, but I will use a slightly narrower term, *phytomass*, which refers to both planktonic organisms (cyanobacteria, coccolithophorids, silicoflagellates, diatoms) and macroscopic aquatic autotrophs (algae, sea grasses), the two categories whose direct harvests by humans are rather limited, as well as to all terrestrial autotrophs (that is, mostly to herbs, trees, and shrubs), whose harvests dominate the photosynthate that is used by humans.

Because of substantial differences in the moisture content of fresh biomass—phytoplankton cells or young plant shoots are more than 95% water, freshly cut wood trunks contain about 50%, mature cereal grain and dry straw have only about 15% water, and some dry seeds have less than 5% moisture—a comparative accounting should be done only in terms of *absolutely dry biomass*. Its values are obtained after desiccating the fresh tissues at 104°C–105°C to constant weight. Dry biomass has the density range of 0.2–0.9 g/cm^3, and about 95% of it is organic matter (although minerals can make up as much as 70% of dry phytomass in some phytoplankton species). Conversions to absolutely dry biomass introduce two possibly significant errors.

First, no conversions done on a scale larger than that of a small and fairly homogeneous plant community can use a single representative figure for the water content of fresh-weight biomass: simplifying assumptions will always be needed to choose a single approximate value, which might easily err by 10%–15%. When I take a closer look at crops, crop residues, and wood harvests I will explain that difficulties remain even at a species level. Second, not all publications make it clear whether the biomass data are reported in terms of fresh or dry matter. Foresters, agronomists, and compilers of wood and crop statistics almost invariably use fresh-weight values, which require conversion to a common dry-mass denominator (g/cm^2 in basic scientific units, more often kg/m^2 in ecological studies and t/ha—tonnes per hectare—in agriculture and forestry).

But there is an even more fundamental complication in assessing the density of phytomass: what to count. "Everything" might be the most obvious answer, but one that cannot be adduced so easily. Destructive sampling (harvesting of all phytomass within a given unit area, reducing it to absolutely dry weight, and adding up the components) is the most reliable way to count everything, but the approach is both time-consuming and labor-intensive, and it has obvious spatial limits. Removing all plants from small, randomly spaced squares is the best practical way to determine the overall standing phytomass of a larger area of a uniform grassland but not of a mature tropical forest, whose diversity of species would require sampling much larger areas; a complete harvest, moreover, would require not only felling large trees but also a particularly difficult removal of all roots, while the sampling of small plots may result in significant over- or underestimates once the results are used to calculate phytomass totals for much larger areas.

Not surprisingly, most phytomass estimates, even for plant communities that are fairly easy to sample, refer only to the *aboveground phytomass* and ignore the *belowground phytomass* of roots. But they should always state that clearly, and if they claim to represent the total phytomass, then they should offer details as to how the root mass was estimated. This is done most often by applying specific correction factors that are usually expressed as *root:shoot ratios*. These ratios are highest in grasslands (commonly around 5), much lower in forests, and have the lowest values (with the obvious exception of root crops) in croplands.

Grasslands are obviously easier to study than forests, whose belowground phytomass is known accurately only from a limited number of sites sampled during one-time investigations and mostly just to the depth of 1–2 m (Jackson et al. 1996). Taproots and branching roots are relatively easy to sample, but accounting for all the fine roots is very difficult: depending on the biome, their mass is between less than 10% and more than 90% of the underground phytomass, and they have a relatively fast turnover (Jackson, Mooney, and Schulze 1997). Different challenges arise when assessing marine phytomass, whose components' brief life span, large fluctuations, and unpredictable massive blooms make it difficult to offer truly representative longer-term (monthly or annual) averages.

Finally, cogent arguments can be made for both drastically reducing and greatly expanding the definition of the Earth's phytomass. Restricting it to the living protoplasm would make the biosphere even less substantial than it is; adding all recently accumulated dead tissues would greatly enlarge its scope. Most of the terrestrial phytomass is locked in structural polymers and in cell walls, tissues that play essential supportive, protective, and conductive roles but that are not alive. Correcting for this fact is not easy. With forests containing roughly 90% of all phytomass, most of this adjustment could be done by estimating the share of living cells in trees—but no simple generalizations are possible.

The radial extent of the cambial zone, the generator of tree growth, is difficult to define because of the gradual transition between phloem (the living tissues) and xylem (which holds most of the dead carbon in a living tree). Some cells in axial and radial parenchyma may remain alive not just for many months but for years and decades, and there are substantial specific differences in the shares of total phytomass made up of fresh leaves, buds, young branches, and rapidly growing fine roots (Reichle 1981; Shigo 1986; Jackson et al. 1996). A very generous allowance would be to assume that no more than 15% of all standing forest phytomass is actually alive. Besides their dead stem wood, all trees and shrubs also have highly variable shares of attached dead branches and roots. Cornwell et al. (2009) put the total phytomass of this coarse dead wood at 36–72 Gt C, or on the order of 10% of all forest phytomass.

In contrast, the standard inclusion of all nonliving structural cells in the total weight of woody phytomass in living plants could be logically extended, first to the inclusion of all dead but still standing trees, shrubs, and grasses (as well as their roots), then to all accumulated surface litter, and eventually even to the huge stores of soil organic carbon, whose mass, unlike the living phytomass, is composed largely of long-lived and not easily decomposed compounds. Arguments in favor of such inclusions are ecologically impeccable: if dead structural polymers in trees must count because they provide indispensable plant services, then litter and soil organic matter should count no less as harbors of soil fauna, as irreplaceable sources of recyclable nutrients needed for plant growth, and as reservoirs of moisture.

Heterotrophs are organisms that can survive and reproduce only by ingesting fully formed organic compounds synthesized by autotrophs, either directly by eating them (as is the case with herbivores and with the vastly more numerous bacterial and fungal detritivores and saprovores) or indirectly by eating other heterotrophs (as carnivores do). The imperatives of energy metabolism mean that the global biomass of heterotrophs is equal to only a small fraction of that of all autotrophs.

Heterotroph sizes range over eight orders of magnitude, from the numerous microbial decomposers to the largest marine mammals (blue whales, *Balaenoptera musculus*, with maximum lengths of more than 30 m).

The term *zoomass* is used much less frequently than phytomass and easily evokes images of thousands of cattle packed in America's beef feedlots or large herds of migrating wildebeests and caribou or multiple V-shaped flocks of high-flying Canada geese—but not of bacterial decomposers or insects. But the total fresh-weight zoomass density of different termite species can add up to more than twice as much (on the order of 100 kg/ha) as the biomass of elephants (Inoue et al. 2001). The term zoomass is used infrequently when referring to zooplankton or to insect biomass and is usually reserved for vertebrates. The abundance of small-scale studies of microbial and fungal biomass contrasts with the still very rare zoomass estimates on large (biome, continental) scales.

Quantifications of invertebrate and vertebrate zoomass have similar limitations: field data make it easy to calculate the resident zoomass of lions or wildebeests within their respective home ranges in some national parks, but there are no reliable accounts of the total zoomass of wild ungulate herbivores or top carnivores in a biome or at the global level. The only possible, and very minor, addition to heterotrophic biomass could be made by including viruses: they have, of course, no intrinsic metabolism, but they exert enormous influence on the survival and productivity of cellular organisms, especially in the ocean, where recent research has shown them to be the most abundant entities containing nucleic acids (Fuhrman 1999; Brum 2005).

Biomass Carbon

An excellent alternative to using absolutely dry mass is to use *carbon* as the common denominator for all phytomass comparisons (I will explain the inevitable complications and uncertainties associated with this choice when taking a closer look at specific harvests). Life's structural unity means that all biomass is made up of the same kinds of carbon-based macromolecular *polymers,* which must be synthesized from a limited number of *monomers* (their simpler subunits): more than 90% of biomass is made up of only about 50 different compounds, including 21 amino acids, 15 monosaccharides, 10 fatty acids, and five nucleotides. Dry biomass is mostly composed of just four kinds of polymers, two of an aperiodic kind (nucleic acids and proteins) and two classes of periodic polymers (carbohydrates and lipids) that act as long-term energy stores and structural components.

Nucleic acids (DNA and RNA) are linear unbranched assemblies of *nucleotides* (pyrimidines and purines), and compared to the other three biopolymers they are present in biomass in only trace amounts (always less than 0.1% of dry-matter mass). They occur in highest concentration in meat and fish and in relatively high concentration in some cereals and vegetables (Imafidon and Sosulski 1990). Their role is qualitative rather than quantitative as they store, transmit, and process genetic information and direct protein synthesis; no less important, they take part in cellular signaling and act as cofactors in enzymatic reactions, and adenosine triphosphate is the principal carrier of biochemical energy in all organisms.

Proteins, composed of *amino acids*, have several fundamental roles: as ubiquitous enzymes they catalyze biochemical reactions; as hormones they stimulate and coordinate specific activities; as antibodies they help to guard organisms against antigens; as transport media (hemoglobin, cytochromes) they shunt molecules among tissues; as storage proteins (milk's casein, eggs' albumin) they act as deposits of amino acids; as structural proteins (keratin, collagen, elastin) they form many longlasting cover (hair, feathers) and connective (tendons, ligaments) tissues; and as contractile proteins (actin, myosin) they enable muscle motion. Proteins are the dominant constituents of all metabolizing heterotrophic tissues, but their presence in phytomass is relatively limited (with the highest concentration in seeds), and they are virtually absent from woody matter, as well as from many vegetables and fruits.

To synthesize their body proteins, humans must ingest nine *essential amino acids*. All food proteins contain these amino acids, but only the proteins of heterotrophic origin have them in ratios that are optimal for human growth and maintenance of metabolizing tissues; plant proteins are always deficient in one essential amino acid (cereals in lysine, legumes in cysteine), and hence a healthy strict vegetarian diet requires a combination of different plant foods. This reality explains the importance of legumes in all traditional (overwhelmingly vegetarian) agricultural societies, as well as the rapid increases in consumption of animal foods (and the concurrent decline in legume intake) that have characterized the universal dietary transition brought about by higher incomes.

Lipids are composed of long chains of *fatty acids* and are used as building blocks of cell membranes and internal organs, in cellular signaling, in the supply of fat-soluble vitamins, and as the most important energy stores (adipose tissue) in heterotrophs: their energy density of 39 kJ/g is unmatched by any other biopolymer. They are present in fairly high concentrations in some plant tissues (particularly in seeds and nuts as polyunsaturated, monounsaturated, and saturated oils) and in even higher amounts in large animal bodies (mostly as subcutaneous and intramuscular saturated solid fats). Humans can synthesize all the fatty acids needed for normal development except linoleic and α-linolenic acid, which fortunately are readily available in plant oils.

Carbohydrates (saccharides) dominate the biosphere's phytomass and are indispensable both because of their essential metabolic functions and because of their structural roles, but they are virtually absent in muscle. Simple sugars— *monosaccharides* (glucose, fructose, galactose, and others) and *disaccharides* (such as sucrose or milk's lactose, with 16.5 kJ/g)—store energy in forms that are readily digestible by heterotrophs, and ribose ($C_5H_{10}O_5$) is the backbone of RNA and, after phosphorylation, a key subunit of adenosine triphosphate. The polymerization of simple sugars produces long-chained *polysaccharides,* whose two key roles are to store energy (in plants as starch, in heterotrophs as glycogen) and to form structures such as the exoskeletons of arthropods with chitin and the stems, stalks, and trunks of plants with cellulose.

Cellulose, a polymer of glucose with the formula of $(C_6H_{10}O_5)_n$, is a linear polysaccharide with a molecular weight usually in the range of 300,000–500,000 daltons, and is the biosphere's most ubiquitous structurally strong macromolecule (its energy content is 17.5 kJ/g). In contrast, monomers making up *hemicelluloses* include most pentoses (five-carbon sugars, including xylose, mannose, galactose, and arabinose) and produce an amorphous and structurally weak biopolymer. The third major constituent of phytomass, *lignin,* is a complex polymer of coniferyl, coumaryl, and synapyl alcohols that links with hemicelluloses, fills cell wall spaces, stiffens plant structures, and, being relatively hydrophobic, helps conduct water inside plants (its energy content is 26.4 kJ/g).

The proportion of lignin in phytomass ranges from as little as 15% in crop residues to just over 50% in nut shells, with softwoods having higher concentrations (>25%) than hardwoods and bark containing more lignin than heartwood. Phytomass containing 25% lignin and 75% cellulose would thus have almost exactly 50% carbon, and this generic value has been used for decades as the standard carbon content of wood, without examining variations between and within common species. Cornwell et al. (2009) analyzed a large set of wood property studies and found median values for lignin, cellulose, and other carbohydrate polymers of, respectively, 26%, 46%, and 21% for leafy trees and 29%, 44%, and 23% for coniferous species, with extractives averaging less than 5% for the former and more than 6% for the latter.

Cellulose has a carbon content of 44.4%, but the value for lignin cannot be reduced to a single figure. Idealized lignin (composed solely of coniferyl units) contains 65% C (Freudenberg and Nash 1968), and wood analyses showed that the biosphere's most abundant aromatic polymer has a carbon content ranging between about 61% and 66% (Wald, Ritchie, and Purves 1947). Lamlom and Savidge (2003) and the USDA (2010) offer the following information for the heartwood of 41 North American trees. The carbon content of absolutely dry wood is between 46% and 50% (average about 48%) for hardwoods (the lowest value is found in birches) and between 47% and 55% (average about 51%) for conifers (with the highest value for redwoods). The differences are due to a higher lignin content of softwoods. In addition, the carbon content of early wood is consistently a few percentage points higher than that of late growth, which contains more cellulose.

Carbon is significantly lower in marine phytomass because of the latter's often high mineral content. Diatoms, silicoflagellates, and radiolarians use $Si(OH)_4$ (silicic acid) to build their intricate opal (hydrated, amorphous biogenic silica, $SiO_2 \cdot 0.4H_2O$) structures, and their dry-matter carbon content can be as low as 20%. Similarly, nannoplanktonic coccolithophores use calcite ($CaCO_3$) to build their exoskeletons. The generic mean of 50% C in phytomass may be acceptable for very large-scale approximations of terrestrial phytomass but will have errors of 4%–5% when applied to specific tree stands, and it will exaggerate the carbon content of crops and their residues (by about 10%) and, even more so, of marine phytoplankton (by as much as 50% or more).

Although this book is concerned with harvesting plants and animals, I should point out (given that bacteria make up the bulk of the biosphere's mass) that particularly large errors can be introduced by converting microbial biomass to carbon equivalents. Traditionally, it was assumed that the dry-matter content of bacterial cells is about 20%, but Bratbak and Dundas (1984) examined three common strains (*B. subtilis*, *E. coli*, and *P. putida*) and found that the real value may be more than twice as high, ranging mostly between 40% and 50%, and hence they recommend converting bacterial volume to carbon content by assuming an average of 0.22 g/cm^3.

Autotrophic and heterotrophic species have been selected for harvest by humans because of exceptionally high proportions of individual biopolymers or because of their specific combinations. Prominent food examples include cereals, selected for their combination of high carbohydrate (typically more than 70%) and a relatively high (7%–15%) protein content; wheat, selected for its uniquely high content of gliadin and glutenin, proteins whose pliability makes leavened breads possible; soybeans, for their unmatched percentage of protein (typically around 35%); sunflower seeds or olives, for their high lipid shares (respectively almost 50% and about 20%); and meat, for its combination of high-quality protein (around 20% of fresh weight) and saturated fat (on the order of 40% in many cuts of beef), whose ingestion provides a feeling of satiety.

2

Biomass Stores: Means and Extremes

In contrast to regularly conducted inventories of wood volumes in commercially exploited forests and the frequent (now satellite-aided) monitoring of crop productivity, accurate quantifications of phytomass in natural ecosystems remain relatively uncommon. But they do support some easily defensible generalizations, illustrate some important exceptions, and help correct some stereotypes. These assessments are usually presented in units of dry matter per unit area (g/m^2 or t/ha) or as the total mass of carbon (g C/m^2 or t C/ha). Our knowledge also remains highly uneven in spatial terms as temperate ecosystems have been studied much more intensively than the tropics. Integrations on biome, continental, and planetary scales have the highest margins of error, and the quantification of heterotrophic biomass is more tenuous than that of phytomass.

Phytomass and Zoomass

Phytomass stores vary greatly among biomes, large areas with similar climates whose global distribution shows distinct latitudinal regularities. The poleward sequence of biomes starts with tropical rain forests, the most diverse terrestrial biome: they fill the entire basin of the Amazon and the Congo, as well as most of Central America and West Africa, and spread across Southeast Asia and Indonesia. Tropical forests give way to tropical grasslands, in Africa from the Atlantic to Sudan and from Angola to Mozambique, in South America both south and north of Amazonia, and in Australia in the northern third of the continent. In turn, tropical grasslands blend into deserts, a biome that is relatively restricted in the Americas but spreads across Africa from Mali to Sudan and then continues across the Arabian Peninsula and Iran to western India and north of the Hindu Kush and Tibetan plateau into the heart of Asia, then all the way to northern China; desert also fills most of Australia's interior.

The tropical rain forests of South America, Africa, and Southeast Asia are the biosphere's largest depositories of phytomass, with the richest plant communities, arrayed in several vertical layers, storing as much as 1,000 t of dry phytomass per hectare. This photograph of Amazonian forest in Peru is available at http://upload.wikimedia.org/wikipedia/commons/f/f6/Amazonian_rainforest_2.JPG.

Temperate grasslands fill the semiarid and arid centers of North America and Asia, while temperate deciduous forests were the natural climax vegetation across most of Atlantic and Eastern Europe, southeastern North America, and parts of northeastern Asia. Extensive boreal forest (taiga) spreads across North America from the Pacific coast to the Atlantic and in Europe from Scandinavia to the Sea of Okhotsk. Treeless tundra is the least biodiverse biome found along the northern fringes of North America and Eurasia. The perpetually dark and cold abyssal, nutrient-poor waters of the open ocean are the aquatic counterpart of desert. Highly diverse marine biomes are limited to relatively small areas of coral reefs and kelp forests, to the regions of upwelling (where rising cold water replenishes surface nutrients) along the western coast of the Americas and Africa, and to estuaries of major rivers.

Before citing any numbers for phytomass densities I must stress several important yet often ignored facts, none of them more critical than our surprisingly limited basis for accurate assessments of global phytomass. Houghton, Hall, and Goetz (2009, 1) put it best: "Our knowledge of the distribution and amount of terrestrial biomass is based almost entirely on ground measurements over an extremely small, and possibly biased sample, with many regions still unmeasured. Our understanding of changes of terrestrial biomass is even more rudimentary." This conclusion is true even when we take into account the latest advances in quantifying carbon in tropical forests (Saatchi et al. 2011), although the creators of many global vegetation models and the even more complex simulations of biosphere-atmosphere interactions want us to believe that their understanding has become deep enough to recreate faithful virtual versions of Earth's biomass *in silico*

Besides the limited basis of actually measured phytomass stocks, we have to reckon with a large variability (spatial and temporal) of phytomass densities. The possibilities of bias resulting from destructive sampling of small plots were noted in the last chapter, but even the best available large-scale mapping of forest ecosystems frequently ends with very different totals (Houghton et al. 2001). The range of phytomass densities spans three orders of magnitude (the difference between the northernmost tundras and giant trees in the rain forest of the Pacific Northwest), and even the difference between averages in cold and hot deserts, on the one hand, and the richest forests on the other is more than 30-fold.

Spatial differences in average forest density are often an order of magnitude even within 1 ha (100 × 100 m) vegetation patches, while even the highest resolution that is now readily available for satellite vegetation imaging is 250 m, lumping all phytomass within 6.25 ha into a single category; the other two monitoring resolutions make a single data point from areas of 500 × 500 m and 1 km × 1 km.

Table 2.1
Ranges of Phytomass Densities of Terrestrial Ecosystems

Biomes and Ecosystems	Phytomass Density (dry matter, t/ha)
Tundra	<5
Grasslands	
Temperate	10–20
Tropical	10–35
Wetlands	20–60
Forests	
Boreal	40–100
Temperate	50–1,000
Richest communities	1,700–3,500
Tropical	100–400
Richest communities	700–1,100

Note: For data sources, see the text.

Temporal differences are even greater: a satellite pass may record the minimum phytomass density in a freshly clear-cut or burned forest, but rapid regrowth might occur in the following years and decades, and new stocking records may be set by the secondary growth.

Most of the biosphere's standing phytomass (nearly 90%) is in forests, and most of the forest phytomass (nearly 75%) is in the tropics, about three-fifths in the equatorial rain forests and the rest in seasonally green formations (table 2.1). Tropical rain forests can be rather monotonous—especially those dominated by the family of dipterocarp trees in Southeast Asia and in the northeastern basin of the Congo River (Connell and Lowman 1989), but most of this biome has an admirably high biodiversity, and its richest communities contain 200–300 species/ha, with maxima in excess of 600 plant species.

The standing phytomass is concentrated mostly in tall, massive, often buttressed trees forming the forest's dense canopy and in a smaller number of emergent trees whose crowns rise above the closed-canopy level. Long woody vines (lianas), aggressively growing strangler species, epiphytic plants (orchids, bromeliads), and relatively bare forest floors (a result of rapid decomposition of the litter) are other notable hallmarks of this biome (Gay 2001; Carson and Schnitzer 2008). Dense tropical rain forests have between 150 and 180 t C/ha of aboveground phytomass, and their total storage (including living, dead, and underground biomass) is commonly between 200 and 250 t C/ha (Keith, Mackey, and Lindenmayer 2009).

Cool, temperate, moist forests store between 250 and 650 t C/ha, while boreal forests have a total phytomass usually no greater than 60–90 t C/ha and an aboveground living component of only 25–60 t C/ha, or 50–120 t/ha: the Russian mean is about 80 t/ha, the European and U.S. averages are around 100 t/ha, and the Canadian means range from just around 40 t/ha in the drier western provinces to about 55 t/ha in eastern Ontario and Quebec (Kurz and Apps 1994; Potter et al. 2008). Actual timber (roundwood) harvests are only a small fraction of the standing aboveground phytomass; in temperate forests they are no higher than 1.5–2 t/ha.

The most extensive ecosystems with the highest recorded stores are in the old-growth formations of western North America, which benefit from abundant moisture in maritime locations (Edmonds 1982). Mature stands of Douglas fir (*Pseudotsuga menziesii*) and noble fir (*Abies procera*) store up to 1,700 t/ha, and the aboveground maxima for the Pacific coastal redwoods (*Sequoia sempervirens*) are around 3,500 t/ha, three times as much as the richest tropical rain forests. These forests also contain the biosphere's most massive organisms, giant sequoias (*Sequoiadendron giganteum*), with a phytomass in excess of 3,000 t and a life span of more than 3,000 years.

Keith, Mackey, and Lindenmayer (2009) found another instance of high carbon density in an Australian (highlands of Victoria) evergreen temperate forest dominated by *Eucalyptus regnans*: its average aboveground carbon content in living tissues is 1,053 t/ha and the total is 1,867 t C/ha (more than 3,700 t/ha) in all (living, roots, and dead) phytomass in stands with trees older than 100 years. Douglas firs, sequoias, and *Eucalyptus regnans* are also the world's tallest trees. Douglas firs can reach 110 m, the tallest coastal redwood (*Sequoia sempervirens*) is nearly 116 m, and the record height for *E. regnans* is about 125 m (Carder 1995; Gymnosperm Database 2011). However, as with every large tree, most of the phytomass of these giants is dead wood. Another remarkable attribute of old-growth forests is that, contrary to the standard belief that they are carbon neutral, they continue to accumulate phytomass (Luyssaert et al. 2008).

Even those ecosystems that have relatively low stores of woody phytomass, such as chaparral (the arid shrubland of southern California) and temperate woodlands, typically store more phytomass per unit area than do the richest grasslands. Tall tropical grasslands can have phytomass densities surpassing 30 t/ha, temperate grasslands store usually around 20 t/ha, and the short-grass tundras average no more than 5 t/ha. Unlike in forests, most of the grassland phytomass is stored belowground: the mean global root:shoot ratio is about 3.7 (Jackson et al. 1996), and in many grasslands it is in excess of 6 (Yang et al. 2010). In contrast, both freshwater and coastal wetlands—whose limited areas have been subject to extensive conversion to cropland, urban, and industrial uses—have phytomass stores commonly in excess of 50 t/ha.

Energy losses along the food webs mean that the heterotrophic biomass must be dominated by tissues of short-lived microorganisms. Prokaryotes—unicellular bacteria and archaea—are abundant not only in soils and near-surface waters but also in extreme environments (undersea sediments, deep ocean thermal vents, and rocks up to several kilometers below the Earth's surface, hot springs, and highly alkaline lakes), and they are also (often symbiotically) active inside plants and animals; nitrogen-fixing bacteria in legumes and gut microbes in ruminants are excellent examples of unusually rewarding symbioses. Biomass totals diminish rapidly at higher trophic levels: the biomass of soil invertebrates is much smaller than that of soil bacteria, and the zoomass of large mammalian herbivores is much larger than that of the carnivores that prey on them. The top predators (large felids, tuna, sharks) have large bodies, but their aggregate zoomass is a very small fraction of the overall biomass of their respective ecosystems.

Compared to quantifications of phytomass, quantifications of heterotrophic biomass are relatively uncommon. Bacteria, both autotrophic and heterotrophic, are present in soils in quantities ranging from 10^1 to 10^3 g/m^2 (Coleman and Crossley 1996). Earthworms (Oligochaeta) are the most conspicuous soil invertebrates, and their biomass usually dominates the total for soil macrofauna; in soils of the temperate zone it usually adds up to just around 5 g/m^2, and in cultivated soils rich in organic matter it can surpass 10 g/m^2 (Edwards and Lofty 1972; Hartenstein 1986). Barros et al. (2002) found similar densities in Amazonian farmland soils (between 1 and 9 g/m^2), but maxima above 100 g/m^2 in some Amazonian pastures. Millipedes (Diplopoda) add only between 1 and 5 g/m^2. The most common range for all soil macrofauna (dominated by earthworms, ants, and termites) is between 5 and 15 g/m^2.

The biomass of ants (Hymenoptera) is usually well below 0.5 g/m^2 (Brian 1978), while typical densities of termites (Isoptera) are an order of magnitude higher. Averages in Amazonian rain forests are 2–2.5 g/m^2. Very similar rates (1–3 g/m^2) prevail in the region's croplands and pastures, and in the annually flooded *várzea* forest termite densities are just 0.1 g/m^2 (Martius 1994; Barros et al. 2002). Values around 5 g/m^2 are typical for the four dominant species in the tropical woodlands of Australia's Queensland (Holt and Easy 1993), and means of 10 g/m^2 were found in the Atlantic forest in northeastern Brazil, in São Paulo state, and in the dry evergreen forest of northeast Thailand (Inoue et al. 2001; Costa-Leonardo, Casarin, and

Table 2.2
Densities of Bacterial and Invertebrate Biomass

Organisms	Biomass Density (g/m^2)
Soil bacteria	10–1,000
Earthworms	
Common densities	5–10
Maxima	>100
Millipedes	1–5
All soil macrofauna	3–15
Ants	<0.5
Termites	
Common densities	1–3
Maxima	5–10
Total soil macrofauna	5–15

Note: For data sources, see the text.

Ferreira 2003; Vasconcellos 2010). Zimmermann et al. (1982) found peaks of more than 4,000 termites in tropical moist forest and wet savanna: assuming 2 mg/termite, this translates to rates between 8 and 10 g/m^2 (table 2.2).

Mass per unit area is not an appropriate measure for flying insects, and any large-scale averages of biomass densities of small ectothermic vertebrates (mainly frogs and snakes) are just statistical artifacts, as their densities are highly variable. But reptiles and amphibians can dominate the vertebrate zoomass in some tropical rain forests, where their overall density may rival that of invertebrates (Reagan and Waide 1996). An abundance of terrestrial mammalian zoomass correlates with individual body mass, but the expected decline in density with increasing body weight is not, as formerly thought (Damuth 1981), a simple log-log linear relationship with the allometric exponent of –0.75. Instead, the variation is distinctly nonlinear: only the populations of intermediate-sized mammals (body masses between 100 g and 100 kg) had allometric exponents close to the expected –0.75, while heavier animals (100 kg to 3,000 kg) had exponents close to zero (Silva and Downing 1995).

This means that among the nearly 1,000 studied mammalian populations, the expected density of animals weighing 10 g would be about 1,000/km^2, while there would be only about 10 weighing 1 kg and one with a body mass of 100 kg—but still nearly one weighing 1,000 kg. This would be only the best-fit values as the actual densities for every mass category range over at least two (and even three) orders of magnitude. Nevertheless, the central tendency is clear: the largest terrestrial herbivores dominate the mammalian zoomass of their respective ecosystems (African elephants commonly account for at least half the total), while the zoomass of the smallest animals, dominated by rodents, will be usually less than 0.2 g/m² (0.2 kg/ ha), a rate confirmed by field studies (Golley et al. 1975). But in suitable environments rodents can reach extraordinarily high rates: the rat zoomass on a tropical floodplain of the Adelaide River in Australia's Northern Territory was as high as 4.7 g/m² (Madsen et al. 2006).

In accounts of mammalian zoomass, it is not surprising that ecologists have paid particular attention to the world's richest assemblages of large herbivores, in Africa's savannas and rain forests, where the mean mass of mammals is significantly greater than in Amazonia (Cristoffer and Peres 2003). Coe, Cumming, and Phillipson (1976) compiled a list for major national parks of East, Central, and South Africa. The highest recorded zoomass means were 19.9 g/m² in Uganda's Ruwenzori National Park (calculated in the 1960s) and 19.2 g/m² in Tanzania's Manyara National Park. Schaller's (1972) estimate for the entire Serengeti was 4.2 g/m², similar to the average for the southern section of South Africa's Kruger National Park, but other published Serengeti means are up to twice as high. In two other African ecosystems dominated by large herbivores, estimated zoomass densities range from 8.5 g/m² for the Simanjiro Plains of northern Tanzania (Kahuranga 1981) to 10.3 g/m² for the wet-season zoomass in Kenya's Maasai Mara National Reserve (Ogutu and Dublin 2002).

Dry-season maxima for the Ngorongoro Crater were 12 g/m² (Hanby et al. 1995), while the total for all carnivores (in one the world's best habitats for large predators to hunt ungulates) was put by Schaller (1972) at less than 0.03 g/m². Large herbivore zoomass is generally less dense in the West African parks: at Arli and Po in the Upper Volta region it was about 1.8 g/m², at Kainji in Nigeria just 1.2 g/m² (Milligan, Ajayi, and Hall 1982), and in Benin's Pendjari 1.1 g/m² (Sinsin et al. 2002). The total zoomass for all large diurnal mammals at different sites of the Lope Reserve in Gabon was between 1 and 6 g/m2 (White 1994), and southern India's tropical forest in the Nagarahole National Park supported 7.6 g/m² of wild herbivores, compared to wild herbivore densities of 4.7–6.3 g/m² in the country's tiger reserves (Karanth and Sunquist 1992).

Hayward, O'Brien, and Kerley (2007) tabulated average densities for all major herbivorous species in ten of South Africa's conservation areas. Converting those densities to biomass by using an average of 75% of adult female body mass (this approximation accounts for the lower weights of subadults and young animals) yields rates of 50–300 kg/km^2 for eland (*Taurotragus oryx*), about 70 kg/km^2 for zebra (*Equus zebra*), and 30–200 kg/km^2 for wildebeest (*Connochaetes gnu*). The African elephant (*Loxodonta africana*), the largest surviving terrestrial mammal, has its highest densities (in excess of 3,000 kg/km^2) on natural grasslands in the eastern part of the continent; the rate can be an order of magnitude lower on transformed grassland and in most parts of West Africa (Sinsin et al. 2002). But even the smaller forest elephant (*L. africana cyclotis*) can have a biomass of 2,000–3,000 kg/km^2 (White 1994; Morgan 2007). In contrast, the biomass density of all small mammals (mostly rodents) in temperate grasslands and forests rarely surpasses 3 kg/ha, and their most common biomass densities are less than 1 kg/ha (French et al. 1976; Smith and Urness 1984).

Finally, the zoomass densities of chimpanzees (*Pan troglodytes*), mammals that are genetically closest to humans: in Tanzania's Gombe Stream National Park there are more than five animals/km^2 (Pusey et al. 2005), but this community, habituated to humans, is now surrounded by densely inhabited and cultivated areas. Its former chimpanzee densities were between 1.29 and 1.93 individuals/km^2, in line with typical counts in the forests of East and Central Africa: 1.45–2.43 and 1.45–1.95 in, respectively, Uganda's Kibale National Park and Budongo Forest Reserve, and 2.2 in Congo's Odzala National Park (Bernstein and Smith 1979; Ghiglieri 1984; Bermejo 1999; Williams et al. 2002; Plumptre and Cox 2006).

Chimpanzee densities are lower in northern Congo (0.33–1.75 individuals/km^2), while a long-term mean in Côte d'Ivoire's Taï National Park was 2.8–3.2 animals/km^2 between 1982 and 1996 (Boesch and Boesch-Achermann 2000); more recent numbers have ranged between 1.03 and 2.14 (Lehmann and Boesch 2003; Devos et al. 2008). There are both interannual and seasonal density variations in all studied populations. The population structure of chimpanzee groups is dominated by smaller females and infants, resulting in an average body mass of only about 25 kg/animal, compared to about 40 kg/adult male and 34 kg/adult female. This results in average zoomass densities ranging from highs of more 100 kg/km^2 to lows of less than 25 kg/km^2.

Inevitably, aggregate mammalian densities decline when expressed as averages for large regions or for biomes: at those scales, only savannas and some tropical forests harbor in excess of 2 g/m^2, while the mammalian zoomass in most equatorial and montane rain forests as well as in temperate woodlands is below 1 g/m^2 (Prins and Reitsma 1989; Plumptre and Harris 1995), and totals for avifaunas usually do not surpass 0.05 g/m^2 (Edmonds 1974; Reagan and Waide 1996). Continental zoomass means are less than 1 g/m^2, and hence their actual precise rates make little difference in counting the planetary biomass: unavoidable errors in estimating the

Table 2.3
Zoomass Densities of Wild Terrestrial Mammals

Organisms	Live Weight (kg/ha)
All large wild mammals	
Ruwenzori National Park (Uganda; 1960s)	199
Lake Manyara Park National Park (Tanzania)	192
Ngorongoro Crater (Tanzania, 1970s)	125
Masai Mara (Kenya)	100
Serengeti (Tanzania)	40–80
Gabon, lowland forest	10–60
African elephants	20–30
Eland	0.5–30
Wildebeests	0.3–20
Chimpanzees	0.2–1
All small mammals	1–3
Rodents	
Common densities	<0.2
Maxima	>4

Note: For comparison, densities of traditional African cattle herding are 6–30 kg/ha and the highest zoomass densities of domesticated animals (Dutch Friesian dairy cows on pasture) are 700–750 kg/ha.

biomass of soil bacteria, soil invertebrates, or the underground biomass of forests yield uncertainties that are orders of magnitude larger. Table 2.3 offers both the typical and the highest recorded densities of terrestrial mammals at a glance.

Estimates of Global Biomass

Realistic quantifications of the Earth's phytomass are relatively recent: the first attempt was made only in 1926 by Vladimir Ivanovich Vernadskii in his first edition of *Biosfera*. He put the total weight of all "green matter" at 10^{20}–10^{21} g, and noted that this enormous mass did not seem excessively large and that the total was of the same order of magnitude as all of the biosphere's living matter (Vernadskii 1926). Vernadskii was correct in his second conclusion, for the heterotrophic biomass adds up to only a small fraction of phytomass, but his phytomass estimate was an enormous exaggeration. Vernadskii realized this error, and during the 1930s he kept revising the total downward, but his last published value was, at 10^{16} g, too low (Vernadskii 1940).

The only other cited pre–World War II estimates were of the right order of magnitude: Noddack's (1937) total of about 300 Gt of carbon (600 Gt of dry biomass) was only a small fraction of Vernadskii's original aggregate. Neither of these estimates was constructed from an aggregation of major vegetation subtypes; the first attempt to do so was made only during the late 1960s, when three Russian biologists reconstructed the Earth's potential terrestrial phytomass of the preagricultural era by aggregating the phytomass of 106 major plant formations (Bazilevich, Rodin, and Rozov 1971). Their total was 2.4 Tt, or about 1.2 Tt C. Shortly afterward Whittaker and Likens (1975) used 14 ecosystem types to calculate global phytomass for the year 1950 at 1.837 Tt (about 920 Gt C), nearly 25% below the Russian estimate for the preagricultural world, a plausible difference given the intervening conversion of forests and grasslands.

A major refinement came when Olson, Watts, and Allison (1983) gathered most of the available information on phytomass stores in various ecosystems and presented their global terrestrial phytomass estimate of $0.5° \times 0.5°$ cells. Recognizing many inherent uncertainties, they opted for a range of 460–660 (mean 560) Gt C rather than for a single figure, and concluded that any value above 800 Gt C should be seen as unrealistically high, while totals well below 560 Gt C would not be all that surprising. Most of the subsequently published terrestrial phytomass totals were based on global carbon cycle models and have included a relatively low value of 486 Gt C (Amthor et al. 1998), and a 60% higher total of 780 Gt C (Post, King, and Wullschleger 1997).

The Pilot Analysis of Global Ecosystems, an attempt to synthesize data from national, regional, and global studies, underscored the continuing uncertainty surrounding phytomass estimations by adopting an excessively broad range of 268–901 Gt C (Matthews et al. 2000), while Saugier, Roy, and Mooney (2001) settled on 652 Gt C and Houghton and Goetz (2008), stressing the continued uncertainties in our understanding of global phytomass, offered a nearly twofold range of 385–650 Gt C. The latest appraisal of tropical forest phytomass ended up with about 250 Gt C (Saatchi et al. 2011); assuming that tropical forests account for 50% of the global total, the terrestrial phytomass would be just 500 Gt C. Table 2.4 compares all of the reviewed estimates of global terrestrial phytomass. It should be noted that differences among these values are much larger than the totals published for the oceanic phytomass. Because phytoplankton has a very short life span—its fast turnover means that its entire stock is consumed every two to six days (Behrenfeld and Falkowski 1997)—its standing mass adds up to only 1–4 Gt C.

Table 2.4
Evolution of Global Estimates of Terrestrial Phytomass

Year	Author	Phytomass (Gt C)
1937	Noddack	300
1966	Bowen	507
1970	Bolin	450
1971	Kovda	1,395
1971	Bazilevich, Rodin, and Rozov[1]	1,200
1972	Duvigneaud	592
1975	Whittaker and Likens[2]	920
1976	Baes et al.	680
1979	Ajtay et al.	560
1979	Bolin et al.	590
1982	Brown and Lugo	500
1983	Olson, Watts, and Allison	559
1984	Matthews	734
1984	Goudriaan and Ketner	594
1987	Esser	657
1990	IPCC	550
1992	Holmén	560
1992	Smith et al.	737
1993	Hall and Scurlock	560
1994	Foley	801
1997	Post, King, and Wullschleger	780
1998	Amthor et al.	486
1998	Field et al.	500
1998	WBGU	466
1999	Potter	651
2000	Matthews et al.	268–901
2001	Roy, Saugier, and Mooney	652
2008	Houghton and Goetz	385–650

Notes: 1. Potential plant cover. 2. Plant cover in 1950.

Estimates of heterotrophic mass are much more uncertain, mainly because of highly variable densities of microbial biomass in soils and the poorly known presence of these prokaryotic organisms in extreme environments, particularly in the uppermost region of the Earth's crust. Additional uncertainties are caused by the enormous variety, variability, and mobility of arthropods. Whitman et al. (1998) estimated that about 26 Gt C are stored in soil bacteria, but the real value may be anywhere between 15 and 50 Gt C. And the uncertainty is even greater as far as the mass of subterranean prokaryotes is concerned.

If we assume, as Gold (1992) did, that microbes fill 1% of all porous spaces in the topmost 5 km of the Earth's crust, the prokaryotic biomass could be as much as 200 Tt C, but if the microbes filled just 0.016% of the available porous space (Whitman et al. 1998), then they would add just 200 Gt C—and the real total may be an order of magnitude smaller. Bacteria in the subsea sediments may hold another 200–300 Gt C (Parkes et al. 1994). Although prokaryotes provide the trophic foundation of all metazoan life, most of their harvests have been always only incidental, as they are present on every surface and inside all phytomass and zoomass tissues. The only notable exception has been the deliberate collection of cyanobacteria—particularly those belonging to the genera *Nostoc*, *Spirulina*, and *Aphanizomenon*—for food in parts of Asia and Africa (Ramírez-Moreno and Olvera-Ramírez 2006).

Fungal biomass in soils also ranges widely, from 10^0 to 10^2 g/m^2 (Bowen 1966; Reagan and Waide 1996), but in comparison to high bacterial counts its global presence is negligible, adding up most likely to only between 2 and 5 Gt C—and the biomass of metazoa (multicellular animals) is even more insignificant. As already shown, it is dominated by invertebrates, with earthworms and termites (inhabiting soils in nearly 70% of all nonglaciated land) being the two largest contributors. Even a conservative mean of 5 g/m^2 in all agricultural and forest soils would yield nearly 500 Mt of earthworms, or about 100 Mt of dry weight and less than 50 Mt C.

Assuming that the global termite estimate of 2.4×10^{17} individuals offered by Zimmerman et al. (1982) is of the right order of magnitude, then multiplying it by an average body mass of 2 mg/worker yields a total live weight of 480 Mt, a dry weight of just over 100 Mt, and roughly 50 Mt C. Estimates of total insect biomass are highly uncertain: after all, we do not even know the correct magnitude of the total number of insect species. Erwin's (1982) often cited estimates of 30–100 million species are almost certainly exaggerated given that the latest attempt to estimate the global total ended up with about 8.7 million for all eukaryotes and 7.8 million for all animals (Mora et al. 2011).

The biosphere's vertebrate zoomass is now dominated by domesticated animals, in particular by seven genera of mammals—cattle (*Bos*), horse and donkey (*Equus*), water buffalo (*Bubalus*), pig (*Sus*), sheep (*Ovis*), goat (*Capra*), and camel (*Camelus*)—and four genera of birds—chicken (*Gallus*), goose (*Anser*), duck (*Anas*), and turkey (*Meleagris*). In the year 2000 their live weight added up to about 600 Mt, with cattle and water buffaloes accounting for nearly two-thirds of the total and pigs for more than 10%. With the water content of empty bodies averaging 55% for bovines and just over 60% for pigs (Garrett and Hinman 1969; Mitchell, Scholz, and Conway 1998), this translates to about 280 Mt of dry weight and 125 Mt C.

The great volume of the inhabited medium, the high mobility of many marine organisms, and the extraordinary patchiness of seafloor heterotrophs make the quantification of oceanic zoomass quite difficult. The greatest challenge is to quantify the smallest and hence the most abundant organisms. To begin with, we do not even know the number of their species to the nearest two or three orders of magnitude. The Census of Marine Microbes began in 2003 with 6,000 kinds of identified species and with expectations of as many as 600,000, but subsequent sampling has brought hundreds of thousands of new microbial forms, and expectations are now for at least 20 million, possibly even billions (Qiu 2010)—and we do not know how many of these organisms are autotrophs and how many are heterotrophic protists. There are also considerable uncertainties about the deep-sea zoomass: it may be "an empire lacking food" (McClain 2010), but sampling has come up with biomass densities 100 times greater than previously reported for depths below 500 m (De Leo et al. 2010).

Moreover, all known zooplankton taxa are short-lived, with typical longevities of up to 10–12 weeks in cold waters but only half as long in the warmest seas (Allan 1976; Laybourn-Parry 1992). This rapid turnover means that at any given time, the standing biomass of marine zooplankton will be only a small fraction of the total mass that is available for capture by species at the next trophic level. For the fish zoomass there are two independent estimates prepared by Wilson et al. (2009). The first one uses a model that integrates fish biomass over 36 × 36 km cells and is based on photosynthetic production, trophic transfer efficiency, and predator body mass ratios: it resulted in 899 Mt (fresh weight) of fish zoomass. The second one, a stratified ecosystem model, yielded a total more than twice as high (2.05 Gt). A highly uneven distribution of this zoomass is best illustrated by the fact that half of all fish biomass is found in only 17% of the world's ocean area (Jennings et al. 2008).

Another recent estimate of marine zoomass reevaluated the amount of Antarctic krill, more than 80 species of tiny (typically just 1–2 cm) shrimplike crustaceans, dominated by *Euphausia superba*, that add up to one of the biosphere's largest sources of easily accessible protein. Krill's maximum distribution covers about 19 Mkm^2, and with a mean total abundance of 8×10^{14} post-larvae its total zoomass is 379 Mt (Atkinson et al. 2009). The authors also estimated that the annual gross postlarval production of 342–536 Mt supports a predator consumption of 128–470 Mt/year by fish, squids, seabirds, and baleen whales.

Quantifications of the global whale zoomass may get even the order of magnitude wrong because the extant numbers of most of the massive species are known only within unhelpfully broad bands. For example, the IUCN (2011) offers best estimates of "hundreds of thousands" for sperm whales (*Physeter macrocephalus*) and any-where between 10,000 and 25,000 for blue whales (*Balaenoptera musculus*), and because we do not have sufficiently detailed understanding of the age and sex com-position of these animals, calculating errors will only increase when the total numbers are multiplied by uncertain body mass averages.

None of this has prevented Christensen (2006) from reconstructing historical (prewhaling) and modern abundances on the global scale (implausibly, down to the nearest 10 or 100 individuals of all species), and Pershing et al. (2010) from using these numbers, in conjunction with age-structured models and average body masses, to calculate the biomasses of eight species or species groups of baleen whales (blue, fin, humpback, sei/Bryde's, gray, right, minke, and bowhead; sperm whales were excluded). The prewhaling zoomass added up to about 103 Mt in 2.56 million animals, while the totals for 2001 were, respectively, about 16 Mt and 880,000 individuals, implying an overall loss of about 85% of the original baleen whale zoomass.

Terrestrial invertebrates are more important for the maintenance of ecosystems than vertebrates, but they are harvested by humans only in relatively small quanti-ties. Insectivory (including the larvae and pupae of some species, with relatively common choices including ants, crickets, and grasshoppers) is still practiced in many traditional societies (while bees and silkworms remain the most important domes-ticated insects), and many snails are also eaten (Menzel and D'Aluisio 1998). But in aggregate, terrestrial invertebrates make only a marginal contribution to global nutrition, whereas marine invertebrates (cephalopods, mollusks, crustaceans) are an important source of dietary protein.

The rate of biomass renewal is expressed either as annual productivity (as mass or energy per unit area) or as overall production (aggregate productivity for a country, an ecosystem, a biome, or the entire Earth). Specific productivities of animals that have traditionally been killed or domesticated for meat, eggs, and dairy products will be reviewed later; in the next chapter I explain why quantifying primary productivity is so challenging. I also define all the relevant measures of this key variable and explain its advantages and limitations in order to avoid misleading conclusions regarding the human impact on the Earth's biota.

3

Biomass Productivities

The biosphere's productivity can be quantified as a cascading series of variables: the most inclusive ones, the two rates at the cascade's top, gross and net primary productivity, cannot be measured directly and can be quantified (far from accurately) only thanks to our improved understanding of photosynthetic processes, the environmental responses of autotrophs, and the properties of plant metabolism. In contrast, the rates at the cascade's end can be readily measured in mass or energy terms (although in practice they are often estimated on the basis of limited field sampling): crop yields (expressed mostly in t/ha) and timber harvests (commonly quantified as m^3/ha) are the most common values of interest. But data for crop and wood yields do not capture the complete biomass productivity and must be adjusted in order to find the totals of initially produced phytomass or zoomass and hence to express more accurately the actual intensity of harvests and to assess the overall impact of human actions on the affected natural ecosystems and on various agro-ecosystems or aquacultures.

Primary Productivity

Gross primary productivity (GPP) is the most inclusive measure, subsuming the total amount of new phytomass that is photosynthesized during a given period of time by all autotrophs. The rate excludes all photorespiratory losses that accompany the conversion of solar radiation to the chemical energy of phytomass, and its global tally can neglect marginal contributions by chemotrophic prokaryotes that form new biomass in the absence of sunlight. A large part of this new photosynthate is not deposited as new plant tissues but is rapidly reoxidized during *autotrophic respiration* (R_A). Plant respiration reduces the flux of fixed carbon, but (unlike the entirely wasteful photorespiration) it cannot be seen as a loss because the

Brazilian sugarcane is the world's most productive field crop, with fresh yields averaging about 75 t/ha and with a carbohydrate (saccharose) yield of 8 t/ha. Photograph of mature crop courtesy of Governo do Estado do Espírito Santo.

reoxidation of fixed carbon energizes autotrophic growth and maintenance (that is, the synthesis of biopolymers from their monomers, the transport of photosynthates within plants, and the repair of diseased or damaged tissues).

Autotrophic respiration thus forms an indispensable metabolic bridge between photosynthesis and plant structure and function (Amthor and Baldocchi 2001; Trumbore 2006). Its relative shares are commonly expressed as the quotient R_A/GPP and are usually lowest for some intensively cultivated crops. In primary production terms, agriculture can be defined as an endeavor aimed at maximizing GPP while minimizing R_A; this quest is achieved primarily through the selection of high-yielding cultivars, adequate fertilization, and, if need be, irrigation. R_A varies widely both among biomes and ecosystems (as a function of location and climate) and during successive stages of plant growth (it generally increases with plant age). Its rates are mostly between 0.3 and 0.65 in grasslands and between 0.55 and 0.75 for boreal and temperate forests; the highest rates (at least 0.75–0.85) have been attributed to tropical rain forests. A mean of 0.5 has often been used as the first-order approximation for estimates on larger spatial scales.

Subtracting R_A from the GPP yields the *net primary productivity* (NPP; NPP = GPP – R_A), the amount of phytomass that becomes available to heterotrophic organisms, be they bacteria, insects, or humans. The NPP of major biomes ranges from negligible amounts in extreme environments (hot or cold deserts) to nearly 1 kg C/m^2 (20 t/ha) in the richest tropical rain forests. Forest NPP is higher than the productivity of grasslands growing in the same environment because trees have a much higher leaf area index (LAI), the upper area of foliage per unit area of ground that captures solar radiation. While grasslands typically have an LAI no higher than 3, and often less than 2, mean values for forest are above 3 even for boreal growth, and are commonly more than 5 for multistory temperate and tropical trees with heavy canopies (Myneni et al. 1997; Scurlock, Asner, and Gower 2001).

Large-scale averages of the terrestrial NPP/GPP ratio have been commonly assumed to be around 0.5, and this mean was confirmed by calculations based on four years of satellite observations: between 2000 and 2003, global terrestrial ecosystems had an NPP/GPP ratio of 0.52 (Zhang et al. 2009). The ratio is generally lower in densely vegetated ecosystems than in sparsely overgrown regions, forests have a lower ratio than shrub and herbaceous ecosystems, and the ratio increases with altitude. Climate exerts a key influence: the ratio has a decreasing trend with a higher precipitation total of up to 2.3 m/year but is static for annual precipitation above that threshold; as for average temperature, the ratio declines with temperatures between –20°C and 10°C and rises with temperatures increasing between –10°C and 20°C.

The standard approach is to consider only the R_A of the phytomass that was photosynthesized during the period in question, but the quantification should also take into account all respiratory losses from preexisting phytomass, a flux that, as Roxburgh et al. (2005) concede, appears to be impossibly difficult to measure. They also offer a systematic and revealing decomposition of the measure. Their definition of the instantaneous rate of change of the total phytomass carbon stock—GPP – R_A – L, where L is the total of nonrespiratory losses, including not only litterfall, root death, and exudation but also herbivory and natural physical disturbances—corresponds to what others call the *net ecosystem productivity* (NEP). That flux, representing all carbon accumulated by ecosystems, has been usually defined as NPP reduced by *heterotrophic respiration* (NEP = NPP − R_H), but it should also include all nonrespiratory carbon losses.

Chapin et al. (2006) argued that the term NEP should be restricted to just the difference between GPP and ecosystem respiration and proposed a new term, net ecosystem carbon balance (NECB), for the net rate of carbon accumulation (or loss) in ecosystems. R_H is influenced by the ubiquitous microbial and insect depredations to the recurrent seasonal grazing by large ungulates and also reflects such irregular episodes of massive insect attacks as invasions of crop-devouring locusts or bark beetles destroying coniferous trees. In bioenergetic terms, agriculture should thus also be seen as an endeavor aimed at minimizing R_H. This quest is accomplished most often through the application of pesticides (dominated by insecticides and fungicides) but also through the use of protective crop cover, the fencing of fields, or even by deploying noisemaking devices to minimize damage done by birds or grazing herbivores. As expected, the components of NEP vary with scale, and at a regional or global level the term *net biome productivity* (NBP) is used as an alternative to NEP.

NPP cannot be measured directly, and the standard method of frequent harvesting of sample plots (restricted by logistics and cost to areas of 10^3 m^2 and hard to do in mature forests) captures reasonably well only the aboveground share of the overall productivity; infrequent sampling would, of course, miss a great deal of litterfall (once the dead phytomass is on the ground it is difficult to assess its exact age). But the sequential harvesting of aboveground phytomass does not capture either the belowground increment or the carbon losses that do not involve respiratory flows. The component of belowground productivity most difficult to measure is the often voluminous but always short-lived fine root turnover (Fahey and Knapp 2007). As expected, overall accuracy is better at smaller scales, particularly where many field observations are available. Luyssaert et al. (2009) found that their model-based account of European (EU-25) forest NPP (annual mean of 520 ± 75 g C/m^2) was within 25% of the results obtained using forest inventories or scaled-up terrestrial observations.

Accounts of agricultural production may be the most reliable source for quantifying actual NPP, but a recent continent-wide analysis of the carbon balance of European (EU-25) croplands showed that even the world's best output and land-use statistics, combined with process-oriented and remote-sensing models, still yielded a surprisingly large range of NPP rates. Even when an identical cropland area was assayed, different production factors resulted in estimates of mean annual NPP of cropland in 25 EU nations as low as 646 and as high as 846 g C/m^2 (Ciais et al.

2010), a 30% difference. Expressed in common yield terms (although, of course, NPP is much larger than actual harvestable yield), this corresponds to an average annual productivity of 13–17 t/ha.

More complete appraisals of CO_2 fluxes can be now derived using gas exchange techniques (eddy covariance and mass balances within the convective boundary layer) that are fairly easily accomplished with small grass or crop plots but are much more difficult with forest growth (they require the erection of tall towers, the use of tethered balloons, or regular sampling with aircraft). But even these techniques are of no help in discriminating the autotrophic (derived from roots) and hetero-trophic (derived from bacteria) components of soil respiration, or in quantifying non-CO_2 losses. Total CO_2 flux methods should yield productivity estimates that are perhaps 20%–50% higher than the standard values (Geider et al. 2001). For example, the NPP of a Brazilian rain forest near Manaus is as high as 15.6 t C/ha, while the total that neglected fine root turnover was nearly 40% lower. And Scurlock, Johnson, and Olson (2002) believe that the harvest-based estimates of grassland NPP may be no more than 50% and perhaps as little as 20% of the real rate. At the same time, NEP values are significantly lower once all nonrespiratory fluxes are taken into account.

The most obvious of these carbon losses is from ubiquitous litterfall: a large part of NPP is continuously discarded in plant litter, and while most of these tissues (buds, flowers, fruits, leaves, twigs, branches) are consumed (some rapidly, others very slowly) by resident heterotrophs, some litter carbon is carried away (dissolved and particulate organic carbon is moved by leaching beyond root level and by leaching and soil erosion into streams and carried away by wind) and does not get a chance to become part of the producing ecosystem's heterotrophic respiration. Not surprisingly, specific information about the partitioning of litter flows among bacterial and fungal decomposers, invertebrate and vertebrate foraging, and losses beyond an ecosystem boundary is not easily available. Meentemeyer et al. (1982) estimated the worldwide annual leaf fall at 35.1 Gt and the total litter production at 54.8 Gt, or about half the global NPP. Studies in various ecosystems have shown ranges of 5–15 t/ha in tropical forests (11 t/ha may be a good mean) and mostly between 4 and 8 t/ha in temperate and boreal biomes, where 4.5–5 t/ha may be a typical loss.

Other carbon losses include those in the form of volatile organic compounds (mainly monoterpenes) emitted in particularly copious volumes by some trees (Fuentes et al. 2000; Tunved et al. 2006; Schurgers et al. 2009), other exudates (sap, resins, waxes), methane produced by methanogenic bacteria, CO from photochemical and thermal oxidation of organic matter, and carbon supplied to root symbionts. Over longer periods of time and on larger scales, the accounts must also include phytomass losses resulting from such natural disturbances as fires and destructive floods, which can cause substantial episodic destruction of plant growth (the effects of droughts should be reflected in a reduced GPP). This means that the commonly used NEP values are only best approximations and that an all-encompassing definition of NEP should explicitly incorporate all of the carbon flows leaving an ecosystem (Randerson et al. 2002). If all of the primary productivity components that cannot be captured by standard aboveground sampling were to add just 5% to the total, there would be no need for concern, but the truncated assessments of NEP result in much larger errors.

For comparative purposes, both the NPP and the NEP are normally expressed in annual terms, although the ecological literature contains references to maximum daily, monthly, and seasonal productivities. Here are a few examples of annual forest NEP (with all values expressed as shares of GPP) calculated from data in Luyssaret et al. (2007): about 13% for a boreal humid evergreen forest, around 22% for temperate evergreen as well as deciduous forests, and just 11% for a tropical humid evergreen forest. Daily means of NPP that are sustainable during the weeks of the most rapid growth are about 20 g/m^2 (200 kg/ha) for the photosynthetically more efficient C_4 plants and 5–15 g/m^2 (average of about 13 g/m^2) for C_3 species. As a general rule, short-term rates should not be extrapolated to obtain yearly fluxes, and there should be explicit references to what is included in any particular rates.

Calculating and Monitoring Global Primary Productivity

The global estimates of NPP made prior to the 1970s are either too low or too high when compared with the most likely range of phytomass production that emerged in the last three decades of the twentieth century with more field studies and better models. Although Liebig (1862) was credited by Lieth (2005) with the first estimate of global NPP, a careful reading of Liebig's original text shows that his was only a conditional example aimed at explaining the cyclical nature of photosynthesis and respiration, not a deliberate estimate of actual phytomass productivity.

As a result, the oldest published NPP estimate was made by Ebermayer (1882). It was based on typical yields of Bavarian forests and crops applied to woodland, grassland, cropland, and barren areas, and it amounted to only 24.5 Gt C/year. Schroeder (1919) and Noddack (1937) estimated even lower totals at, respectively, 16 and 15 Gt C/year, while Deevey (1960) suggested as much as 82 Gt C/year as

the best value of terrestrial NPP. Research into mass and energy flows in the biosphere expanded during the 1960s and 1970 (Eckardt 1968; Lieth and Whittaker
1975), and thanks to better regression models linking NPP and key climatic variables, estimates for both terrestrial and marine NPP began to fall within a much
more constrained range during the 1970s: the lowest estimate was 45 Gt C/year
(Lieth 1973), the highest, by Ajtay et al.(1979), was 60 Gt C/year, and perhaps the
most widely cited total, by Whittaker and Likens (1975), put the terrestrial NPP at
55 Gt C/year.

All subsequent global estimates of NPP have benefited from satellite monitoring
and from better models of primary productivity. Remote sensing of photosynthetic
productivity is based on the phenomenon that chlorophyll, the principal plant
pigment, reflects less than 20% of the longest wavelengths of visible light but about
60% of near infrared radiation. These differences in reflectance can be used in relatively detailed ecosystem mapping and can be converted into a normalized difference
vegetation index (NDVI). Multispectral Scanning Systems (MSS) on the first Landsat
satellite, launched in July 1972, recorded reflected radiation in four different spectral
bands between 0.5 and 1.1 μm, with a resolution of 80 m. Later Landsat satellites
had the Thematic Mapper, with a resolution of 30 m, and the French SPOT satellite,
launched in 1985, revealed details at 10–20 m.

The high cost of Landsat and SPOT imagery made the Advanced Very High
Resolution Radiometer (AVHRR, installed on the National Oceanic and Atmospheric Administration's polar-orbiting satellites) the tool of choice for monitoring
global NPP. AVHRR's maximum resolution of 1–4 km sufficed to capture large-
scale patterns of vegetation coverage and to detect seasonal variability (Gutman and
Ignatov 1995). The first NDVI calculations used reflectances in the visible band
(0.58–0.68 μm) and the near infrared band (0.73–1.1 μm); calculations for reflectances in the thermal infrared band (11 and 12 μm) and compensations for changing
illumination conditions, surface slope, and viewing aspect were added later (Gutman
et al. 1995).

At the same time, the results of a growing number of field NPP studies were
collated in a standardized format thanks to the Global Primary Production Data
Initiative based at the Oak Ridge National Laboratory in Tennessee. The database
was begun in 1994 and now contains data on 65 intensively studied sites (mainly
grasslands and tropical and boreal forests, with georeferenced climate and site
characteristics data) compiled between 1930 and 1996 (ORNL 2010). NPP totals
published during the last two decades of the twentieth century had a range of
48–65 Gt C/year, and improvements in satellite monitoring and expanding primary

production databases have helped narrow the most likely range to between 55 and 60 Gt C/yr.

The first modern attempt to estimate marine NPP (inadvisably extrapolating from a small number of questionable values) was Noddack's (1937) total of 25.4 Gt C/year for the open ocean and 3.2 Gt C/year for coastal seas. This was followed by Riley's (1944) wide range of 44–208 Gt C/year based on just seven western Atlantic sites. Koblents-Mishke et al. (1968) used data from 7,000 ocean sites to come up with a range very similar to Noddack's (27–32 Gt C/year, excluding all benthic production). De Vooys (1979) put the annual marine NPP at 46 Gt C/year. Estimates of marine NPP had also benefited from satellite monitoring.

Data from the Coastal Zone Color Scanner (CZCS) made it possible to monitor global patterns of phytoplankton growth, most notably the seasonally high productivities in major upwelling regions off the western coast of Africa and South America. Two satellite-based models of marine NPP came up with very similar results: 37–46 Gt C/year by Antoine et al. (1996) and 47.5 Gt C/year by Behrenfeld and Falkowski (1997). There is a strong seasonal variation, with the monthly average ranging from 3.8 to 4.6 Gt C/year and longer-term anomalies driven by El Niño/La Niña fluctuations. The Pacific Ocean accounts for slightly more than 40% of the total, and the marine NPP of the two hemispheres is roughly equal because the smaller expanse of the northern ocean has about 60% higher productivity per unit area.

At the century's end, the Potsdam NPP Model Intercomparison meeting provided a comprehensive survey of global phytomass productivity estimates based on 16 models that were compared using standardized input variables (Cramer et al. 1999). After two extreme values were excluded, the results ranged from 44.3 to 66.3 Gt C/year with the mean of 54.1 Gt C/year. This translates to roughly 89–132 Gt (mean, 108 Gt) of dry phytomass, or, assuming an average energy density close to 20 GJ/t, to about 2 ZJ. Two conceptually similar models of terrestrial and marine NPP resulted in a combined global total of 104.9 Gt C/year, with 56.4 Gt C/year on land and 48.5 Gt C/year in the ocean (Field et al. 1998; Geider et al. 2001). The most likely planetary total of satellite-based modeling of NPP was thus between 100 and 110 Gt C/year or 200–220 Gt/year of dry phytomass and at least 3.5 ZJ of energy (an annual flux of about 110 TW).

A superior tool for monitoring global NPP became available with the launching of the *Terra* satellite in December 1999 (*Aqua*, devoted to ocean studies, was launched in May 2002). The satellite carried a Moderate Resolution Imaging Spectroradiometer (MODIS, resolution of 0.25–1 km) designed to track reflected radiation in 36 discrete spectral bands. The combination of measurements of canopy reflectance, information on biome type, the fraction of photosynthetically active radiation that is absorbed by vegetation (a variable that changes with plant growth and senescence), and daily surface climate conditions made it possible to quantify continuous fluxes of terrestrial GPP and to make spatially explicit GPP/NPP estimates that could be compared on seasonal and annual bases, and hence made it possible, for the first time, to discern any longer-term trends in global primary productivity (Heinsch et al. 2003; Running et al. 2004).

In 2000, during the first full year of MODIS operation, the GPP and NPP totals were, respectively, 108.42 and 56.06 Gt C/year (NTSG 2006), and the means for the next two years were 109.29 and 56.02 Gt C/year (Zhao et al. 2005). A MODIS-based GPP has been routinely calculated for eight-day periods since February 2000, and archived information is available either as global and regional maps with 1 km resolution or as data sets (USGS 2010). Matching the 1 km resolution of the sensors with plot-scale measurements on the ground makes the validation of MODIS-derived NPP values difficult. Comparisons of the MODIS-derived GPP with the GPP calculated from tower-based measurements of CO_2 for forest and grassy ecosystems showed that the satellite-based values tended to overestimate GPP at sites with low productivity and underestimate the fluxes at high-productivity sites (Zhao et al. 2005; Turner et al. 2006).

Beer et al. (2010) approached the problem by combining eddy covariance flux data (high-frequency vertical wind vector and CO_2 and H_2O concentration measurements obtained using a using 3-D anemometer and an infrared gas analyzer mounted on platforms high above the vegetation canopy) and process-oriented biosphere models to quantify terrestrial gross CO_2 uptake, its global distribution, and its covariation with climate. They found that the GPP over more than 40% of the vegetated land is associated with precipitation (up to 70% in grasslands, shrublands, and croplands; only 30% of GPP variability in tropical and boreal forests, where temperature plays a greater role) and that tropical ecosystems (forests and savannas) account for 60% of total GPP of 123 ± 8 Gt C/year.

When Ito (2011) surveyed the relevant literature from 1862 to 2011, he found 251 estimates of total terrestrial NPP, whose mean, standard deviation, and median were respectively 56.2, 14.3, and 56.4 Gt C/year. His original intent was to see whether the estimates of global NPP were converging, but he found that even the estimates published after 2000 had substantial uncertainty (coefficient of variation ±15%). This uncertainty has only increased with the publication of a new paper on interannual variability in the oxygen isotopes of atmospheric CO_2 driven by El Niño (Welp et al. 2011). The paper showed how the decay time of the El Niño anomaly can be used to constrain the global estimate of GPP, and found a shorter cycling time of CO_2 with respect to the terrestrial biosphere and oceans than did previous estimates. The authors concluded that the fast response is most plausibly accounted for by revising the global GPP upward to the range of 150–175 Gt C/year.

That is an increase of 25%–45% above what Cuntz (2011) calls carbon's "gold standard" of global GPP, around 120 Gt C/year; he also notes that gold does not tarnish easily—and a closer look at the new value shows how some small shifts in assumptions can bring the total closer to the prevailing standard.

Most notably, Welp at al. (2011) assume that 43% of all CO_2 molecules entering through the stomata are eventually fixed, a rate that is highly dependent on other assumptions and on the global partitioning of photosynthesis between the more efficient C_4 species (whose fixation rate is about 60%) and the less efficient but dominant C_3 plants. Reducing the overall fixation rate lowers the global GPP estimate, and at 34% it would be 120 Gt C/year. Consequently, the only solid conclusion we can make is that despite all the recent advances, our understanding of global primary productivity will have a substantial margin of uncertainty (at least ±15%) for years to come.

As for the spatial and temporal variability of the NPP, satellite observations have confirmed many well-known phenomena (for example, during the first four years of monitoring, the global NPP exhibited annual fluctuations up to nearly 4%), but they have also brought some unexpected results. The NPP reaches its annual terrestrial maxima in excess of 1 kg C/m^2 in the equatorial zone: in the Amazon basin and in the regions located immediately north and south of it, in the Congo basin, in parts of eastern and western Africa, and across the Indonesian archipelago. High productivities also prevail in the highlands of Central America and in the wettest parts of Southeast Asia. A secondary NPP maximum is in the temperate mid-latitudes of the Northern Hemisphere: most of Europe, eastern North America, and China average 400–600 g C/m^2, while most of Asia's interior and Australia's interior, as well Siberia and western North America, have means below 300 g C/m^2.

4

Phytomass Harvests

Modern phytomass harvests fit mostly into four distinct categories. Food harvests have been transformed as humans have evolved from simple foragers collecting edible plants, hunting animals, and catching aquatic species to agriculturalists relying first on extensive shifting cultivation and later on intensive methods of farming, including large-scale domestication of animals and worldwide fishing efforts. Phytomass use as fuel was relatively limited in all foraging societies, but it increased with a sedentary existence and with the use of wood and charcoal in the production of metals. Because the evolution of agriculture also involved the domestication of animals, the third major purpose of biomass harvest has been to secure feed for these mammals and birds.

I hasten to add that the order of harvest purposes just presented is valid only in terms of its evolutionary appearance. In many late nineteenth-century societies fuelwood was mostly displaced by coal, but large quantities of feed were needed for draft animals as agricultural activities continued to rely heavily on draft horses and cattle. This animal labor was in turn displaced by machines, and wood became a marginal fuel in all modern affluent societies, but the importance of animal feed has only increased: metabolic imperatives mean that the annual total of feed phytomass that is required to produce an abundance of meat, eggs, and dairy products now greatly surpasses that of global food harvests and is also greater than the aggregate of woody biomass used for fuel in all affluent economies.

The final, less homogeneous category subsumes all uses of plants as raw materials. The earliest uses of phytomass in crafting simple tools go far back into our hominin past, but once again, only sedentary habitation resulted in a voluminous need for construction timber (in maritime societies also for building ships) and wood for agricultural implements (hoes, plows, harrows), household utensils, furniture, and an enormous variety of increasingly sophisticated manufacturing tools

Traditional cereal cultivars yielded much more straw than do modern varieties (and much more straw than grain). Joachim Patinir's 1514 *Rest on the Flight to Egypt* (now hanging in Madrid's Prado) illustrates that reality: a scene on the far right of the painting shows a peasant harvesting wheat, with the grain ears well above his head and all but hiding armed men carrying halberds.

(the process that culminated in the intricate designs of the early modern era) and conveyances (wheelbarrows, carriages). Other important uses of phytomass as raw material include the use of fibrous plants (cotton, flax, hemp, agave) to make cloth (as well as ropes), as a source of colorants and medicines, and the pulping of woody biomass to produce paper.

The number of plants ever harvested by humans for food, fiber, medicinal and ornamental uses, and animal feed runs into the thousands, but only about 50 species have accounted for the bulk of all harvested phytomass during the millennia of preindustrial agriculture, and this number was reduced by large-scale intensive cultivation and modern dietary preferences to fewer than 20 dominant species. A similar simplification has affected wood harvests: large-scale monoculture tree plantings, begun in the late eighteenth century, and modern afforestation favor a small number of fast-growing species, now increasingly grown on intensively managed tree plantations.

Crops and Their Residues

Food harvests in temperate and subtropical latitudes of the Old World have been traditionally dominated by half a dozen major cereal grains (wheat, rice, barley, rye, oats, buckwheat), leguminous grains (beans, peas, lentils, soybeans), and oil seeds (olives, rapeseed, linseed, sunflowers, peanuts), supplemented by a large variety of leafy and root vegetables, fruits, and nuts. Premodern diets in the Americas shared a heavy reliance on corn and beans, and three of the hemisphere's major crops, potatoes, tomatoes, and peppers, became worldwide favorites within a few centuries after their introduction to Europe, Asia, and Africa. The large-scale production of sugar crops (sugarcane and sugar beets) is of an even more recent origin.

Crop harvests are reported in terms of fresh phytomass, whose water content varies from less than 10% for some seeds to more than 90% for many vegetables; staple grains are marketed with a moisture content of less than 15%, while tubers have more than 70%. Depending on the time and method of harvesting, water content varies appreciably even for the same cultivar of staple grains. The moisture content of cereal grains at harvest ranges between 11% and 25%, with optima between 14% and 17%. The long-term storage of grains requires that their moisture be reduced to less than 13.5%, but actual water shares are often 1%–2% above or below that mean.

Virtually any food phytomass can also serve as feed for some domesticated species (particularly omnivorous pigs), while healthy ruminant nutrition requires cellulosic roughage, which has been always provided by cereal straws. Concentrate feed (cereal and leguminous grains) was traditionally used only sparingly, during periods of heavy draft work, and became a norm in animal feeding only with the emergence of a more intensive cropping that created food surpluses and hence allowed a significant proportion of farmland to be devoted to forage crops (mixed hay, alfalfa, clover, vetch) or feed grain (now most notably corn and soybeans) and tuber (potato, cassava, turnip) crops.

A complete quantification of crop phytomass productivity must also account for all belowground growth and for the residual phytomass that is either removed from fields, to be used as bedding and feed (then often recycled) or as raw material (removed from short-term carbon cycling), or directly returned to soil (plowed in). Shares of belowground phytomass range from less than 10% for many vegetables to between 20% and 25% for cereal grains, about 35% for corn, and well over 50% for root crops (about 60% for potatoes). The largest amount of residual phytomass is harvested as cereal and leguminous straws and as sugarcane stalks; chopped corn stover and tuber and vegetable vines are usually left behind and are recycled back into the soil.

While crops are harvested for their carbohydrate, lipid, or protein content, the composition of crop residues is dominated by cellulose, hemicellulose, and lignin (Barreveld 1989). Cellulose accounts for 30%–50% of residual phytomass, but it can make up as much as 61% of rice stumps (Mukhopadhyay and Nandi 1979). Hemicellulose accounts for up 25%–30% of residual dry phytomass, while lignin's share is 10%–20%. When converting crop residues to carbon equivalents, 45% has been the most often used generic value: although very close to the shares prevalent in most cereal straws it is still a simplification, mainly because of differences in mineral content. Shares of alkaline elements (Ca, Mg, K) are not that different among crop residues, but oats have only about 1% of silica, winter wheat more than 3%, and some rice cultivars more than 13%, and their total ash content may reach about 20% (Antongiovanni and Sargentini 1991). The carbon content of rice straw and corn stover is thus no higher than 40%, and that of wheat is about 45% (Demirbaş 2003; Sawyer and Mallarino 2007).

Harvested residues have low density, with cereal straws rating just 50–100 kg/m^3 (compared to more than 500 kg/m^3 for wood); uncompressed corn stover weighs just between 21 and 111 kg/m^3, and sugarcane bagasse leaving the final mill averages about 120 kg/m^3 (Smil 1999b; Lam et al. 2008). Crop residues traditionally had a wide variety of competing uses. They were a major (even the only) source of household energy (inefficiently burned in open fires or simple stoves) and a common construction material (particularly in straw-clay mixtures for bricks and roof thatching), as well as the main source of bedding for domestic animals and indispensable feed for domesticated ruminants. Recycling of crop residues was an important source of soil organic matter and plant nutrients, and straw was also an excellent substrate for cultivating mushrooms and a raw material for making paper. While some of these uses have virtually disappeared (thatching, fuel) in modern settings, some remain as important as ever, none more so than feeding ruminants.

Unlike other domesticated animals, ruminants can digest cellulose because the microorganisms in their rumen produce the requisite enzymes; indeed, to maintain normal rumen activity, at least one-seventh of their normal dry-weight diet should be in cellulosic roughages (NRC 1996). Some residue feeding is still done as traditional stubble-grazing of harvested grain fields, but in all modern economies most feed straw is consumed as a part of chopped residue mixes (Bath et al. 1997). In modern feeding straw is often made more palatable and more nutritious by alkali treatment (soaking or spraying with 1.5%–2% NaOH solution) and nitrogen enrichment with ammonia or urea (Schiere and de Wit 1995). The combination of treated straw and protein-rich food processing wastes (oil cakes) can replace hay or silage and make it possible to feed beef or dairy cattle without claiming farmland for concentrate and roughage crops. This option has the highest appeal in land-scarce Asian countries.

The excellent water absorption capacity of cereal straw (equal to two to three times their mass for uncut residues, at least three times for chopped material) makes it also the preferred material for animal bedding. Besides keeping the confined animals clean and comfortable, bedding residues make manures easier to handle and limit the leaching loss of absorbed nutrients; where straw is plentiful, about 250 kg are used for each 1 t of excrement. Recycling of this nitrogen-rich material, often after composting, remains a key component of organic agriculture (Kristiansen, Taji, and Reganold 2006). An even more important use of crop residues is achieved through direct recycling, in which cereal stubble, chopped corn stover, and legume and tuber vines are left on the ground to protect against wind and water erosion and are incorporated into soil, maintaining its organic content and increasing its water retention capacity.

Even in some modern settings straw continues to be used as fuel and construction material. Denmark has been a leader in straw burning: about 1.4 Mt of wheat straw (of the total annual output of about 6 Mt) are burned annually in small household ovens and for centralized district heating and electricity generation (Stenkjaer 2009). Clean shredded straw (most often wheat or rice) is used to make boards by compressing the material (it can also be fused with internal resins) and bonding it to external paper or sandwiching it between laminated strand lumber. Stramit International and Agriboard Industries are the leading makers. There is also a fringe interest in straw-bale buildings as a form of frugal architecture (Steen et al. 1994; Strawbale 2010).

Straw pulps are similar to short hardwood fibers, but their low density, the high collection and transportation costs of crop residues, and their relatively high silica content make more tree-free paper unlikely. Another marginal use of crop residues has been as a feedstock in biogas generation (Smil 1993; Demirbaş 2009), and residual phytomass can be also tapped as feedstocks for extracting organic compounds such as furfural (used as a selective solvent in crude oil refining and in bonded phenolic products) from pentosan-rich corn cobs, rice hulls, or sugarcane bagasse, or directly from corn stover and cereal straws (Staniforth 1979; Di Blasi, Branca, and Galgano 2010). Yet another common use of cereal straws is in mushroom cultivation. Wheat straw, often mixed with horse manure and hay, is used for the cultivation of *Agaricus bisporus* (white button mushroom), while *Volvariella volvacea* (straw mushroom) is grown in wetted rice straw or straw mixed with cotton waste (de Carvalho, Sales-Campos, and de Andrade 2010).

The availability of crop residues is usually calculated by using residue:crop (that is, most often, straw:grain, or S:G) multipliers. These values are usually given as ratios of the fresh weight of the harvested plant parts to the dry matter of the crop's residual mass, and for modern cereal cultivars they are typically around 1.0, ranging mostly between 0.7 and 1.5. Residue yields can be also determined as the inverse value of a crop's harvest index (the ratio of crop yield to the crop's total aboveground phytomass), but in this case the total is somewhat larger than the one obtained from the S:G ratio because the latter measure does not include stubble. Harvest indices of major crops have been increasing, which means that S:G ratios have been declining (for more details, see the next chapter).

Forest Phytomass

When the harvests of forest phytomass are considered from a global perspective, the key distinction is between wood to be used as a construction material (roundwood in forestry statistics, which becomes sawn wood, commonly timber or lumber in everyday parlance) or pulped to make paper and woody biomass collected and cut for fuel (fuelwood or firewood), whose oven-dry energy density is about 19 MJ/kg (Francescato et al. 2008). This distinction does not apply in affluent countries, where all but a small amount of fuelwood is a part of the roundwood harvest that is conducted by standard mechanized commercial methods of tree felling and transportation to processing facilities.

But the distinction is important in all of those low-income countries of Asia, Africa, and Latin America where wood (or charcoal) is either the dominant or the single most important source of energy for household cooking (and heating) as well as for many local artisanal manufactures: in all these cases the term woody phytomass is much more fitting than forest phytomass. Most of it is collected by household members, usually by women and children, who bring home head loads of up to 25 kg, some of it bought in local markets; this wood has not been harvested by mechanized felling, and much of it does not come from forests but from any woody phytomass, including bushes, tree groves, and commercial plantations (rubber, coconut), as well as roadside and backyard trees. These sources may account for more than half of all wood burned by rural households. Surveys by the Regional

Wood Energy Development Programme in Asia found that during the late 1990s, nonforest fuelwood provided more than 80% of all woody phytomass burned in Bangladesh, Pakistan, and Sri Lanka (RWEDP 1997).

The second and much more obvious distinction is that any part of a tree can be burned by households or local industries, including nonwoody phytomass (dry leaves, palm fronds), bark, small twigs (suitable for kindling), and roots, whereas the phytomass that is commercially harvested and traded as roundwood is limited to stems. The standard American definition of roundwood restricts it to stems with diameter greater than 12.7 cm (5 inches) at breast height above a 30 cm stump and up to 10 cm diameter outside bark at the top. The same definition is also used for growing stock, and in both cases it excludes live, sound, 10 cm diameter trees of poor form (these cull trees typically amount to 5%–6% of the total live tree volume). As a result, a great deal of phytomass either remains unharvested (roots and stumps, which will gradually decompose or resprout) or is trimmed from the bole: in conventional chain-saw felling, all branches and the treetop are cut off and left to decay on the forest floor (usually where it was produced, sometimes collected in windrows).

The bole used to be bucked into 1.2–2.5 m lengths for easier removal (dragged to the nearest road by horse teams) but is now almost always removed uncut, mostly by draggers and skyline assemblies, while particularly valuable logs may be lifted by helicopters. Bark is stripped from timber only at a processing site and is often used, together with sawdust, to generate heat and electricity. Depending on species and age the bole is no more than 50%–60% of tree mass, while the typical shares for branches and leaves or needles and for the stump and roots are, respectively, about a quarter and a fifth of the total phytomass in temperate and boreal forests.

While the quantification of crop harvests is done in mass terms (t/ha), foresters have traditionally preferred to express the yields in volumes (m³/ha). This requires knowledge of specific conversion factors, and, because of a rather wide range of specific wood gravities, using a single approximate value (a practice common for first-order estimates) can cause a substantial error (USDA 2010). The densest temperate zone hardwood (white oak) has an absolutely dry density of 0.77 g/cm³, other temperate hardwoods (birches, maples, oaks, walnuts) average mostly between 0.6 and 0.7 g/cm³, and the commercially most important softwoods (spruces, firs, redwoods) have densities between 0.45 and 0.5 g/cm³, while some pines and poplars have densities as low as 0.35 g/cm³; both extremes are tropical woods, 0.11 g/cm³ for balsa (*Ochroma pyramidale*) and up to 1.3 g/cm³ for *lignum vitae* (genus *Guaiacum*).

Another complication in converting wood to a common mass denominator is its highly variable water content. The water content of freshly felled roundwood is almost never less than 40% and can be well above 100% (when measured on an oven-dry basis). Poor people who collect their own fuelwood do so because they obviously do not have any tools (saws or axes) to cut down trees, but there is a major advantage to this limitation: fallen or dead branches and twigs are much drier than green stemwood and can be used immediately for cooking or heating, while wood with a water content above 67% will not even ignite. Green stemwood must be split into smaller pieces and air-dried, often for long periods of time, before it reaches its minimum equilibrium moisture of no less than 20%–25%. Moreover, twigs and branches are easily carried bundled as head loads.

A simple formula is used to convert volumes of freshly cut roundwood to oven-dry weight: oven-dry weight = specific density × 1,000 (specific density of m³ of water) × (1 + wood moisture/100), and tables list these values according to wood's water content and specific gravity (Glass and Zelinka 2010). Green heartwood has a water content as low as 35% in some pine and fir species and as high as 80%–90% in some hardwoods, but most values will be between 40% and 60% by weight. With a moisture content of about 40% and a specific gravity of 0.4, a freshly felled cubic meter of coniferous wood will have an oven-dry weight of about 560 kg, and 1 m³ of hardwood with a specific gravity of 0.6 and a moisture content of 60% will have a dry weight of 960 kg. The mean of these two common combinations is 760 kg/m³, while wood with a specific gravity of 0.5 and a moisture content of 50% has an almost identical mass of 750 kg/m³. I will use this value later in the book when converting roundwood volume to mass.

Inevitable errors in converting fresh volumes of commercial roundwood to the common denominator of absolutely dry phytomass are further compounded by uncertainties regarding the actually harvested volume, as illegal logging has been common in several major wood-producing countries, including Russia, Brazil, China, Indonesia, Malaysia, Congo, Gabon, and Ghana (Global Timber 2011). The challenge is even greater when trying to account for wood used by low-income households. No statistics are kept regarding the volumes or weights of woody phytomass collected by families in rural areas. Short-term studies of actual collecting activities have to reckon with different sizes of typical head loads or animal loads, and if fuelwood is stacked near a house for future use, then (depending on the species, size of wood pieces, and the tightness of stacking) 1 m³ can have a mass of 310–460 kg.

To get fairly accurate accounts of the actual standing forest phytomass is even more challenging. Such accounts are available only for those affluent countries that conduct periodic forest inventories and refer to standing forest phytomass only in its narrowest definition, the growing stock of timber. One standard way to ascertain this resource is by measuring first the diameter of growing trees (all of them in a small area, otherwise by sampling) at breast height (1.4 m above the ground), using either a special tape calibrated in π (giving directly the diameter reading) or calipers; the next step is to find the average height of major species that make up the forest and then to use the two measures in formulas designed to express the roundwood volume (West 2009).

Foresters quantify the growing stock of timber in volume measures; in temperate regions it can go as high as 90–100 m³/ha, but the rates are considerably lower when averaged over larger areas. The U.S. data show a growing stock volume of less than 40 m³/ha in both the northern and southern regions and more than 60 m³/ha in the West, with the annual increment averaging almost 3% of the standing stock and with tree mortality amounting to about 0.7%/year (USDA 2001). Species-rich tropical rain forests with closed canopies have a much higher growing stock, up to about 180 m³/ha. Broader measures of forest resources—that is, data on all aboveground phytomass and the all-encompassing total of all above- and belowground tissues on a national or a biome scale—remain a matter of uncertain estimates.

5

Zoomass Harvests

Terrestrial trophic pyramids are invariably broadly based, with phytomass (primary producers) being commonly 20 times more abundant than the mass of herbivores (primary consumers), and the zoomass in the highest trophic level (this may be the third level in the simplest ecosystems, and only a few terrestrial communities go beyond the fifth level) may be equal to just 0.001% of the phytomass. These realities limit the mass of heterotrophs that could be profitably hunted (or collected): many small herbivores (insects, rodents) are not worth the effort, while the largest carnivores are too scarce and too dangerous to hunt. The situation is reversed in the ocean as phytoplankton's short life span limits the total amount of standing autotrophic biomass and results in inverted trophic pyramids, with the aggregate heterotrophic biomass being commonly twice, and up to four times, as large as the biomass of primary producers and with some fish and marine mammals being fairly massive.

There are no universally valid rules as far as energy transfers between successive trophic levels are concerned. Lindeman's (1942) pioneering work on conversion efficiencies in Wisconsin's Lake Mendota found that the primary consumers incorporated nearly 9%, secondary consumers a bit over 5%, and tertiary consumers about 13% of all energy that was available at the previous trophic level—and this led to the often cited conclusion that typical energy transfers between successive trophic levels (ecological efficiencies) average about 10%. Subsequent research disproved this simplistic conclusion, as studies in aquatic ecosystems showed transfer efficiencies as high as 40% and as low as 2%, but the 10% rule is still frequently invoked.

Endothermy exacts its high metabolic cost, as only 1%–3% of ingested phytomass is converted to new zoomass by birds and mammals; in contrast, many ectotherms invest roughly an order of magnitude more of their energy intake (10%–25%)

In 1557, Pieter van der Heyden engraved Pieter Brueghel the Elder's illustration of the proverb "big fish eat little fish." This picaresque image illustrates both a long tradition of harvesting marine zoomass and the increasing sizes of species at the top of the trophic pyramid.

to produce new zoomass (Golley 1968; Humphreys 1979; Currie and Fritz 1993). With respective means at 2% and 15%, an even split of the consumed phytomass (assuming 10% of 55 Gt C of the global net primary production, or NPP) would result in net herbivore production of some 470 Mt C/year—while Whittaker and Likens (1973) calculated what they considered to be a high estimate of about 370 Mt C/year. Global meat production (carcass weight) in 2010 was about 290 Mt; reverse application of typical live weight/dressed carcass factors (Smil 2000) yields at least 450 Mt of slaughtered live weight, or at least 90 Mt C. The annual slaughter of domestic animals would thus account for 20–25% of net secondary production.

Animal domestication in general, and modern mass-scale commercial breeding and production in particular, pose a different set of energetic challenges. Little can be done about changing the metabolic efficiencies of individual species (for example, pigs are inherently more efficient converters of feed than cattle), but higher productivities can be achieved by supplying adequate amounts of optimally selected feeds. As a result, modern varieties of meat-producing heterotrophs mature faster and reach higher slaughter weights sooner than the traditional breeds and require less feed per unit of edible mass than their unimproved predecessors. But these improvements could not prevent a massive increase in the total biomass required to feed the domesticated animals: because of enormous increases in their numbers, better feeding efficiencies could only reduce the rate of increase in the overall claim.

Hunting and Domestication

Hunting opportunities are fundamentally determined by the efficiency of energy transfer between trophic levels. Shares of the NPP that are actually consumed by herbivores are only 1%–2% in deciduous temperate forests, may surpass 25% in temperate meadows and wetlands, and reach maxima of 50%–60% in rich tropical grasslands; in the ocean they can peak at more than 95% in some patches of phytoplankton (Crawley 1983; Valiela 1984; Chapin, Matson, and Mooney 2002). Terrestrial rates between 5% and 10% are perhaps most common, implying annual transfers of roughly 3–6 Gt C. But when the calculation is restricted to vertebrates, the primary targets of hunting, it is mostly just around 1% (around 0.5 Gt C). Only part of this ingested energy is digested: for herbivores (whose diet contains a high share of digestion-resistant phytomass), this transfer amounts generally to less than 30%, while for carnivores (eating high-lipid, high-protein tissues) it may surpass 90%. The next step is to see what share of digested energy is neither respired nor spent on reproduction and becomes available for growth of new tissues.

Ectotherms have an inherent advantage (many nonsocial insects channel more than 40% of ingested energy into growth), for endotherms have to spend a great deal of ingested energy on maintaining high core body temperature: only about 3% of ingested energy ends up in growth in large mammals, and for small mammals and birds the rate is even lower, 1–2% (Humphreys 1979). Multiplying the three rates—exploitation, assimilation, and production—yields overall trophic efficiencies that are invariably well below 10% for most hunted vertebrates, and mostly from a fraction of 1% to a few percent for the most commonly targeted herbivores. But this may be largely irrelevant as far as the human claims on wild zoomass were concerned. Herbivorous mammals and birds have been always the most important target of hunting, and Hairston, Smith, and Slobodkin (1960) showed that the numbers of herbivores are most often limited by predators rather than by the energy available for transfer and that only the predators are resource-limited. This reality put natural limits on the avian and mammalian zoomass available for hunting, and the size of hunted animals was a result of particular environmental and trophic outcomes.

Grasslands have mammalian herbivores ranging from small rodents to the largest terrestrial mammals, and the dominant ungulates are larger than their predators: killing one of those animals provides a large amount of meat. In contrast, in tropical rain forests most of the vertebrate grazers are small and folivorous (mostly monkeys) and hence present much less rewarding targets for hunts, and their numbers are ultimately controlled by larger predators (leopards, eagles). And, as already noted, in marine ecosystems many herbivores are much larger than the planktonic autotrophs, and secondary and tertiary consumers are larger still, presenting rewarding opportunities for killing and capture.

Not surprisingly, the most commonly hunted animals were those that combined fair size with relatively high productivity and high territorial density. Wild pigs (adult body mass of up to 90 kg), deer, and antelopes (the smallest animals at less than 25 kg, the large male eland well over 600 kg) are perfect examples of such species. Before the domestication of dogs, persistence hunting (made possible by the superior endurance running and cooling abilities of our species) was an efficient way to bring down these animals (Liebenberg 2006). With simple tools (spears, bows and arrows) they could be hunted with much less effort. In contrast, hunting of the largest herbivores required cooperative effort and often some elaborate strategies: perhaps most notably, herds of buffaloes were chased over cliffs, their bodies smashed by the fall. Head-Smashed-In buffalo jump near Fort McLeod, Alberta, is one of UNESCO's world heritage sites.

In all affluent countries, hunting has been transformed from a quest for food to an often expensive pastime, but the killing of wild animals for food goes on in many countries in Asia, Africa, and Latin America. China's omnivory creates particular problems, as the traditional preference for exotic foods has been combined with rising disposable incomes, resulting in a growing demand for shark fins, sea cucumbers, turtles, and even tigers. The African bush meat trade is also very strong, and the chunks of fresh or roasted meat offered for sale are not only from the abundant rodents and small grazers but also from gorillas and chimpanzees.

Since animals selected for domestication had to combine a fair body size with a relatively high productivity, the third important factor was their demand for feed, in respect to both quantity and quality. No large carnivorous animal could be reared profitably for meat production. Omnivorous pigs had yet another advantage: their basal metabolic rate is slower than that of sheep, goats, or cattle, and hence they convert feed into muscle and fat with higher efficiency. On the other hand, ruminants can digest cellulosic tissue that is indigestible by other animals and hence are able to survive entirely by grazing. But in places with longer winters and plenty of snow this advantage is lost, and much time and effort go into harvesting, drying, and storing hay needed for the months of winter feeding when the animals remain penned in.

Adults of the smallest mammals that were domesticated for meat—guinea pigs and rabbits—weigh, respectively, about 1 kg and at least 2 (even more than 3) kg per adult. The largest ones—cattle and water buffaloes, domesticated initially for draft and for milking—have typical slaughter weights as little as 250–300 kg in low-income countries but more than 500 kg with optimized feeding, with maximum bull weights approaching or even slightly surpassing 1 t. Similarly, horses domesticated for transport and draft (but also eaten in many countries of the Atlantic and Central Europe, Latin America, and Japan) have body masses ranging from less than 400 kg to more than 700 kg for the heaviest kinds. The range for domesticated birds is narrower, from pigeons (adult body weight less than 500 g) to turkeys (adults in the wild up to about 10 kg, twice as much for penned-in nonflying breeds).

Feed supply constrained the availability of animal foods in all traditional societies. While some dairy cultures sustained a modest but fairly widespread consumption of dairy products, only the richest strata of societies in the preindustrial world enjoyed meat dishes. Typical diets were nearly meatless even in the industrializing Europe of the early nineteenth century, and overwhelmingly vegetarian diets were the norm in rural China and India even during the 1980s. Only the combination of rising food crop yields (freeing more land for feed crops), higher disposable incomes, and a truly worldwide trade in animal foods has brought modern meaty diets to all of the world's affluent countries, and now also to such rapid modernizers as South Korea, China, and Brazil.

Fishing and Marine Mammals

The inverted trophic structure of marine ecosystems makes them a poor environment for phytomass harvests: primary producers are too small, too short-lived, and too dispersed and their biomass fluctuates too much (and too unpredictably) for them to be relied on as a source of food or feed. The only exceptions are algal macrophyta growing in the intertidal zone, particularly the seaweeds belonging to the genus *Porphyra* (Japanese *nori*, originally collected in the wild, now cultivated on nets) and the genus *Laminaria* (kelp, Japanese *konbu*, a large-leafed brownish plant from cold coastal waters with a naturally very high content of monosodium glutamate). These and a few other seaweeds have traditionally been collected in a limited number of countries in East and Southeast Asia, Atlantic Europe, Canada,

Hawaii, and New Zealand, but only in Japan have they become a notable part of everyday diet (Smil and Kobayashi 2012).

In contrast, coastal hunting and the collection of marine species (fish, mollusks, other invertebrates, and fatty marine mammals) provided the energetic basis for some of the first permanent human settlements: species-rich coastal waters, regular migrations of some anadromous fishes (particularly salmon and eels) and whales, and the possibility of collecting invertebrates inhabiting the intertidal zone (mussels, clams) made it possible to secure an exceptionally nutritious diet without changing abodes. Long-distance (open ocean) fishing was a relatively late development, pioneered in Atlantic Europe at the beginning of the early modern era, when improved ship designs made it possible to venture far offshore in search of fish or whales.

Harvests of marine heterotrophic zoomass in both coastal and open ocean waters spanned an enormous size spectrum and a large variety of species, traditionally ranging from whitebait (young fish of many species whose length is generally less than 5 cm) to sperm whales, and eventually (after the introduction of steam-powered whaling ships and large harpoons during the 1860s) even the largest living mammal, the blue whale. Krill is now the smallest marine heterotrophic species that is harvested commercially on a large scale by netting in the Antarctic waters. Its capture began in 1961, when the Soviet ships caught just 4 t of it, but by early 1980 the harvest had approached and one year even surpassed 500,000 t before it fell to around 100,000 t a year and then recovered a bit (FAO 2011a).

Fish species that are most commonly caught by nets (purse seines, used to encircle fish schools and then gather the fish by closing the netting's bottom, or wall-like vertically hanging gill nets in which swimming fish get entangled) range from sardines to cod, salmon, and smaller tuna; different kinds of nets are also used to catch shrimp and the two most commonly sought marine invertebrates, octopus and squid. Commercial catches of the largest deepwater carnivorous fish (bluefin tuna and swordfish) rely on lines many kilometers long (the longest ones are up to 80 km) carrying smaller lines baited at regular intervals with hooks. The capture of ocean heterotrophs goes under the collective term of fishing, but a significant share of this effort involves harvests of invertebrate species. In global terms, the mass of crustaceans (dominated by shrimp and crabs) and mollusks (mostly mussels, clams, and oysters) has recently added up to more than 15% of all capture in ocean waters, and in value terms their share has surpassed 20% of the total (FAO 2011a).

Perhaps the most notable fact regarding the modern fishery is that its gradual expansion has made it truly global. Long-distance fishery became more common between the world wars, thanks to the diffusion of inexpensive diesel motors. It expanded rapidly after the early 1950s and reached a plateau by the late 1980s. No fishing ground is now too remote, no latitude's weather is too forbidding to prevent the large-scale exploitation of almost every available fish species, including the diadromous salmons, smelt, and eels, and pelagic (those living in the water column, ranging from sardines and still abundant herring to the most endangered tunas) and demersal fish (living on or near the ocean bottom: cod, haddock, halibut).

This exploitation has led to an increasing investment in marine aquaculture. Raising freshwater fish in ponds, lakes, and reservoirs has a long history in East Asia and parts of Europe, and it continues to dominate the worldwide aquaculture, accounting for more than half of the annual fish output. Marine aquaculture was traditionally a much less important activity, concentrated on cultivating seaweeds and culturing oysters and clams: its foremost practitioners included Japan and Hawaii. In terms of fresh weight, seaweeds still dominate the global aquacultural output, but the practice now includes the production of more than 50 species of fish and marine invertebrates. Leading products in terms of zoomass and value are, respectively, mollusks (mainly the Pacific oyster, *Crassostrea gigas*) and crustaceans (shrimps and prawns), and diadromous species (salmon) top the marine fish list. Even such large carnivores as bluefin tuna are now aquacultured (caught young in the wild and fed to slaughter weight in cages). In turn, this large-scale production of carnivores has led to a rising demand for other marine zoomass (above all squid and sardines) for feeding.

The only notable retreat from a mass killing of marine mammals has been the worldwide moratorium on commercial whale hunting imposed in 1982 by the International Whaling Commission with a starting date in 1986 (Kalland and Moeran 1992). This ban has worked, but not perfectly, as several countries (Japan, Norway, Iceland) continue to conduct small-scale whale hunts, in Japan's case under the dubious cover of "research" whaling that kills mostly minke, as well as a few sei and sperm whales, and has been producing annually between 1,000 and 2,000 t of whale meat (Ishihara and Yoshii 2003; ICR 2010).

6

Land Cover and Productivity Changes

For a few million years of their early evolution, hominins made a limited claim on the biosphere, as they foraged for food and ate it raw. This began to change with the controlled use of fire for cooking and defense against predatory animals. Inevitably, this led to accidental fires, whose unchecked progress added to the phytomass destruction that had been taking place naturally by fires ignited by lightning. Eventually, humans began to use fire for other than immediate food and security needs, above all to clear forested land for shifting cultivation and later to open up new areas for permanent cropping. Other important uses of fire were to suppress any encroachment of shrubs and trees on grasslands used for grazing domesticated herds, and to open up land for new settlements.

Ever since that time (this means for close to 10,000 years in some of the oldest continuously inhabited regions with domesticated animals), the large-scale impact of human activities on the biosphere has gone beyond the conversion of forests, grasslands, and wetlands to crop fields, beyond the expansion and management of grazing lands, and beyond deforestation driven by the demand for fuelwood, metallurgical charcoal, and timber. These diverse actions have caused many second-order changes, seen not only in the directly affected ecosystems but also in distant places, at regional and even global levels. Soil erosion (and the concomitant loss of accumulated nutrients and reduction of an eroded site's potential primary productivity) and the emission of greenhouse gases are two ubiquitous examples of such effects.

Fire and Land Conversions

Fire has always been the least labor-intensive way to clear a forest in the absence of good tools (the axes and saws needed to cut down large trees) and strong draft animals (to remove the logs). Preindustrial societies could not control deliberately

Forest fires, both wild and deliberately set, have been one of the most important factors in changing the global stocks of phytomass. This Envirosat image shows plumes of massive forest and peat fires in central Russia east of Moscow during the unusually dry summer of 2010. A high-resolution image can be downloaded at http://esamultimedia.esa.int/images/ EarthObservation/images_of_the_week/forest_fires_MoscowMER_FR_20100729 _43977.jpg. Photograph courtesy of European Space Agency, Paris.

set fires, and hence the results of burning varied from a fairly complete combustion to an only partially burned forest. Intensive fires burning moderately dense growth of generally young trees consumed more than 80% of all standing aboveground phytomass and left behind a relatively level surface (covered by ash, a concentrated source of mineral nutrients to support the newly planted crops) that could be prepared for planting by hoeing or even by using draft animals. In contrast, fires that swept rapidly through a mature growth would obliterate the understory plants and smaller trees but leave some large trunks nearly intact, or heavily seared but still standing.

The resulting rough surface with large stumps and charred logs could be cultivated only by hoeing and irregular planting, mostly of tuber crops. These realities make it very difficult to partition phytomass carbon lost in forest fires, particularly in the tropics: a major share is obviously emitted as CO_2, but inefficient combustion and smoldering fires also produce CO. Moreover, under some conditions a significant share of phytomass carbon turns into biochar (charcoal produced at low temperature), a material that can remain intact in soils for several hundred to a few thousand years, providing a long-term sequestration of the element while improving soil quality and enhancing primary productivity. Anthropogenic Amazonian *terra preta* (black earth) soils, created by surprisingly dense late prehistoric (600–1200 CE) settlements scattered throughout the region (Heckenberger et al. 2003), are the best-known example of this counterintuitive outcome when wood burning results, at least partially, in the prolonged storage of carbon.

Nobody in a modern economy is now converting forestland to cropland by burning, and shifting cultivation relying on periodic burning today produces only a very small share of the worldwide food output. But during the last two decades of the twentieth century satellite monitoring illustrated the extent of fires used to clear forested land to create new fields and pastures and for transportation, settlements, and industrial uses throughout Latin America and Southeast Asia, particularly in the Brazilian Amazon and parts of Indonesia (Levine 1991; Bowman et al. 2009; Lauk and Erb 2009). Satellite images also show fires set to stimulate pasture regeneration during the late summer months across the Sahel, from Senegal to Eritrea, and in an even broader swath of land extending across the southern part of the continent from Angola to Mozambique (Barbosa, Stroppiana, and Grégoire 1999; Roberts, Wooster, and Lagoudakis 2009). Grasslands, too, are often burned in Central America, parts of India, and Southeast Asia.

As decades of burning were reducing the area of tropical rain forests, an opposite trend in land-cover change began to accelerate in many affluent countries in the temperate zone. Higher crop yields and the consolidation of farming in the most suitable regions led to a widespread abandonment of marginal cropland and its gradual reforestation. This trend has been most notable in eastern North America and in more than half of all European countries. The conversion of suburban agricultural land to residential, commercial, and industrial uses and the growth of transportation corridors have been affecting primary production all around the world, particularly in the rapidly urbanizing societies of Asia and Latin America. But not everything is lost: lawns and city parks maintain, and in patches can even surpass, previous primary productivity.

A curiously neglected consequence of unplanned or poorly planned urban and industrial expansion has been the fragmentation of land use, a worldwide phenomenon with a substantial aggregate impact on primary productivity. This expansion often creates isolated (and often poorly accessible) patches of farmland among new buildings and roads. Some of those patches continue to be cultivated for a while, but most are soon abandoned to weeds. Other land near industrial plants or mines has to be abandoned because of high levels of soil contamination. A new industry devoted to the restoration of this land (often relying on bioremediation using plants whose metabolism concentrates soil pollutants in roots or leaves or enhances beneficial microbial activity) has been successful in returning this land to plant cover.

Effects on Biomass Productivity

Deforestation, grassland burning, and permanent cropping have had a number of obvious environmental consequences for the affected ecosystems. In most cases, the primary productivity of new plant cover has been lower than that of the displaced natural ecosystems. Where high-yielding crops displaced short grasses the productivity may be higher, but the nature of standing phytomass has changed profoundly: unlike crops, natural grasslands have most of their phytomass underground to boost water storage and minimize soil erosion. The consequences have included changes in the micro- and mesoclimate resulting from alterations in the albedo (the share of incoming solar radiation reflected by surfaces), evapotranspiration, and soil moisture retention; these alterations in turn have affected primary productivity, either by changing its overall rate or by altering the specific composition of plant communities.

Where the conversions have resulted in higher soil erosion rates—burning down a closed-canopy forest could increase the typical erosion rate, particularly on a sloping land and in a rainy climate, by several orders of magnitude—the effects have extended further afield. Eroded soil, carrying particulate organic matter and dissolved nitrogen, is deposited downstream in river channels and can eventually change the stream course or bring more extensive flooding, while periodic flooding in lowlands can accelerate the deposition of rich alluvial soil suitable for intensive farming. A common assumption was that these changes affected productivity only on a local or regional scale and that they could not have a larger, even global, impact.

That assumption was challenged by Ruddiman (2005), who argued that the anthropogenic influences on global climate (and hence on primary productivity) began with the adoption of shifting agriculture by Neolithic cultivators and intensified with subsequent permanent cropping and animal husbandry. These effects resulted not only from CO_2 emissions from land-cover changes but also from the release of CH_4 from flooded soils and ruminant herds and N_2O emissions from manures. Ruddiman (2005) concluded that despite their relatively small numbers, the first farmers eventually hijacked the entire global climate system, and that humans have remained in control ever since. His hyperbolic statement ("in control"?) focuses attention on the long history of the second-order effects of human actions on the biosphere's productivity.

These effects have been magnified since the middle of the nineteenth century, first with large-scale industrialization, energized by the combustion of coal, then with the rapid post-1950 global economic growth, increasing dependence on hydrocarbons, a nearly universally high (Africa being the only exception) reliance on synthetic nitrogenous fertilizers, increasing acreage devoted to paddy fields, and expanded dairy and beef production. Other anthropogenic emissions affecting primary productivity include airborne pollutants (in the form of photochemical smog, black carbon, and sulfur and nitrogen oxides) that can affect sensitive crops as well as natural plant communities.

The accidental or deliberate introduction of pathogens can have devastating effects on plant productivity, with some impacts extending to a large regional or even continental scale. After its introduction on Asian nursery stock at the beginning of the twentieth century, chestnut blight fungus (*Cryphonectria parasitica*) destroyed the American chestnut, formerly a dominant species of the North American continent's eastern forests (Freinkel 2007). Another Asian pathogen, an ascomycete fungus of *Ophiostoma* genus spread by bark beetles, has devastated first European and then American elms (*Ulmus americana*). The European infestation began in 1910, the first affected trees were found in the United States in 1928, and by 1981 the slow westward spread of the fungus had reached Saskatchewan, leaving Alberta and British Columbia as the only sizable unaffected areas (Hubbes 1999).

The latest dangerous accidental Asian invader is a small sap-sucking insect (*Adelges tsugae*, woolly adelgid) that threatens the very survival of the two eastern North American hemlock species, *Tsuga canadensis* and *Tsuga caroliniana* (Nuckolls et al. 2009). A reverse infestation is now afflicting Chinese forests as *Dendroctonus valens*, the red turpentine beetle native to North America, where it colonizes dead and dying trees, is destroying millions of growing pines (Youngsteadt 2012). *Phytophthora ramorum*, an aggressive and unpredictably advancing fungus, has caused sudden oak death in California and Oregon and since 2003 has been destroying larch plantations in Britain (Brasier and Webber 2010).

The deliberate introduction and subsequent diffusion of invasive plant species can not only alter primary productivity, it can also change the very appearance of affected ecosystems: kudzu vine (*Pueraria lobata*) is perhaps the most disturbing example of this undesirable effect. This prolific vine was brought to the United States from Japan in 1876 as an ornamental and forage plant, and during the 1930s its planting was promoted by the Soil Conservation Service to reduce soil erosion on farms. The vine is now an uncontrollable invader (often entirely smothering even large trees) throughout the U.S. Southeast, from east Texas to Virginia (Webster, Jenkins, and Jose 2006).

II
History of the Harvests: From Foraging to Globalization

Iulius, Augustus, nec non et Iunius Aestas ·   AESTAS Adolet Fontis imago))   Frugiferas aruis fert Aestas torrida meßeis ·

In 1570 Pieter van der Heyden engraved Pieter Brueghel the Elder's *Summer*, an exemplification of millennia of traditional Old World agriculture dominated by the cultivation of cereals.

The predecessor species belonging to our genus—starting with *Homo habilis,* who appeared nearly 2.5 million years ago—spent all of their evolution as simple heterotrophs. Our species, *Homo sapiens,* has spent no less than 95% of its evolution (assuming it evolved by about 200,000 years ago) in a similarly simple foraging mode. All of these hominins gathered edible phytomass (tubers, stalks, leaves, fruits, grains, nuts), collected small heterotrophs (mushrooms, insects, small invertebrates, mollusks), hunted a variety of animals (mainly herbivorous species), and caught fish and aquatic mammals. Given the length of these experiences and the variety of environments they eventually occupied, it is not surprising that humans developed a remarkable range of food acquisition strategies, from snaring animals to spear fishing to elaborate group hunts to using other animals as lures or actual hunters (cormorants, birds of prey, dogs) as a way to secure a large variety of foods, from truffles to fish.

Small numbers of prehistoric foragers and their low population densities (some maritime societies were the only exception) would indicate limited harvests of the biosphere's productivity. Although all prehistoric population estimates are highly uncertain, the totals may have been no higher than 125,000 people half a million years ago and as low as 15,000 (or perhaps even less than 10,000) following the population bottleneck caused by the Toba mega-eruption 74,000 years ago (Ambrose 1998; Harpending et al. 1998; Hawks et al. 2000). Subsequent expansion brought the global total to no more than a few million people before the emergence of the first sedentary agricultural societies some 10,000 years ago.

Low population densities were the case in most places for most of the time, though prehistoric hunters have been also credited with being the prime agents of the late Paleolithic extinction of megaherbivores in North America, Eurasia, and Australia, a claim I examine in some detail. But it was the hominin use of fire, rather than foraging for food or the hunting of large herbivores, that was responsible for the most extensive preagricultural human impact on the biosphere. The epochal shift from foraging to farming was a complex and protracted process: there was no Neolithic "agricultural revolution." In many places the process ceased unfolding in its earliest stages and for millennia had not progressed beyond the two extensive food production ways of pastoralism and shifting cultivation.

And, contrary to a common claim, subsistence foraging was not quickly discarded because of agriculture. The emergence of sedentary farming was characterized by the prolonged coexistence of cropping and foraging. Foraging on land (ranging from mushroom picking to the seasonal hunting of mammals and birds) has never disappeared, even in the most affluent societies, and it continues to play an important role in many low-income societies, particularly in Africa. And modern societies have been engaged in ocean foraging—dominated by commercial catches of scores of fish species, as well as marine invertebrates and large mammals—with unprecedented intensity as formerly low-volume and largely coastal fishing practices were transformed into mass-scale and truly planetwide hunts.

Agricultural practices and the domestication of animals evolved independently in several regions and were followed by the gradual intensification of traditional cultivation that could not prevent a relatively high frequency of malnutrition and recurrent famines. Even during the early modern era, roughly the sixteenth to the eighteenth centuries, neither was uncommon in parts of Europe and China, economically the most advanced regions of the world. Some traditional agricultures achieved remarkably high productivities and were able to support unprecedented population densities; the three Asian regions come first to mind: China's Guangdong province, Japan's Kantō Plain, and Java.

But truly revolutionary production gains came only during the twentieth century with the massive deployment of intensive inputs that began with the introduction of synthetic (nitrogenous and phosphatic) fertilizers and field machinery powered by internal combustion engines and intensified after World War II with the adoption of new cultivars, the application of herbicides and pesticides, and a wider use of various forms of irrigation. Modern agriculture concentrated the impacts associated with food production in areas that satisfied at least some, if not all, preconditions for annual or permanent cropping, and higher yields and improved productivities in animal husbandry have reduced the amount of land, water, or nitrogen needed to produce adequate diets: the great post-1850 agricultural revolution drastically cut the amount of land needed to feed an average person.

But the global population growth began to accelerate after 1850, and surged after 1950. The increase was driven not only by better nutrition but also by epochal improvements in health care and the overall standard of living produced by a new industrial-urban civilization energized by fossil fuels, using new mechanical prime movers and turning out massive amounts of affordable material possessions, thanks to better organized manufacturing. As a result, crop cultivation now claims a larger share of ice-free surface than at any time in history, and the effect of rising population numbers has been potentiated by a worldwide dietary transition whose main feature has been a higher consumption of meat, eggs, and dairy products. Metabolic imperatives cause unavoidable and, in the case of large ruminants, also very high feed-to-meat conversion losses, and as a result, most of the crops produced (or imported) by affluent countries are not for human consumption but are feed crops for animals.

Long-term trends in biomass consumption for fuel and as raw material have resembled those of food production. Notable improvements in the typical efficiencies of wood and straw combustion (as well as in wood conversion to charcoal) and better ways of harvesting and using timber came only after millennia of no or very slow change, but these gains did not lower the overall need for woody biomass or crop residues. The opposite has been true: the combination of larger post-1850 populations (after 1950 particularly in low-income countries, where many people had no access to modern fuels) and a much higher per capita consumption of all raw materials resulted in an unprecedented demand for fuelwood, timber, and pulpwood.

7

The Evolution of Foraging

A reliable quantification of the resulting impact on the biosphere is impossible, as we lack any realistic assessments of total populations engaged in specific foraging activities, but there can be no doubt that in parts of several biomes (in the richest tropical rain forests and in boreal forests), it remained marginal for millennia. The relative richness of the coastal aquatic zoomass meant that some maritime regions were among the most intensively exploited environments. Cold and hot deserts and their neighboring transitional zones to more vegetated environments were at the other end of the diversity spectrum: a paucity of both collectible plants and larger species whose hunting could supply enough meat for small bands of foragers kept the population densities very low, and with no possibility of converting such lands to permanent farming, these areas were also among the last places where foraging persisted into the twentieth century and could be studied by modern anthropologists.

That the continuous gathering of preferred plant species and the hunting of favorite large, slowly reproducing animals could exert notable local and regional effects on the exploited ecosystems is self-evident. But the idea that hunting alone— by relatively small populations equipped with simple wooden and stone weapons— could be the only, or at least the primary, reason for a relatively rapid extinction of large terrestrial mammals at the end of the Pleistocene remains controversial. While it is easy to advance arguments in favor of this interpretation and to construct theoretical models claiming to prove the case, it is another matter to produce convincing evidence and to explain the many realities that do not conform to that interpretation. In contrast, there is no doubt about the extensive and recurrent impact of deliberately or accidentally set fires on the biosphere's productivity.

The extent of foraging in traditional agricultural societies was highly time- and place-dependent: rewarding opportunities for hunting animals and collecting wild plants were rather rapidly exhausted on the treeless alluvial plains that became the

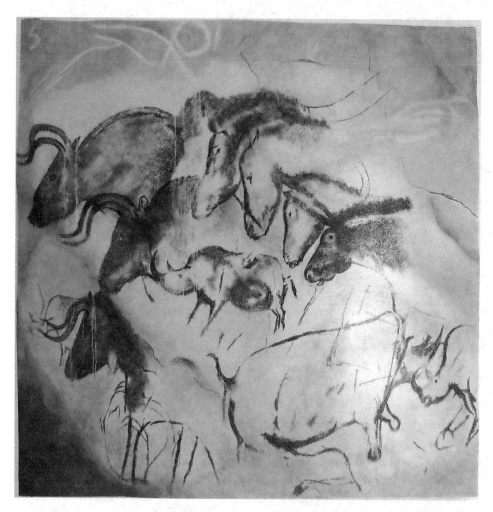

Marvelous paintings of horses, bison, and rhinos from the Chauvet Cave in southern France date from about 30,000 years ago and remind us that prehistoric foragers did much more than just collect edible and combustible phytomass. They also hunted animals, including many species of megafauna. This image, in the public domain, can be downloaded at http://upload.wikimedia.org/wikipedia/commons/d/d1/Chauvet_cave%2C_paintings.JPG.

productive cores of several ancient civilizations. In other environments, however, particularly those with mixtures of forested and cultivated areas, the seasonal pursuit of animals and plant collecting coexisted for millennia with annual cropping, and some of these practices are still alive (if on a scale that makes only a marginal contribution to the overall food supply) even in some of the world's most affluent postindustrial societies.

Plants, Meat, and Fire

Our evolutionary heritage is clear: all extant large primates (chimpanzees, gorillas, and orangutans) are omnivores whose diet is dominated by a variety of phytomass tissues but also contains some heterotrophic biomass, ranging from ants and termites (sometime cleverly collected by sticks or blade tools) to small invertebrates and mammals. Archaeological records and food choices observed among those foraging societies that survived into the twentieth century indicate a very large range of collected plant tissues, from dwarf willow leaves and the contents of caribou stomachs in the Arctic to a profusion of roots, tubers, leaves, stalks, seeds, nuts, fruits, and berries in the tropics.

Tropical gatherers collected commonly as many as 50–100 different species (Scudder 1976), and similarly large inventories of wild plants are still exploited by many African societies today. In Uganda's Bulamogi county, the list contains 105 species belonging to 77 genera (Tabuti, Dhillion, and Lye 2004), while the Luo of western Kenya gather 72 different vegetables (Ogoye-Ndegwa and Aagaard-Hansen 2003). Such diversification of plant food sources is to be expected given the combination of a relatively low energy density of most plant tissues (particularly leaves), difficult accessibility (whether tubers or fruits in high tree canopies), the seasonal occurrence of fruits, seeds, and nuts, and short- and long-term fluctuations in plant productivity.

But equally unsurprising is the fact that a much smaller number of species, often less than a handful, usually accounted for most of the food energy derived from plant collection: nuts and seeds (plant tissues with the highest energy density, commonly above 20 MJ/kg and as much as 25 MJ/kg) and starchy tubers (with moderately high energy density, up to about 4 MJ/kg, but bulky and easy to collect) were usually dominant. Prehistoric foraging for plants could, at worst, damage the exploited species in a very limited and temporary way: population densities were too low to have more extensive impact, and human foraging was of secondary importance when compared to the pressure exerted by large numbers of herbivores and seasonally extensive forest fires. This means that the most important human impact on preagricultural ecosystems came first because of the use of fire and later (when weapons and hunting tactics allowed it) from the hunting of megaherbivores.

The importance of fire in human evolution is easily exaggerated, but it cannot be denied. The date of the first controlled use of fire will always remain elusive: the long-term preservation of any convincing evidence in the open is highly unlikely, and evidence of the earliest fires inside an occupied cave was destroyed by generations of later use. The earliest date for an incontrovertibly proven controlled use of fire use has been receding: in the early 1990s Goudsblom's (1992) most likely timing was about 250,000 years ago, but by the early 2000s the date had been pushed as far back as 790,000 years ago (Goren-Inbar et al. 2004). The fossil record suggests that the consumption of some cooked food could go as far back as 1.9 million years ago, but unequivocal evidence for the widespread use of fire goes only to the Middle Paleolithic (300,000–200,000 years ago), the period when *Homo sapiens sapiens* displaced the European Neanderthals (Bar-Yosef 2002; Karkanas et al. 2007).

Not long afterward, people also began to use fire as an engineering tool: as early as 164,000 years ago early modern humans were heat treating stones to improve their flaking properties (Brown et al. 2009), and Mellars (2006) has suggested that the controlled burning of vegetation may have been done in South Africa as early as 55,000 years ago. But we cannot reconstruct any meaningful record of such controlled burning, and the heat treating of stones could not have translated into a major demand for wood. This leaves cooking, an activity that has always been seen as an important component of human evolution but that was elevated by Wrangham (2009) to having a "monstrous" effect on our ancestors.

Wrangham argues that cooking enormously increased the range of foods hominins could eat (undoubtedly true), as well as the food quality (a debatable proposition, as there is no easy way to compare easier digestion with the loss of various nutrients), and that the adoption of cooking resulted in many fundamental physical and behavioral changes ranging from smaller teeth and a less voluminous digestive tract to the necessity of defending stealable patches of accumulated food (which in turn promoted protective female-male bonds) and eventually led to complex socialization, sedentism, and "self-domestication."

Such claims and counterclaims are easy to make, but the impact of universally adopted frequent cooking on the biosphere is impossible to quantify beyond some plausible indications. All prehistoric cooking was done with simple open fires, but that still leaves a number of possibilities: food can be cooked directly on slow-burning pieces of wood, buried in hot embers, placed on hot rocks, suspended in or above flames on sticks, encased in a tough skin or a thin clay covering, placed in simple stone ovens, or cooked with hot stones added to leather pouches filled with water, and, of course, it can be consumed in states ranging from only lightly singed to much overcooked.

Cooking methods have very different energy conversion efficiencies, defined in this case as the quotient of the heat transferred to plant or animal tissues to cook them to a desired consistency and the heat content (chemical energy) of the phytomass needed to maintain a fire during the cooking period. Experiments have established actual efficiencies ranging from only a few percent to as much as 10%. A plausible set of deliberately generous assumptions (to set a likely high limit) indicates that the annual consumption of wood (or other phytomass) for cooking was 100–150 kg per capita (box 7.1).

Even if the outlined calculation errs by as much as two orders of magnitude, there can be no doubt that Paleolithic cooking fires had a negligible impact on the biosphere's phytomass stores. More important, even if we had near-perfect information about typical cooking practices and frequencies, we still could not quantify the amount of woody phytomass used in prehistoric campsites with any reasonable degree of accuracy because campfires were set and maintained not just for cooking but also for defense against large predatory animals, protection against insects, and the provision of heat and light, and many of those fires led to accidental grassland or forest fires.

Hominin consumption of the heterotrophic zoomass began with collecting insects, small invertebrates, and mollusks and scavenging the carcasses of animals (Domínguez-Rodrigo 2002). Larger animals could be killed only with larger weapons, and throwing spears date back to 400,000-380,000 years ago (Thieme 1997), predating the emergence of early modern humans, which is now dated to around 195,000–150,000 years ago (Trinkhaus 2005). The cooperative hunting of large animals (fallow deer, aurochs, horses, wild pigs, asses) and subsequent meat sharing is convincingly demonstrated by tool marks on animal bones found at Qesem Cave in Israel during the period 400,000–200,000 years ago (Stiner, Barkai, and Gopher 2009). In contrast, the first bows and arrows appear in the archaeological record only about 25,000 years ago

The antiquity of the human quest for meat is not at all surprising: decades of field studies of chimpanzees, our closest primate ancestors, have shown that not only the common chimpanzees (*Pan troglodytes*) but also the smaller bonobos (*Pan paniscus*) are fairly regular hunters and consumers of meat. Small monkeys (particularly the red colobus) are the species most often targeted by common chimpanzees,

Box 7.1
Open-Fire Cooking of Meat: Wood Consumption

Realistic assumptions needed to set plausible maxima of phytomass consumption for late Paleolithic cooking include the following: an average daily food energy intake of 10 MJ (that is, about 2,500 kcal, adequate for an adult and hence exaggerating the average food need for an entire population), the meat share of this consumption at 80% (or 8 MJ, too generous for most cases, particularly when the meat would have to come from smaller animals), the food energy density of a large animal carcass of 8–10 MJ/kg (typical for mammoths, but the rate is only about 5–6 MJ/kg for such large ungulates as eland or elks), and an average ambient temperature of 20°C in warm climates and a 10°C mean temperature in northern latitudes (needed to determine energy needs to cook meat to desired doneness).

Furthermore, I assume that the temperature of meat when done would be 80°C (77°C is generally considered sufficient for well-done meat), that the heat capacity of this meat is about 3 kJ/kg°C (relatively high because meat is mostly water, whose specific heat capacity is 4.18 kJ/kg), that a typical cooking efficiency of an open fire is just 5% (with the most likely range between 2 and 8%), and that the average energy density of wood is 15 MJ/kg (this is the lower heating value, accounting for the fact that a substantial amount of energy is needed to evaporate the water present in freshly harvested wood). These assumptions result in a daily per capita intake of close to 1 kg of mammoth (and about 1.5 kg of large ungulate) meat, and to cook it without any energy losses would require roughly 200–300 kJ; with a 5% open-fire efficiency, it would take about 4–6 MJ of wood, or 1.5–2.2 GJ in a year, an equivalent of 100–150 kg of (some fresh and some air-dried) wood.

If the same fuel demand were to apply to every individual of some 200,000 people who lived 20,000 years ago, the overall need would add up to 20,000–30,000 t of woody phytomass a year (the generous assumptions made to calculate this total would also take care of the fuel needed to cook the plant portion of the average diet, assumed here to be only 20% of the total intake). This means that even if we assume a relatively large late Pleistocene population, the aggregate demand for cooking fuel amounted to only a negligible share of terrestrial primary productivity and to an even smaller share of the standing woody phytomass: its estimated preagricultural total (see chapter 1) is more than 100 million times larger.

while bonobos eat small antelopes (duikers), flying squirrels, and other small primates (Hohmann and Fruth 2008). Chimpanzee hunts are sometime cooperative and meat is often shared, either to reinforce social bonds within a group or, more often, to attract females (Teleki 1973; Boesch 1994; Stanford 1999).

The evolution of cooperative hunting (and subsequent meat sharing) is perhaps best explained by the fact that it maximizes an individual's chance to get high-quality protein and essential micronutrients in meat (Tennie, Gilby, and Mundry 2009). And the exchange of meat for sex is not an infrequent occurrence, but is done on a long-term basis (Gomes and Boesch 2009). These findings led anthropologists to expect a similar importance of skilled and cooperative hunting to derive the nutritional benefits of meat and improve the probabilities of reproductive success in the evolution of our species. More frequent meat eating helps explain significant differences in height and body mass (a roughly 40%–50% gain) between *Homo habilis* and *Homo erectus* (McHenry and Coffing 2000), and lower meat consumption in sedentary agricultural societies was a factor in their common health problems.

Moreover, Aiello and Wheeler (1995) argue that the only way humans could support a larger brain size without increasing total metabolic rate was to reduce the size of another internal organ. This would have been difficult to accomplish with liver, heart, or kidneys, but the size of a metabolically expensive gastrointestinal tract could be reduced with a better, more meaty diet and would free more food energy to support larger brains. Meat consumption remained limited in all traditional agricultural societies, but higher meat consumption has been a universal marker of rising incomes; meat eating is still widely seen as a sign of affluence, and meat sharing is still used to create personal bonds in most cultures.

Lean meat is, of course, a virtually pure protein of the highest quality (unlike all plant proteins, it has the correct proportions of all essential amino acids needed for human growth), and as such, it has a rather low energy density. The meat of wild herbivores, be they small lagomorphs (rabbits, hares) or large ungulates (antelopes, elks, bison), is very lean: fresh carcasses are more than 70% water, 20%–23% protein, and less than 10%(some species less than 1%) fat (subcutaneous and intestinal), resulting in an overall food energy density of only 4.8–6.0 MJ/kg (compared to 15 MJ/kg for grains and more than 16 MJ/kg for fatty meat). That is why even a relative abundance of this lean meat could leave people feeling hungry and craving high-energy-density, satiety-inducing fat. Moreover, prolonged eating of nothing but lean meat causes acute malnutrition known as rabbit starvation (*mal de caribou* of the early French explorers in Canada) that brings nausea, diarrhea and eventual death (Speth and Spielmann 1983). This helps to explain why Hayden (1981) concluded that the high regard for meat in foraging societies is nothing but a misconception created by ethnographers, and that the real quest was for fat.

There is no need to go to this extreme in order to confirm a common preference for killing fatty animals, including underground mammals (antbears, badgers, porcupines) and, once the adequate weapons and hunting strategies were in place, such large herbivores as mammoths, aurochs, bison, eland, or elks. And eating high-fat foods was a metabolic imperative in the Arctic (Cachel 1997), a requirement that was easier to satisfy thanks to the abundant insulating fat in marine mammals (seals, whales). But the environment often hindered the quest for fatty meat or made very little of it available. Contrary to the common perception, species-rich tropical forests were an inferior place to hunt.

Bailey et al. (1989) suggested that perhaps no foragers could have ever lived in tropical rain forests independent of cultivated foods. They modified this conclusion in the light of archaeological evidence from Malaysia and conceded that environments with high densities of sago and pigs could be exceptions (Bailey and Headland 1991). But it is obvious that in contrast to grasslands and open woodlands, hunting in tropical rain forests is much more difficult because most of the resident zoomass is relatively small, mostly folivorous (and hence arboreal and inaccessible), and often nocturnal. Not surprisingly, Sillitoe (2002) found both gathering and hunting in a tropical rain forest of the Papua New Guinea highlands costly (with people expending up to four times more energy on hunting than they gained in food) and concluded that it could not have been a prominent component of the highland subsistence: proto-horticultural practices and shifting farming were needed to secure adequate food.

Killing the largest terrestrial herbivores with simple wooden and stone weapons entailed high risks. Such weapons were available since the Late Paleolithic—in 1948 a nearly complete spear from the last interglacial period (125,000–115,000 years ago) was found (inside an elephant skeleton) in Germany, and in 1996 throwing spears found at Schöningen were dated to 380,000–400,000 years ago (Thieme 1997)—but a high rate of success would have been impossible without cooperative hunting strategies. Many anthropologists and students of the Quaternary environments believe such hunting caused a relatively sudden demise of the late Paleolithic megafauna.

Hunting and Megafauna

The term *megafauna* conjures enormous, long-extinct dinosauria, but when used in the context of late Pleistocene extinctions (they took place during the last millennia of that geological epoch, which lasted from about 2.58 million years to 12,000 years before the present) it describes all heterotrophs with an adult body mass of more than 40–50 kg. The extant megafauna includes such still abundant species as deer, moose, and wildebeests, as well as such endangered mammals as gorillas and tigers. The extinct megafauna, whose abundance and widespread distribution overlapped with the evolution of *Homo sapiens*, included large herbivores with an average body mass well in excess of 1 t (1,000 kg). Besides the woolly mammoth (*Mammuthus primigenius*) and the woolly rhinoceros (*Coelodonta antiquitatis*), the two iconic animals of the period, the megafauna included the straight-tusked elephant (*Paleo-loxodon antiquus*), the giant deer (*Megaloceros giganteus*), the steppe bison (*Bison priscus*), and the auroch (*Bos primigenius*), as well cave-dwelling lions (*Panthera leo*) and bears (*Ursus spelaeus*).

The list of extinct megafauna in the Americas also includes the mastodon (*Mammut americanum*), giant ground sloths (*Megatherium*), beavers and polar bears, camelids, horses, and saber-toothed felids (*Smilodon, Xenosmilus*). Australia had giant herbivorous marsupials (the wombat-like *Diprotodon* was the largest), as well as a carnivorous marsupial lion (*Thylacoleo carnifex*). Some of these animals (aurochs, lions) survived in much diminished numbers into the earliest part of the Holocene, which began 12,000 years before the present: even for mammoths, the latest survival dates in their refugium on the North Russian Plain are about 9,800 ^{14}C years, or approximately 11,100 years before the present (Stuart et al. 2002).

But all of them were gone by the time humans began their transition from foraging to a sedentary existence. Once radiocarbon dating allowed a fairly reliable determination of ancient biomass samples preserved in a variety of sediments, the idea that hunting caused the great extinction, as first argued by Owen (1861), became quantitatively supportable: a relatively sudden and, in geological terms, a very brief period of megafaunal extinction during the late Pleistocene appeared to coincide with the arrival of humans in Australia and North America. Roberts et al. (2001) concluded that all of Australia's land mammals, reptiles, and birds with a body mass of more than 100 kg became extinct across the entire continent around 46,400 years ago, or as rapidly as within 1,000 years of human arrival. In North America nearly half of all late Pleistocene extinctions (affecting 16 genera of large mammals) took place between 12,000 and 10,000 years before the present, while the arrival of the first skilled hunters from Asia (the Clovis people) is dated to between 13,200 and 12,800 years ago (Thomas et al. 2008).

Beginning in the late 1950s, Martin began to argue that the rapid late Pleistocene megafaunal extinctions in North America were caused solely by hunting (Martin 1958, 1967, 1990, 2005). His "overkill hypothesis" posited "blitzkrieg" extinction, a swift and massive killing of large animals by an advancing wave of migrating hunters. The rarity of megafaunal kill sites (just 14 of them have been found in North America) is explained by an extremely low probability of the bones being preserved in unprotected upland environments (Fiedel and Haynes 2004). If humans were solely responsible for this rapid and massive extinction, it would have been the most remarkable case of preagricultural destruction of biomass, comparable only to the effects of deliberately used fire.

The principal arguments for the overkill hypothesis are the well-known effects of even relatively low but persistent rates of killing of slow-breeding animals that give birth to a single offspring after a long gestation (8 months for elks, 11 months for horses, and 22 months for elephants). Fiedel (2005) also argues that dogs introduced during the Paleoindian colonization of the Americas had more than one role to play in megafaunal extinction. Genetic evidence indicates that wolves were not domesticated first in sedentary Mesolithic societies of the Middle East but less than 16,300 years ago from several hundred wolves in southern China (Pang et al. 2009).

The two main competing explanations of megafaunal extinction invoke various environmental hypotheses, ranging from catastrophic encounters with extraterrestrial bodies and rapid changes in temperature to gradual fragmentation or a complete loss of preferred habitats, usually attributable to climate change, and a hypervirulent paleodisease. Ascribing the extinctions to a "hyperdisease" (MacPhee and Marx 1997; Lyons et al. 2004) is not convincing. Such a pathogen would have to survive in a carrier species without harming it; it would have to be sufficiently virulent to infect large numbers of animals; it would have to cause very high (more than 50%) death rates; and the epizootic would have to able to infect species as different as mammoths and tigers without ever causing a universal epizootic. No virus, bacterium, or fungus can meet all of these requirements. And the proponents of the overkill hypothesis also dismiss climate change as the main cause of extinction because such a major species loss did not take place during one of the many similar climatic reversals that marked the Pleistocene.

After more than half a century of debate, the "overkill" controversy remains unresolved. Brook and Bowman's (2004) simulations of megafaunal extinction on a medium-sized land mass found that median times to extinction were less than 800 years, confirming that the late Pleistocene human colonizations had "almost certainly" triggered a "blitzkrieg" of killing. But Waters and Stafford (2007) revised the Clovis time range to a surprisingly brief span between 13,300 and 12,800 years ago that overlaps non-Clovis sites in both Americas and implies that humans inhabited America before Clovis. This conclusion was buttressed by Thomas et al. (2008).

They used genetic profiling of coprolites from Oregon to demonstrate a human presence in North America at least 1,000 years before the Clovis complex, a dating that would extend the blitzkrieg period to more than 2,000 years. The latest, and a very convincing, demonstration of pre-Clovis hunting comes from a projectile point made of mastodon bone and embedded in a rib of a disarticulated mastodon found in Washington state: Waters et al. (2011) dated the point to 13,800 years ago, proving that hunters were killing proboscidean megafauna at least two millennia before Clovis. Even before these latest findings, Grayson and Meltzer (2003) offered their requiem for North American overkill—but Fiedel and Haynes (2004) called their conclusions "a premature burial."

Arguments often hinge on the accuracy of stratigraphic interpretation, geochemistry, and radiocarbon dating. Australia's megafaunal extinctions are a perfect illustration of these disagreements. Miller et al. (2005) analyzed isotopic tracers of diet preserved in eggshell and teeth to conclude that the megafaunal habitat, a mosaic of drought-adapted trees, shrubs, and grasslands, was converted by fires set by humans to today's fire-adapted desert scrub shortly after the continent's human colonization, some 50,000 years ago. This blitzkrieg story was questioned by Trueman et al. (2005), whose dating of Cuddie Springs deposits demonstrated a prolonged coexistence (for a minimum of 15,000 years) of humans and the now extinct megafauna. But Gillespie (2008) claimed that the latest dating shows that the Australian evidence fits the overkill pattern, and Grün et al. (2009) explained that Trueman et al. (2005) were misled by a flawed geochemical argument and that a correct redating of the Cuddie Springs fossils puts their age at 51,000–40,000 years before the present, or within the interval that matches the last megafaunal survivals in other parts of Australia.

On the other hand, closer examinations of some of the seemingly convincing overkill explanations show that they often err by generalizing on the basis of limited (or questionable) evidence or by relying too much on theoretical models based on questionable assumptions. Barnosky (2008) identified "megafauna biomass tradeoff" (including humans as a megafauna species) as a driver of Pleistocene (and future) extinctions, and even quantified the process for the past 100,000 years. His conclusion: until about 12,000 years ago, the loss of animal megafauna largely matched the growth of human biomass, and after the megafaunal crash the global system gradually recovered to its precrash level, but with the megafaunal biomass concentrated in a single species, *Homo sapiens*.

Given our poor understanding of both zoomass densities and hominin numbers during the past 100,000 or even 10,000 years such an exercise is merely a speculation translated into values whose errors must remain unknown. What is clear in this specific case is that Barnosky (2008) had substantially underestimated the total mass of megafauna during the past millennium while overestimating the mass of humanity (by assuming average body weight of 67 kg during the past 500 years). Similarly uncertain is the attempt by Doughty and Field (2010) to estimate the total mass of vegetation that was liberated by the extinction of Pleistocene megaherbivores and then gradually appropriated by humans: their calculation shows about 1.4 Gt C (put with an unlikely accuracy at 2.5% of the global terrestrial net primary production, or NPP) and they estimated that humans began to harvest that much phytomass only sometime during the seventeenth century.

Alroy (2001) limited his simulation to the Pleistocene-Holocene time as he assumed slow human population growth rates, random hunting, and low maximum hunting effort to model megafaunal extinction; his prediction of its median occurrence was about 1,200 years after the initial invasion by humans, a period roughly corresponding to the dates for the earliest Clovis artifacts and the demise of the North American megafauna. But how would his simulation fare when taking into account the pre-Clovis migrants, and how applicable would it be in Africa, where our species arose, where capable hunters were killing animals for millennia before the settling of Americas—and where no megafaunal extinctions took place during the late Pleistocene or, for that matter, for nearly another 10,000 years afterward? Paying attention to specifics (focusing on individual species, continents, and regions) thus leads to a more complex understanding that does not fit any simplistic preconceived model.

Some large species never made it into the late Pleistocene to be hunted down by humans. The straight-tusked elephant was extinct in Europe perhaps as early as 40,000–50,000 years ago, and no later than about 33,000 years (Stuart 2005). Some 25,000 years of genetic decline preceded the extinction of the cave bear: neither climate change nor hunting alone was responsible (Stiller et al. 2010). Other species were under stress long before the late Pleistocene: Webb (2008) concluded that Australia's increasing mid- to late Quaternary aridity had changed the distribution of large mammals and reduced their numbers and, as a result, by the time of human colonization the Australian megafauna had reached a precarious stage that made it vulnerable to any level of hunting.

Ancient DNA analyses exclude humans as the drivers of musk ox decline in Greenland (Campos et al. 2010), and the latest research on megafaunal extinctions in Eurasia shows a remarkable degree of consensus, with little inclination to see hunting as a decisive factor. Pushkina and Raia (2007) found that Eurasian hunters generally selected the most abundant species and that the continent's late Pleistocene extinctions were driven primarily by climatic changes; only after the ensuing habitat fragmentation and reduction of ranges and numbers did human interference, be it by hunting or by other activities, became important.

As for the demise of Eurasia's largest herbivore, people and mammoths had coexisted for millennia. There is no evidence of any large-scale hunting of the species in sparsely populated regions of Siberia, and the gradually unfolding extinctions (no signs of any blitzkrieg) are best attributed to environmental changes (Reumer 2007; Stuart and Lister 2007; Kuzmin 2009). The latest support for this conclusion comes from Allen et al. (2010). They used a global climate model to simulate paleoclimates during the last glacial stage and early Holocene (42,000–10,000 years before the present) and concluded that the principal cause of megafaunal extinctions was the replacement of open and largely treeless glacial stage vegetation, whose mesophilous herbs had a higher capacity to support large herbivores.

The extinction of mammoths and of other species of the *Mammuthus-Coelodonta* faunal complex was driven primarily by sometimes dramatic shifts of dominant ecosystems, above all the loss of open, grass-dominated landscapes (tundra or steppe vegetation) and their displacement by tree-dominated formations with mixed conifer-broadleaf forests in the mid-latitudes, coniferous forests further north, and peat bogs and tundra in the Arctic. Natural habitat fragmentation created isolated populations, which then underwent insular evolution that was sometimes marked— as in the case of the dwarf mammoth of Wrangel Island—by diminished stature (Guthrie 2004), and human predation could have hastened the demise of some of these isolated and vulnerable species, surviving in fragmented refugia.

Even so, some animals previously thought to have become extinct by the late Pleistocene made it far beyond the Pleistocene-Holocene transition. Stuart et al. (2004) showed that the giant deer (Irish elk) survived in western Siberia more than three millennia after its previously accepted extinction date, to around 7,700 years ago. There were at least two Holocene mammoth refugia, in Siberia (Kuzmin 2009) and on Wrangel Island in the high Arctic, where small-sized mammoths survived well into the Holocene, with the youngest radiocarbon-dated tusk as recent as 2,038 BCE (Vartanyan et al. 1995), contemporary with ancient Egypt's Tenth Dynasty. A new dating of an American mastodon from northern Indiana shows that the species survived beyond the Clovis period into the late Younger Dryas time (Woodman and Athfield 2009).

And new radiocarbon dates from Alaska and the Yukon Territory (AK-YT) show that steppe bison (*Bison priscus*) and wapiti (*Cervus canadensis*), the two species of megafauna that survived into the Holocene, became more numerous before and during the human colonization of the region and before the extinction of mammoths and horses. There are no signs of any overkill or blitzkrieg but clear links between climate change and extinctions and human colonization: indeed, "the humans might have been not so much riding down the demise of the Pleistocene mammoth steppe as they were being carried into AK-YT on a unique tide of resource abundance" that was created by the Pleistocene-Holocene transition (Guthrie 2006, 209).

New research tends to view late Pleistocene extinctions as a result of multiple causes, including both natural (above all climate and vegetation change) and anthropogenic (mainly selective hunting and fire) factors (Yule 2009). For example, in their state-of-the-debate review, Koch and Barnosky (2006) move away from explaining the extinctions simply on the basis of a sweeping wave of hunting that began immediately after human colonization. They conclude that the anthropogenic extinction was a combination of direct effects (hunting) and indirect impacts (competition for resources, habitat alteration), and that the outcome could have been less severe in the absence of concurrent climate change in many affected regions.

The latest two analyses of megafaunal disappearance also concluded that the observed pattern of extinction was not due to a single factor but to the combination of climate change and human action. Lorenzen et al. (2011) used ancient DNA, models of species distribution and the human fossil record to show that individual species responded differently to climatic shifts, habitat changes, and human encroachment: climate change alone can explain the demise of Eurasian musk ox and woolly rhinoceros, while the combination of changing climate and intensified hunting was behind other extinctions. Prescott et al. (2012) modeled the process of extinction on five continental landmasses and concluded that the outcome is best explained by combining human arrival and climatic variables.

Unfortunately, these new, nuanced interpretations stressing continental specifics and avoiding single-factor attributions make it even more difficult to estimate the extent to which hunting affected the late Pleistocene zoomass disappearance. Even the simplest models of extinction due to overkill hinge on concatenated assumptions based on inadequate knowledge. For the estimates of densities and territorial extent of megafaunal species we have to rely on a combination of sparse archaeological records, general metabolic expectations, and analogs with modern large-sized animals. And much less is known about actual hunting preferences (which species were deliberately targeted, which ones were killed in opportunistic manner), about the specific composition of regional or continental kills, and about the rates of hunting success.

This has not prevented Smith, Elliott, and Lyons (2010) from assuming that about 100 million North American herbivores were eradicated by hunting. But their estimate was made without any reference to the total size of the continent's human population. Thanks to genetic mapping, our estimates of prehistoric populations have become better constrained, but the specific numbers of people that inhabited a given region at any particular prehistoric time—and much more so the fractions of these populations that were highly dependent on hunting large animals more than 10,000 years ago—remain only poorly constrained guesses. And if hunting was only a contributory factor, then we would have to apportion the relative importance of individual causes whose contribution hardly remained static for hundreds or thousands of years of the period under consideration: this is a task that our knowledge makes impossible, unless it is a pure guess. Nevertheless, some order-of-magnitude calculations give an indication of plausible impacts (box 7.2).

As the Pleistocene diet was definitely far from purely carnivorous, and as a non-negligible share of animal food came not only from smaller megafaunal species but also from small animals, the total number of large mammals killed annually by hunters could have been less than half that number—or twice that number if the body mass composition of kills was bimodal (including larger numbers of smaller animals). For comparison, the recent slaughter of cattle, the world's most common domesticated megafaunal species, has been about 300 million animals annually; with an average body mass of some 350 kg, this amounts to just over 100 Mt of fresh weight and about 60 Mt of carcass weight.

But even if late Pleistocene hunting had been an important (or, for the sake of argument, dominant) cause of megafaunal extinctions, its overall impact on the productivity of the biosphere would have been overwhelmingly positive. The disappearance of so many species of megafauna, including herbivores more massive than the surviving elephants (mostly between 5 and 8 t for adult mammoths, vs. 3–5 t for most adult African elephants), had profound consequences for the composition and primary productivity of affected ecosystems. As attested by studies of extant African megafauna, both grazers (animals that eat grasses and forbs) and browsers (those that eat primarily woody plants) profoundly affect plant structure, with elephant browsing having a particularly strong negative impact on the total tree cover, and this effect appears to increase synergistically in the presence of fire (Guldemond and Van Aarde 2008; Holdo, Holt, and Fryxell 2009).

Pleistocene megaherbivore browsers, feeding on woody phytomass and invulnerable to predation, created open, grassy landscapes by destroying trees and suppressing their regrowth (Zimov et al. 1995). The removal of trees by massive browsers

Box 7.2
Late Pleistocene Hunting: Megafauna and Meat Consumption

A late Pleistocene population of at least two million people would translate (assuming an average body mass of 40–45 kg per capita) into roughly 100,000 t of fresh-weight anthropomass. As already noted (see box 7.1), even if we assume very high food energy needs for these (sometimes highly mobile) foragers, the average per capita rate (including all children and adults) could not have been higher than about 10 MJ/day. To arrive at a (perhaps unrealistically) high limit on the overall kill would be to assume that, as argued by Mosimann and Martin (1975), the late Pleistocene foragers ate mostly meat, that about 80% of their food needs came from megafaunal species (and the rest mostly from smaller herbivores), and that the live weight of killed animals was 2.5 times the weight of their edible tissues, whose average energy density was as high as 10 MJ/kg.

These assumptions would require an annual kill of nearly 2 Mt (fresh weight) of megafaunal zoomass, approximately 20 times higher than the global anthropomass at that time. If all of the killed animals were mammoths (an unrealistic assumption made in order to get the minimum number of slaughtered animals), then the annual kill— assuming a mixture of *Mammut americanum, Mammuthus primigenius,* and *M. meridionalis* (weighing 4–9 t) and a larger *Mammuthus imperator* weighing more than 10 t (Christiansen 2004)—would have amounted to no less than 250,000 animals and as many as 400,000 animals. Assuming, more realistically, that the largest animals dominated the mixture of hunted megafaunal species, whose average weighted body mass was on the order of 1 t/head, then the total annual kill could easily have been on the order of two million large mammals.

The annual energy requirements of those killed animals could be approximated fairly well either by using standard metabolic equations (Dong et al. 2006) or by calculating their free-ranging metabolic rates from the best available allometric equations (Nagy 2005) and then assuming 11 MJ/kg of dry matter as the average energy density of the phytomass grazed by large herbivores. For a grazer weighing 1 t, the need would between 100 and 125 MJ/day, or an equivalent of about 9–11 kg of dry-matter phytomass. Two million large herbivores would thus have consumed annually 7–8 Mt of dry phytomass, a mass two orders of magnitude larger than that of all wood used for cooking but still less than 0.001% of all terrestrial phytomass in place during the late Pleistocene period. The total phytomass consumption would be smaller for the same overall zoomass if the average mass per head were larger (that is, if most killed species were mammoths rather than large bison or aurochs) because specific metabolism declines with body mass.

provided ideal conditions for smaller grazers, which lived constantly in a landscape of fear (Riginos and Grace 2008) and minimized their vulnerability to carnivores by feeding in the open, where they could better detect the approach of predators. In turn, grazers reduced fire hazard by preventing the accumulation of dry phytomass, and their wastes intensified nutrient recycling. Nonglaciated temperate Pleistocene ecosystems were thus mostly covered by grasses, forbs, and sedges. As a result of megafaunal extinction, these mammoth steppes (relatively diverse in terms of herbaceous species and supporting a high diversity of grazers and associated predators) were largely replaced by mossy or shrub tundra and coniferous taiga at higher latitudes, and by deciduous forests further south.

In many areas the former savanna or parkland was replaced by transitory communities of spruce and broadleaved trees that have no modern counterpart (Johnson 2009a). These forest ecosystems were less diverse, in terms of both plant and animal species, than the open landscapes they replaced, and their primary productivity (typically no more than 10 t/ha) was only moderately higher than that of the grasslands that used to grow in the same environment. But the new forests stored an order of magnitude more phytomass than did the grasslands: even in a relatively sparse temperate or boreal forest there would be about 100 t/ha, and typical rates would be around 200 t/ha, while even rich grasslands would store only on the order of 20 t/ha.

And with a much higher accumulation of phytomass and often thick litter layer, the new forests were also much more prone to fires than the open landscapes that preceded them (Johnson 2009b). But it would be simplistic to claim that the elimination of megafauna led directly to a more flammable world. Marlon et al. (2009) used charcoal and pollen records to appraise the changes in North America's fire regimes during the last glacial-interglacial transition of 15,000–10,000 years ago and found that the timing of large increases in fire activity was not coincident either with the changes in human population density or with the extinction of the megafauna.

Obviously, this generalization (based on 35 samples for the entire continent) leaves room for attributing some of the increase to human actions, but again, any such attribution would be an indefensible guess, and we must conclude that there is no way to quantify the net impact of the most likely combination of anthropogenically driven megafaunal extinction, vegetation change, and increased fire frequency. Perhaps the only defensible conclusion is that the net impact of human actions on the Earth's prehistoric biomass—contributing to the megafaunal extinction (hence reducing the global terrestrial zoomass stores) that led to the expansion of forests (hence increasing global phytomass stores) and to an increased frequency of fires (hence periodically reducing land phytomass)—could actually have been positive. If that was the case, then it was only during the millennia of nomadic and agricultural expansion that human actions led to significant reductions of the biosphere's mass.

Hunting and Collecting in Preindustrial and Modern Societies

Hunting and gathering retained an important place in all early agricultural societies. Excavations at Çatalhöyük (a large Neolithic settlement on the Konya Plain established about 7200 BCE) confirmed the dominance of grains and wild plants, but they also showed that those early farmers hunted large (aurochs) and small (foxes, badgers, hares) animals (Atalay and Hastorf 2006). Tell Abu Hureyra, in northern Syria, where excavations by Legge and Rowley-Conwy (1987) found that hunting remained a critical source of food for 1,000 years after the beginning of plant domestication, was hardly an exception. Predynastic Egyptians (going back to about 5000 BCE) complemented their dependence on emmer wheat, barley, and sheep by hunting ducks, geese, antelopes, wild pigs, crocodiles, and elephants and gathering herbs and roots (Hartmann 1923), and hunting and the gathering of wild plants were common activities in all of China's ancient dynasties (Chang 1977).

In Europe's agricultural areas adjacent to forests or interspersed with many groves and shrubs, the hunting of smaller game (partridges, peasants, hares, rabbits, deer) and the collecting of leaves, fruits, nuts, berries, herbs, roots, and mushrooms for food as well as for medicinal uses persisted not only through the centuries of cropping intensification but right into the modern era. Some of these activities still remain a part of village- and urban-based foraging even in the world's most industrialized and densely populated societies. Mushroom hunting remains very popular not only in Europe (including all nations in the eastern and central part of the continent, as well as in Italy and France) but also in Japan, as attested by the numerous mushroom guides (Consigny 2010).

Hunting in rich countries belongs to three very different categories. Some species are hunted primarily to supply game meat to hunters' families or to commercial markets: Toscana's wild boar (*cinghiale*, much of it made into salami), Central European deer, and British game birds (grouse, pheasants, partridges) are all excellent examples of the first practice. In North America, big game hunted for meat includes deer, elk, moose, and caribou. Others mammals, including large African savanna herbivores and North American bears, moose, and elk, are killed primarily as trophy animals. And in most low-income societies of Asia, Africa, and Latin America, foraging provides an important food quality boost, particularly in terms of high-quality animal protein, vitamins, and minerals. While the hunting and collecting of temperate zone animals have usually been limited to birds and mammals (frogs and snails are the two notable exceptions), tropical foraging includes as well many insects, their larvae, snakes, and lizards. During the rainy season such wild foods could make up about half the total food consumed by farmers in Thailand's central northeast (Price 1997).

There are no reliable larger-scale quantifications of these harvests: food acquired for direct domestic consumption is obtained by opportunistic collecting and hunting, and all but small shares of wild (and often illegally killed) meat (fresh, dried, or smoked) and plants are sold in local markets, also beyond the reach of any statistical monitoring. A rare estimate for Ghana put the total annual consumption of wild meat at no less than 385,000 t during the 1990s (Ntiamoa-Baidu 1998); that would prorate to a bit over 20 kg/capita, or at least twice the average supply of all other meat (beef, pork, goat, poultry) attributed to the country by FAO's food balance sheets (FAO 2011b). Usually the only information we have refers to sales or consumption rates in specific markets or regions during the limited periods of time when these food supplies were studied by outsiders. One of the first studies of this kind was Assibey's (1974) account of wild meat offered for sale at a single market in Ghana's capital: at least 160 t of meat during a period of 18 months.

A review by de Vos (1977) found that in many areas of Africa and Latin America wild meat supplied at least 20% of all animal protein, and nearly 20 years later Njiforti (1996) confirmed that conclusion for northern Cameroon, where the respondents to his study consumed annually almost 6 kg of bush meat (North African porcupine and guinea fowl being the preferred species), representing 24% of the animal protein intake in the region. Hunting in West and Central Africa is strongly market-oriented: many urban residents are recent arrivals with a traditional taste for wild meat, and besides many professional hunters there are many villagers who hunt to supplement their limited income: Kümpel et al. (2010) found that in rural Equatorial Guinea, about 60% of poor to middle-income households were engaged in hunting for urban markets. Bifarin, Ajibola, and Fadiyimu (2008) found that in the forests of Nigeria's Ondo state, hunters averaged more than 140 kg of big game (buffalo, bushbucks, and duikers) every month and nearly 90 kg of smaller species (mainly cane and African giant rats).

Illegal hunting is common even in the relatively best policed national parks, including the Serengeti and partially protected areas surrounding the park

(Nyahongo et al. 2005; Holman, Muya, and Røskaft 2007). Surveys by Ndibalema and Songorwa (2008) indicated that eland and topi are the preferred species of these illegal hunts, but their relative rarity means that buffaloes and wildebeests are killed most often. In some villages adjacent to the park the annual consumption of wild meat was as much as 36 kg per capita, almost as much as Japan's recent per capita meat consumption (Smil and Kobayashi 2012), which is explained by the low cost (with wild meat selling for only about a quarter the price of commercially marketed meat). Fa and Brown (2009) summarized the recent evidence of hunting in Africa's tropical moist forests and confirmed that ungulates and rodents make up most of the consumed zoomass, and that both these classes of mammals are subject to unprecedented rates of killing, but because in most cases even the basic understanding of their ecology is missing, it is impossible to formulate any sensible hunting quotas.

African hunting has received by far the greatest amount of international attention because of the killing of primates (and their orphaned juveniles) in the central and western part of the continent and because of continuing illegal hunting of elephants and rhinos. But I have argued that it is the proverbial omnivory of 1.3 billion Chinese and the prestige bestowed by that ancient culture on exotic dishes and medicines that is the single most important food-related threat to the diversity of zoomass in Asia and beyond. On the one hand, China's remarkable omnivory is a very efficient way of food consumption, as hardly anything gets wasted: pig's skin and chicken feet, silk moth pupae and carp eyes are all eaten; on the other hand, indiscriminate eating means that everything that moves (from snakes to dogs, monkeys to owls) and much that does not (abalones, sea cucumbers) gets consumed.

This was not much of a problem during the decades of Maoist impoverishment, but since the 1980s China's rising incomes have created a rapidly expanding demand for wild meat and other supposed delicacies, which are often featured during banquets given by the country's nouveaux riches or by its famously corrupt officials. The extent of these activities was indicated by a rare Chinese estimate that put a highly conservative annual value spent on eating, drinking, and traveling at public expense at more than 100 billion yuan in 1992 (Wu 1996). By 2010 China's average household consumption expenditure had quadrupled compared to the 1992 level (NBSC 2010), an increase that makes it is easy to imagine how much worse these biosphere-depleting excesses have become.

Some of China's food choices condemned in the West are merely a matter of cultural conditioning; there is surely no rational argument why killing young lambs should be preferable to killing young dogs. But a potentially enormous demand for wild species is a different matter, as drastically reduced numbers of snakes, turtles, and frogs have become the norm throughout southern China (consequently, Chinese reports refer to increased numbers of rodents that were previously held in check by snakes) and as illegal imports of these animals from the Southeast Asia (Vietnam, the Philippines, Indonesia) supply the local demand for these species.

Other surveys illustrate high levels of illegal wildlife trade in one of the planet's most biodiverse regions, east of the Himalayas and in the forested parts of Tibet: this large-scale illegal trade involves both the neighboring countries (particularly Myanmar and Nepal) and activities within China (Li et al. 2000). Not surprisingly, a 1999 survey found that 26% of all wild animal dishes served in restaurants contained species on China's endangered list (Smil 2004). As a result, in January 2002 the China Wildlife Conservation Association asked professional chefs to sign a declaration stating they would not prepare any meals containing endangered species (*BBC News* 2002). But many of those species captured in the wild are now bred commercially, some in astonishing quantities: half of China's commercial turtle farms whose offices had responded to a survey by Shi et al. (2008) sell more than 300 million turtles (not only the common softshell *Pelodiscus sinensis* but also critically endangered species) annually, which means that the actual figure must be much higher.

And the now global search for traditional therapies (a far too polite a term for many superstitions) only adds to China's (and to a lesser extent also to Taiwan's, South Korea's, and Japan's) destructive dietary choices. The potential extent of the Chinese demand is truly stunning: Li et al. (2000) reported that in Yunnan province, more than 6,100 species of plants and nearly 400 species of animals are used in traditional medicines. Species in high demand include not only the leaves, roots, and flowers of plants that are in no danger of extinction (such as garlic, ginger, honeysuckle, magnolia, or peony) but also body parts of such endangered mammals as bears (galls harvested from illegally killed animals from as far as the Canadian Rockies) and the Alpine musk deer (*Moschus chrysogaster*). As a result, an analysis of threats to vertebrate species in China and the United States showed a striking difference: in the United States, habitat destruction was the leading cause, whereas in China it was overexploitation by rural populations and the widespread trade in wildlife products (Li and Wilcove 2005).

This ongoing search for fictional health benefits adds constant and cumulatively destructive pressure to the quest for meat, hides, furs, or plumage, and it has made the hunting of wild animals one of the two main reasons for the extinction of hundreds of species since the beginning of the early modern era (the other one being the loss of habitat due to deforestation and grassland and wetland conversion). In 2010 the most reliable global count of animal species that have become extinct since 1500 surpassed 700 (IUCN 2011), but there is no doubt that the actual total is substantially higher. Most of the disappearances have been among rodents and birds, particularly the smaller ones and the inhabitants of smaller islands.

The list also includes more than a dozen large mammals, above all a number of grazers, among them African quagga (*Equus quagga*), the red and Arabian gazelles (*Gazella arabica*), tarpan (*Equus ferus*), the Portuguese ibex (*Capra pyrenica lusitanica*), and the Syrian wild ass (*Equus hemionus hemippus*); among large carnivores the most notable subspecies losses were those of the Bali, Java, and Caspian tigers (*Panthera tigris balica, sondaica* and *virgata*) and the Barbary and Cape lions (*Panthere leo leo* and *melanochaitus*). In absolute terms, no post-1500 extinction of a vertebrate species surpasses the fate of America's passenger pigeon (*Ectopistes migratorius*). Flocks of this species were composed of tens to hundreds of millions birds, with the total continental count in the billions. John James Audubon recalled how in 1813 on the banks of Ohio, "the air was literally filled with Pigeons; the light of noon-day was obscured as by an eclipse, the dung fell in spots, not unlike melting flakes of snow; and the continued buzz of wings had a tendency to lull my senses to repose."

And Mark Twain's memoirs recall their antebellum surfeit: "I can remember the pigeon season, when birds would come in millions, and cover the trees and by their weight break down the branches. . . . They were clubbed to death with sticks; guns were not necessary" (Twain 2010). By the time the last surviving passenger pigeon died in 1914 it was several decades since the massive flocks had disappeared. The slowly declining numbers of the birds plummeted once the pigeons became massively hunted to supply inexpensive meat: during the 1870s some counties in Michigan were shipping east every year a million or more of the birds, giant flocks were gone by the 1880s, and small flocks survived into the 1890s; the last wild flock of some 250,000 birds was annihilated in 1896 (Eckert 1965). Even if we assume that "only" two billion birds were killed between 1840 and 1890, that slaughter (with an average live weight of 350 g/bird) would have amounted to 700,000 t of avian zoomass.

But when measured in terms of the total wild zoomass destroyed by human actions, none of the post-1500 extinctions has involved such an open and deliberate mass slaughter as did the virtual elimination of the American bison (*Bison bison*) in North America before 1900 (I will deal with the nineteenth- and twentieth-century killing of whales in the next section) and the drastic reduction of the once large and wide-ranging African elephant herds. The bison's demise has been described and documented in detail (McHugh 1972; Branch 1997; Isenberg 2000). Less appreciated is a remarkable comeback of the species. About half a million animals now live in the United States and Canada, mostly in private enclosures or national parks, and bison meat is a small part of the commercial food supply (the nineteenth-century killings were primarily for hides, not for meat).

The retreat of African elephants has been brought about by a combination of intensified traditional hunting (whose effectiveness was greatly potentiated by the use of rifles), an expanding ivory trade (with, once again, China and Japan being traditionally the largest importers, both for various ornamental carvings and for such mundane objects as hankō, personal seals ubiquitous in Japan), and the loss of habitat owing to the expansion of cultivated land. The best historical evidence, based on the ivory exports from the continent, indicates substantial pre-1914 fluctuations and a period of recent high losses starting during the 1970s. The best approximations put the twentieth-century elephant slaughter at roughly twice the level of the nineteenth-century bison killings when compared in terms of freshweight zoomass (box 7.3).

Compared to the slaughter of bison and elephants, tiger killings have involved much smaller zoomass totals: their target is a top carnivore, whose abundance is inherently much lower than that of large herbivores. Common densities are around 1 elephant and up to 10 bison/km^2 (Gates et al. 2005), compared to as few as 1–2 tigers/100 km^2 (Kawanishi and Sunquist 2004). India's tiger numbers were sharply reduced by frivolous hunting (by the European colonial rulers or members of India's many princely families) that was one of the hallmarks of the British rule. A conservative estimate is that 65,000 Indian tigers were hunted down between 1875 and 1925, with the peak of some 17,000 killed during the 1880s (Rangarajan 1998).

Even before the nineteenth century the surviving Chinese tigers were relegated to the fringes of the country's territory, and now China's large illegal trade in tiger products is blamed for a rapid post-1990 decline in India's tiger numbers: some of the country's famous national parks now do not have a single tiger, or none can be found even after extensive night camera searches (Bhardwaj 2006; Datta, Anand, and Naniwadekar 2008). This affluence-driven destruction may be the end. At the beginning of the twenty-first century the country had fewer than 4,000 tigers, a tenth of the total count in 1900 (Bhardwaj 2006). Commonly weak enforcement and the existence of global illegal trade networks that are well financed and often

Box 7.3
Mass Killings: Bison and Elephants

Bison were once present from the Atlantic coast to California and from northern parts of Mexico to northern Saskatchewan and Alberta, their numbers kept well below the carrying capacity maximum by native hunting. As the imported epidemics reduced the native populations, bison numbers rose after 1700, and the continental total reached as many as 50 million animals during the late eighteenth century. A hundred years later, only a few hundred buffaloes escaped the mass slaughter that began during the 1840s and culminated in the 1870s (Branch 1929; Isenberg 2000). Even if we assume, rather conservatively, an average body mass of 500 kg/animal, the killing of some 40 to 50 million animals during the nineteenth century would have amounted to 20–25 Mt of (fresh-weight) zoomass.

The maximum carrying capacity of African elephants was about 27 million animals at the beginning of the nineteenth century, and the actual number of elephants might have been as high as 75% of that value (Milner-Gulland and Beddington 1993). In contrast, the latest continent-wide summary put their definite number at about 470,000 in 2006, with another 160,000 animal in probable and possible categories and 50,000 in a speculative class (Blanc et al. 2007). The best available reconstructions of the ivory trade (Parker 1979; Luxmoore et al. 1989) show a relatively steady flow of around 100 t/year until about 1860, followed by a rise to more than 500 t/year just after 1900; the rapid plunge induced by World War I was followed by a brief rise, then another (World War II) slump, but the annual rate then rose steadily, to peak at more than 900 t by the late 1980s.

Integration of these fluctuating harvests yields about 55,000 t of ivory harvested during the nineteenth century and at least 40,000 t removed during the twentieth century. Assuming an average of 1.8 tusks per elephant is an easy part of the conversion, but selecting an average weight of tusks is associated with a much larger error. The largest and the most sought-after tusks used to weigh commonly more than 50 kg (102 kg is the record), but with the progressing slaughter, the average mass of tusks has been declining (Milner-Gulland and Mace 1991). Assuming, conservatively, 7 kg for the nineteenth century and 6 kg for the twentieth century results in cumulative slaughters of, respectively, about 14 and 12 million elephants. Assuming further that average body mass is 75% of the adult female weight of about 3.5 t brings the twentieth-century total of slaughtered elephant zoomass to around 35 Mt of live weight.

ready to resort to violence provide little assurance that the large-scale destruction of targeted mammals, birds, and reptiles will be reduced anytime soon.

The overexploitation, and eventual extinction, of wild animal species (above all of herbivorous megafauna, primates, and large carnivores) receives incomparably more attention than the damage done by foraging for food and medicinal plants, but their collection remains an important addition to crop-based diets in many low-income countries. Their contribution is mainly in terms of micronutrients (vitamins and minerals) obtained by consuming leaves, seeds, tubers, fruits, and nuts. Among 114 species of trees and shrubs in the Sahelian zone of Africa 23 kinds are of great importance and another 46 of a limited importance in diets; in rural Tanzania, 80% of all leafy green vegetables are consumed, in West Africa, 24 of 165 studied plant species are regularly eaten; and on the continent as a whole, more than 500 different wild plants are consumed as food (Fleuret 1979; Ogoye-Ndegwa and Aagaard-Hansen 2003; Shackleton 2003; Tabuti, Dhillion, and Lye 2004).

One wild plant food is also an important source of two macronutrients that are commonly in short supply in crop-based diets. In West Africa, the beans of a leguminous tree, *Parkia biglobosa,* are boiled and fermented, and the dried concentrate (*dawadawa*), which is commonly added to stews, contains 20%–50% protein (with lysine content similar to that of egg) and 30%–40% lipids (60% being unsaturated fatty acids), as well as iron and calcium (Campbell-Platt 1980). The calcium-, protein-, and lipid-rich leaves of the baobab (*Adansonia digitata*) are a favorite across large parts of western Africa (Lykke, Mertz, and Ganaba 2002; Chadare et al. 2009), and nutrient-rich butter is extracted from shea nuts (*Butyrospermum parkii*). India's variant of *dawadawa* is made from the unripe pods of *jant* (*Prosopis cineraria*) in the country's western semidesert areas (Gupta, Gandhi, and Tan 1974). Even in Switzerland, one of Europe's most affluent countries, collected foods supplied a staple well into the twentieth century: wild chestnuts in the canton of Ticino (much as in parts of northern Italy) were dried and milled to produce a flour (as well as marmalade and beer).

Aquatic Harvests

Fishing is an inaccurate but standard term for activities that include a great deal more than just the capture of the many species of bony fish living in freshwaters (streams, lakes, ponds), in shallow coastal ocean waters, and in pelagic (near the sea surface or in the water column) or demersal (on the seabed) habitats. Marine fishing also includes the capture of numerous cephalopods (many species of squid, cuttlefish, and octopus), crustaceans (in mass terms, their catch is dominated by shrimp and prawns, and it also includes crabs, lobsters, and even krill, tiny marine crustaceans that are the main food of whalebone whales), and mollusks (including clams, mussels, oysters, and scallops).

Many of these invertebrates are also captured (or now increasingly cultivated) in freshwaters, and marine harvests also include the killing of many species of aquatic mammals, including in the past just about every species of (now largely protected) whales, dolphins, seals, and manatees. The most recent worldwide statistics show the total annual landings of more than 4 Mt of cephalopods, more than 5 Mt of crustaceans, and about 7 Mt of mollusks, adding up to about 20% of the total marine catch (FAO 2011a). And marine harvests also include harvests of algal phytomass (recently less than 2 Mt of fresh weight) that is dried and sometimes processed before eating.

Archaeological finds belie the claims that effective fishing methods were a relatively late mode of foraging and that marine fishing was a particularly unappealing choice because of the unpredictability of fish runs and the high energy cost of searching for rich stocks and capturing them in laboriously made nets (Ryder 1969). Rick, Erlandson, and Vellanoweth (2001) summarized the post-1980 finds of fish remains (dating to the terminal Pleistocene and the early Holocene, mostly between 12,000 and 7,000 years before the present) in Pacific coastal communities in Oregon, California, Ecuador, Peru, and Chile that indicate fairly intensive exploitation of up to 26 taxa of fish living primarily in bays, estuaries, and kelp beds and ranging from sea bass to tuna and from lingcod to halibut; the fish were caught by hook and line, as well as by spears, traps, and nets. For one midden site in California, Daisy Cave on San Miguel Island, they were able to estimate that fish provided 50%–65% of all edible meat during the two millennia between 10,500 and 8,500 years before the present.

The heavy reliance of early Holocene coastal populations on ocean foods is also attested by numerous European finds, particularly in the Baltic. Denmark's soils contain vast amounts of fish bone, especially from the Mesolithic Atlantic period, which lasted from 7000 to 3900 BCE and was marked by higher sea temperatures than today (Enghoff et al. 2007). In more inhospitable environments it was necessary to target fatty mammals. In Alaska as well as in Patagonia, mussels and limpets would not have sufficed, and the costal populations exploited sea lions and fur seals (Yesner 2004). And marine zoomass provided the energetic basis for some of the most complex prehistoric societies in the Pacific Northwest (Sassaman 2004). Their sedentism, exceptionally high population density, and ensuing social complexity were made possible by the seasonal abundance of fatty salmons (cod's energy density is just 3.2 MJ/kg, Chinook salmon rates 9.1 MJ/kg) and their preservation by smoking.

But no species was more rewarding to hunt than bowhead whale, one of the world's most massive mammals: even the immature two-year-old animals that were most commonly killed by the Inuit hunters using boats and spears weighed nearly 12 t. Moreover, their tissues had an extraordinary food energy density: at about 36 MJ/kg their blubber, and at 22 MJ/kg their *mukluk* (skin and blubber), were at least twice and up to three times fatter than the fattiest fish, and the tissue could be easily preserved in permafrost dugouts. Sheehan (1985) calculated that the coastal settlements would enjoy food surplus even when assuming a minimum baleen whale harvest near the shore during three to four months of seasonal migration together with a subsidiary fishing and killing of beluga whales, walruses, and seals.

Fishing did not lose its importance with the emergence of ancient complex civilizations but its impact remained spatially circumscribed, confined for millennia to the most heavily exploited coastal waters. Localized overfishing of some preferred species certainly took place in some coastal areas of the Mediterranean during Roman times, but many medieval references to an astonishing abundance of fish attest to the presence of large stocks of herring, mackerel, cod, salmon, pilchards, and sprat. Although there were some temporary reductions as a result of local overfishing or changing water temperature, abundant fish stocks in Europe's Atlantic waters lasted well into the nineteenth century (Roberts 2007). At the very beginning of the modern era, Europeans also discovered fish stocks that became the focus of the world's first long-distance fishery that, remarkably, lasted for nearly five subsequent centuries.

This could happen only after suitable ships became available, and it is not at all surprising that Spanish and Portuguese sailors, the pioneers of the European voyages to Africa, the Americas, and Asia, were also the first practitioners of fishing far from home ports. The Grand Banks of Newfoundland, where the sea appeared to swarm with cod, were discovered by a British expedition (John Cabot in 1496), but Portuguese, Basque, and French ships had opened up the fishery by 1517. Two centuries later the banks were visited annually by hundreds of European ships, but abundant cod stocks were present even during the second half of the twentieth century.

Only then, after decades of overfishing by Western European, Russian, and Japanese fleets outside Canadian waters and by excessive numbers of Canadian vessels, did that great ancient fishery (and tens of thousands of associated jobs) collapse (Bavington 2010). In 1992, the Canadian government imposed a moratorium on cod catches, hoping for an eventual stock recovery, but a generation later there are no signs of any comeback. But even such dire lessons remain stubbornly unlearned: after that stock collapse European cod fishing, particularly in the North Sea, continued with an intensity surpassing the precollapse rates in Canadian waters (Cook, Sinclair, and Stefansson1997).

Pre-nineteenth-century coastal whaling is another ancient practice that had almost no effect on the numbers of migrating whales that entered shallower waters: besides the Alaskans, its practitioners included Basques, Normans, and Scandinavians, as well as the Japanese. Because Japan's ancient Buddhist bans on killing and eating animals did not apply to sea mammals, whale (*kujira*) meat from animals killed in coastal waters was a welcome source of protein and fat from the very beginning of the country's history (Shiba 1986). All estimates of original whale populations before the beginning of large-scale, long-distance whaling are highly uncertain, and the best new models, based on mitochondrial DNA sequence variation, may not bring any more clarity.

Roman and Palumbi (2003) used this technique to put the total numbers of prewhaling North Atlantic whale species at 240,000 for humpbacks (360,000 for fin whales, 265,000 for minke whales). This figure was in stark contrast to the previously estimated total of 20,000 North Atlantic humpbacks, and the new estimate was immediately questioned. Smith and Reeves (2002) used historical records to estimate the total removal of 29,000 humpback whales between 1664 and 2001; added to the extant total of 11,000 individuals in the late 1990s, that would indicate a pre-exploitation total of about 40,000 animals, and historical removals would have to have been roughly six times the reconstructed rate to conform to Roman and Palumbi's (2003) estimate (Smith and Reeves 2004).

Commercially well-organized long-distance whaling began gradually during the seventeenth century, with Nantucket, Massachusetts, emerging as its center (Starbuck 1989). When the North Atlantic whaling grounds showed signs of growing depletion the vessels moved into the South Atlantic, and soon afterward they were active throughout all but the southernmost circumpolar waters in the Pacific, as well as in the Indian Ocean. The world's first truly global industry had been born, its practices immortalized in a great American novel (Melville 1851). Sperm whales (*Physeter macrocephalus*) were the primary target of this great open-boat hunt, which began in 1712 and whose annual catches rose to more than 5,000 animals during the 1830s (Best 1983). Whitehead (2002) used modern visual surveys and a population model that incorporated uncertainties regarding population and catch data to reconstruct a global historical trajectory for sperm whales and concluded that the animal's status needs considerable revision.

White's prewhaling total was about 1.1 million whales (672,000–1,512,000), and the population was still at about 70% of its original level in 1880, when open-boat whaling was about to cease. Contrary to a common impression, open-boat hunting did not have a destructive impact on global sperm whale numbers. Modern whaling was made more lethal with the use of harpoon guns on steamship bows (common by the 1870s); after World War II long-distance expeditions by whaling vessels extended their reach to the last unexploited abodes in the Antarctic Ocean, and the adoption of diesel engines made it possible to hunt even the fastest whales (sei, minke). Japanese vessels began fishing in the Antarctic in 1934, and by 1939 they were killing nearly 10,000 whales, yielding about 45,000 t of meat. Whale meat became much more important after the end of World War II when it emerged as a leading source of scarce animal protein: in 1947 it was the only meat served by Japanese schools for lunch, and it remained popular until the early 1960s.

Japan remained the leading whaling nation after World War II, but its effort peaked in 1965, with nearly 26,000 killed animals (about 70% from the Antarctic seas); a decade later the killing was down to about 13,000 a year, and just over 5,000 whales were taken in 1980, two years before the International Whaling Commission finally decided to end commercial whaling starting in 1986 (Kalland and Moeran 1992). And yet whale meat sales have continued: while Japan observes the ban on commercial fishing, it has been conducting as already noted small-scale but persistent "research" hunts of mostly minke (but also sei and sperm whales), directed by a government-supported Institute of Cetacean Research, which yield 1,000–2,000 t of whale meat every year. Norway and Iceland are the only other countries that have been defying the whale-hunting moratorium.

The post-1970 decline and then the end of commercial whaling came just in time to prevent what might have become an irreparable overexploitation of several species, including the sperm whale. Between the 1830s and the 1920s, worldwide sperm whale hunting kept on declining, but then modern hunting led to rapidly increasing catches that culminated in the annual killing of more than 20,000 animals during the 1960s and in a significant population decline (Best 1983). Whitehead's (2002) models indicate that by 1999, more than a decade after commercial whaling ended, the global population of sperm whales was at 32% (85% confidence interval of 19%–62%) of the prewhaling total, and other species, above all the right whale and the blue whale, suffered even greater declines: by the year 2000 there were fewer than 8,000 right whales and perhaps only 2,300 blue whales (IWC 2010), totals that represent perhaps less than 10% and around 1% of the respective prewhaling populations.

In contrast to whaling, virtually no effective restrictions have been put on most of the world's commercial fishing even as fishing capabilities and typical vessel sizes had undergone a great expansion. The change began during the late nineteenth century with the introduction of steam-powered vessels. Unlike their wind- and tide-driven predecessors, whose operation was limited by those natural energy flows, the new vessels could use destructive trawls (nets dragged across the sea bottom, first introduced in England during the fourteenth century) continuously and they could also deploy much larger nets. Clear signs of the advancing depletion of fish stocks in the most heavily trawled seas were seen in a matter of decades.

The adoption of diesel engines (a superior prime mover compared to steam engines) after World War I further increased the power and operating radius of typical fishing vessels, and the effects of both these factors were multiplied again by the post–World War II introduction of large factory ships with on-board processing and refrigeration. At the same time, fishing vessels also became equipped with sonar and began to deploy monstrous drift nets (some up to 50 km long). All of these technical advances have made it easier to satisfy the rising demand for marine foods and led to what Pauly (2009) labeled a threefold expansion: first, the areal extension of long-distance commercial fishing to tropical oceans and then to the circum-Antarctic waters; second, the depth expansion as longline fleets and trawlers reached the demersal waters; and third, the concurrent taxonomic expansion as fisheries began to target previously spurned and unfamiliar taxa. After four decades this far-reaching, high-tech vacuuming of oceans had predictable results: most of the major fishing areas have become overexploited.

Signs of this trend have included the extension of fisheries well beyond the traditionally exploited waters, the capture of previously ignored, lower-value species, and rising prices. The widening range of operation is best illustrated by the history of Japanese fishing. During the occupation period it was restricted to only about 40% of the area in the eastern Pacific that it accessed before the war, but by 1960 Japan's fishing activities had extended to the Pacific and Indian Ocean waters surrounding Australia, to the Central Atlantic, and to the extreme latitudes of the Pacific: the Bering Sea in the north and east of New Zealand in the south (Swartz 2000).

Perhaps the best example of a massive exploitation of formerly ignored stocks is the fishing for Alaska pollock (*Theragra chalcogramma*), whose abundance attracted fleets of many nations to its rather limited habitat in the northernmost Pacific Ocean centered on the Bering Sea. Most of the catch has been consumed as imitation shrimp and crab meat (Japanese *surimi*). Pollock catches rose exponentially during the 1960s and the early 1970s but peaked in 1986 at less than 7 Mt/year, and by the beginning of the twentieth century they were below 3 Mt/year. Even so, they are the world's second largest single-species fishery, surpassed only by catches of the Peruvian anchovy (FAO 2011a).

As for the prices, Atlantic and Pacific bluefin tuna *(Thunnus thynnus* and *T. orientalis*, Japanese *maguro)* have led the way. The exploitation of these species began with catches of small fish for canning, but the market soon shifted to catching large (maximum weight in excess of 600 kg) specimens for export to Japan, where the meat is the highest-prized variety for sushi and sashimi. Japanese prices rose from just over 10 cents/kg in the late 1960s to about $2.5/kg by 1975, to nearly $30/kg for the fattiest fish by the late 1980s; the record price paid for a giant fish was $213/kg in 1991 (Buck 1995). That record has been repeatedly surpassed, and another large (232 kg) bluefin caught in Japanese waters was auctioned off for $775/ kg in January 2010 (Buerk 2010). Two years later even that price seemed modest, as a 269-kg bluefin was sold in January 2012 for $ 2,737/kg (BBC News 2012).

8

Crops and Animals

The most important force driving the evolution from foraging to crop cultivation and the domestication of animals is clear: gathering and hunting cannot support population densities higher than about one person per square kilometer, even in benign environments with abundant standing biomass. As the numbers of early Holocene humans began to increase, they gradually turned to more intensive ways of food procurement, to what are—in comparison with later traditional agricultures, and even more so with modern farming—still rather extensive ways of growing crops and rearing animals for meat and milk. Given the enormous variety of environments and dominant food sources, there could be no typical population density for foraging: the actual rates ranged over nearly two orders of magnitude, from just one person per 100 km^2 (or 0.01/km^2) in both densely wooded rainy tropics and on cold boreal plains to 0.1/km^2 for hunters on tropical grasslands, all the way to more than 1/km^2 in coastal societies relying on a rich supply (permanent or seasonal) of marine foods (Smil 2008).

The largest forager data set, presented by Marlowe (2005), allows the following generalizations. As expected, population densities increase with primary biomass but level off when the phytomass reaches about 30 kg/m^2: Marlowe's mean for the entire sample of 340 societies is 0.25 people/km^2, with a median of 0.11 people/km^2 (table 8.1). The highest rates (in excess of 1 person/km^2) were among the foragers along the northwestern Pacific coast of North America, but their exceptional densities reflected the seasonal abundance of anadromous salmon rather than high habitat richness. Food shares contributed by gathering, hunting, and fishing correlate significantly with latitude: the latter two activities were dominant in high northern latitudes and important in their southern counterparts, while gathering was the most important contribution in the tropics and subtropics.

In contrast, the population densities of pastoralists were almost always above 1 person/km^2 even in arid grasslands, and shifting cultivation commonly supported

For centuries, rice has been the world's most important staple grain crop. This photograph shows the mechanized harvesting and drying of sheaves on wooden supports in a field near Kyōtō in the early 1990s. Photograph by V. Smil.

more than 10 people/km^2, easily 100 times as many as could be supported in the same environments with foraging. This transition to more intensive modes of food procurement was a matter of complex evolution, and any notion of agricultural revolution (Childe 1951) is an indefensible intellectual construct. For example, wild cereals were collected and processed and the resulting flour was used in baking in Israel during the Upper Paleolithic (19,500 BCE), at least 12,000 years before the domestication of cereals (Piperno et al. 2004). In contrast, Mexico's Tehuacán Valley had no permanent settlements but thousands of years of crop cultivation (Bray 1977), and many settled agricultural societies retained significant elements of foraging for millennia. Clearly, there was no orderly lock-step progression of cultivation and sedentism, no sharp divide between foraging and incipient cultivation.

The cumulative impact of animal domestication and shifting and permanent crop cultivation began to transform ecosystems not only on local or small regional scales but eventually across large areas of some key biomes. As already noted, Ruddiman (2005) has argued that humans actually took control of the climate

Table 8.1
Density of Human Populations

	Population Density (people/km^2)	Live Weight (kg/ha)
Foraging	0.01–>1	0.005–0.5
Pastoralism	1–2	0.5–1
Shifting cultivation	20–30	9–14
Traditional farming		
Predynastic Egypt	100–110	45–50
Medieval England	150	75
Global mean in 1900	200	100
Chinese mean in 1900	400	180
Modern agriculture		
Global mean in 2000	400	200
Chinese mean in 2000	900	410
Jiangsu province in 2000	1,400	630

Note: All densities for foragers and pastoralists are per unit of exploited of land; all densities for traditional and modern agriculture are per unit of arable land.

through these actions. That conclusion is debatable, but there can be no doubt about the large-scale and often lasting environmental transformations brought about by pastoralism and cropping. And these changes affected not only fertile river valleys, coastal lowlands, or forests near growing population centers but even Amazonia, a region that was considered until recently the very epitome of undisturbed nature.

The transition to permanent cropping involved a complex mix of natural and social factors. The climate was too cold and CO_2 levels were too low during the late Paleolithic, but the subsequent warming, as Richerson, Boyd, and Bettinger (2001) put it, made agriculture mandatory in the Neolithic. After all, between 10,000 and 5,000 years before the present, agriculture had evolved independently in at least seven locations on three continents (Armelagos and Harper 2005). At the same time, the importance of cultural factors is undeniable, as sedentary farming fostered association and more complex social relations, promoted individual ownership and the accumulation of material possessions, and made both defense and aggression easier—but Orme (1977) went too far when he argued that food production as an end in itself may have been relatively unimportant.

Traditional low-yield cropping was the dominant subsistence mode in all preindustrial societies. Its practice shared some notable universal commonalities: grains as principal staple crops in all extratropical climates, the rotation of cereals and legumes, cattle used predominantly as draft animals rather than as sources of meat. But there were also many unique cropping sequences, some involving rotations of more than a dozen crops; differences in the domestication of animals (dairying cultures vs. societies abstaining from milk consumption); and distinct diets. Even the least intensive forms of cropping were able to support more than 100 people/km², and the best Asian and European practices could feed more than 500 people/km², that is, more than 5 people/ha (Smil 2008; table 8.1).

But even the best traditional productivities did not suffice to meet the post-1850 food demand driven by expanding populations, rapid rates of industrialization and urbanization, and higher incomes. Modern high-yield cropping has been a recent creation: its beginnings go back only to the last decades of the nineteenth century and its greatest returns have come only since the 1950s. Superior new cultivars can take advantage of new agronomic practices that combine high rates of fertilization, applications of herbicides and pesticides, supplementary irrigation, and the near total mechanization of field and processing tasks and that can provide excellent diets by supporting more than 10 people/ha. Moreover, these practices take advantage of comparative environmental advantages and increase the shares of national diets coming from imports (it is not uncommon to import more than 20% of all food energy). They also made it possible to consume unprecedented amounts of animal foods.

Shifting Cultivation and Pastoralism

Dissimilar as they appear to be, these two most extensive ways of food production are both suited only for low population densities because they both share an intermittent use of land. Shifting cultivation (swidden or slash-and-burn farming) and pastoralism could be also seen as intermediate evolutionary steps between foraging and permanent sedentary farming: they share extensive land use with the former but orderly practices (regular planting of crops, seasonal use of best pastures) with the latter. Of course, their environments differ greatly: shifting agriculture is best suited for forests and woodlands, while animal grazing converts grassy phytomass (indigestible by humans) to high-quality animal foods (milk, blood, meat) and hence enables people to survive in regions too arid for regular cropping.

As a result, the population densities of some nomadic pastoralists were as low as those of gatherers and hunters, while shifting agriculture could support population densities an order of magnitude higher than even the relatively best-off (sedentary maritime) foraging societies. But these extensive food production ways shared yet another important commonality: in areas with low population densities they proved to be extremely long-lived and resilient to change, while in other regions they gradually morphed into various combinations of shifting and permanent farming, or what might be best called seminomadic agropastoralism with some foraging still present. Rising population densities brought a rather rapid end to these modes of subsistence, mainly because of shorter soil regeneration cycles in shifting cultivation and the overgrazing of many pastures.

Shifting cultivation begins by removing natural vegetation (preferably a secondary growth, which is usually easier to clear away than a primary forest), often by burning or (once better tools became available) by felling large trees first and then slashing younger growth before setting woody piles afire, or simply by slashing a secondary growth without burning. The resulting patches ranged from fairly level surfaces that were often turned into densely planted gardenlike plots (interplanting and multilayered polycultures were common) to roughly surfaced small fields containing remnants of large burned trees where crops were grown in mounds or clumps. Regardless of the clearing method, shifting cultivation shared the underlying strategy of exploiting the nutrients accumulated in plants and soil.

Phytomass burning recycles all mineral nutrients (K, Ca, Mg), and while nearly all nitrogen in the burned phytomass is lost to the atmosphere as NO_x, a great deal of this key macronutrient remains in soil, and additional supply comes from the mineralization of decaying (slashed or uprooted) vegetation, so that crops can be grown for a few seasons without any other nutrient inputs. The staples of shifting cultivation included grains (rice, corn, millet), roots (sweet potatoes, cassava, yams), and squashes and various legumes (beans, peanuts), and the most diverse plantings included one or even two scores of edible, fibrous, and medicinal plants. Such plots required a considerable amount of maintenance (weeding, hoeing, pruning, thinning), and they also needed guarding or fencing, while many fields with staple crops were left in a semiwild state (subject to considerable damage by pests and animals) until the harvest.

Relatively large areas of forest were needed in places where just one or two years of cultivation were followed by long regeneration periods of more than 10 and even as many as 60 years; more intensive cultivation with a few crop years followed by only slightly longer (4–6 years) fallow periods were eventually quite common, and in many locations higher population densities compressed the cycle to the shortest fallow spans possible, with a crop or two followed by a single year of regeneration. Australia was the only continent where this practice was never important, while in large parts of Asia, Africa, and the Americas it not only preceded permanent crop-ping but has also coexisted with it, often literally side by side. Shifting cultivation was not only the foundation of ancient Mesoamerican cultures, it left its mark on relatively large areas of the Amazon.

Those regions are readily identified by their soils: *terra preta* (or, more precisely, *terra preta do índio*) can be up to 2 m deep; it contains relatively large amount of charred phytomass residues as well as elevated levels of organic matter from crop residues, human waste and bones (often in excess of 13% of total organic matter, that is, more than 5% C); and it was created by centuries of shifting cultivation, mostly between 500 BCE and 1000 CE (Glaser 2007; Junqueira, Shepard, and Clement 2010). Lehmann et al. (2003) estimated that *terra preta* soils cover no more than 0.3% of the low-lying parts of the Amazon basin, but other estimates are up to an order of magnitude higher.

In Asia, permanently cropped valleys and shifting cultivation in neighboring hills and mountains have coexisted for millennia, and in the southeastern part of the continent the forest clearing for cultivation has survived into the twenty-first century (Allan 1965; Watters 1971; Okigbo 1984). Swidden cultivation was also common in the temperate and boreal zones of preindustrial North America: Canada's Hurons were among the northernmost practitioners, cultivating corn and beans for at least 5 and up to 12 years before clearing a new patch of forest and letting the land regenerate for 35–60 years (Heidenreich 1971). With a typical corn harvest of 1.5 t/ha and a 50-year rotation cycle, they could support between 10 and 20 people/ha.

Shifting cultivation was not uncommon in medieval Europe—even in the Île-de-France region around Paris it was practiced until the early twelfth century (Duby 1968)—and it survived in parts of Central and northern Europe and Siberia not only into the nineteenth but even to the middle of the twentieth century. In Ger-many's Ruhrgebiet, tree fallows (harvested for bark and charcoal) alternated with rye until the 1860s (Darby 1956); rye was also used in Siberian shifting cultivation until the 1930s; and in Finland of the late 1940s birch and pines (cut and sold as logs) were followed by alder, whose burning prepared the soil for turnips. Other crops used in European shifting cultivation in Scandinavia, Central and Eastern Europe, and northern Russia (in some cases until the 1950s) included barley, flax, millet, oats, and wheat, and fallow periods in the sub-Arctic forests were as long as 40 years.

Net energy returns of shifting cultivation (quotients of harvested edible phyto-mass and energy consumption above basal metabolism) are as low as six- to tenfold and as high as 20- to 30-fold in various tropical environments, with rates between

15 and 20 being perhaps the most common and with maxima above 50 for some crops (Smil 2008). Average per capita requirements for shifting cultivation range from as much as 10 to less than 2 ha of land in fallow and under the crops, and the area actually cultivated is most commonly between 5% and 20% of those totals. This translates to population densities as low as $10/km^2$ and mostly between 20 and $40/km^2$, or an order of magnitude above the rates of $1–2/km^2$ typical of pastoral societies.

Specific claims made by shifting cultivation on the biosphere's phytomass vary widely, with the difference due mainly to the kind of cleared vegetation and to the length of the fallow period. Short fallows of less than five years, removal of relatively thin secondary (or later) growth, and a single year of cropping may claim only a small fraction of phytomass that is destroyed by burning primary growth, followed by a longer (more than 10 years) fallow period. Detailed monitoring of a Javanese short-rotation system—one year of mixed vegetables followed by a year of cassava followed by four years of bamboo fallow—showed the importance of the regenerative period: at its end, organic matter in the top 25 cm of soil had increased by about 7 t/ha, an outcome well justifying a traditional saying that "without bamboo, the land dies" (Christanty, Mailly, and Kimmins 1996). Two specific examples based on data obtained by field studies illustrate the overall impact of short- and long-rotation practices (box 8.1).

Even if the population densities supportable by shifting cultivation were to cluster rather tightly around 20–30 people/km^2 and even if we knew approximate totals of people engaged in these practices in the past, we still could not calculate the overall claims on the biosphere because the standing phytomass destroyed by felling or slashing and burning can vary by a factor of four. Shifting cultivation still survives in some tropical and subtropical regions, and some of its practitioners—including the Hanunoo of the Philippines (Conklin 1957), the Kekchi Maya of Guatemala (Carter 1969), the Tsembaga of New Guinea (Rappaport 1968), the Yukpa of Colombia (Ruddle1974), the Iban of Sarawak (Freeman 1980), the Kantu' of Kalimantan (Dove 1985), and the swidden cultivators of India's Manipur state (Shah 2005) and China's Yunnan province (Yin 2001)—were studied in considerable detail by modern anthropologists.

Perhaps the most revealing conclusion is that they found fairly similar and relatively high net energy returns (the quotient of the food energy in harvested crops and the energy spent in their cultivation) ranging between 15 and 25. But McGrath (1987) correctly pointed out that what has traditionally been seen as a highly productive system when assessed in terms of energy outputs and inputs is really an

Box 8.1
Shifting Cultivation: Phytomass Removal and Harvests

In Central America *milpa*, a plot of land devoted to the shifting cultivation of corn supplemented by half a dozen (interplanted and planted in sequence) fruit and vegetable species, is managed by a short cycle of four years of forest fallow and one year of crops (Carter 1969). An average corn yield of just over 1 t/ha (15 GJ/t) means that (assuming a 20% preharvest loss to wild animals and pests) the edible yield is about 11 GJ/ha planted. The short rotation entails clearing relatively thin secondary growth; even if it stores no more than 50 t/ha or (at 16 GJ/t of dry matter and a 90% removal rate after slashing and burning), the total is about 2.9 TJ of woody phytomass from the four fallowed hectares. More than 99.5% of the total phytomass removed from the site on a five-year fallow/crop cycle is thus the cleared woody matter, and with an average annual food requirement of about 10 GJ per capita, this cultivation will need about 4.5 ha per person, supporting about 20 people/km².

A very different shifting cultivation was studied by Rappaport (1968) in New Guinea: a long (15-year) fallow and a highly productive two-year gardenlike polyculture of more than 20 root (taro, yam, cassava), vegetable, and fruit species. The planted area is thus only about 12% of all managed land, and with forest phytomass stores at 150 t/ha and an 80% removal rate, every cycle would destroy 1,800 t of woody matter, or about 29 TJ. Even if the annual edible harvests were as high as 25 GJ/ha, in two years they would still amount to less than 0.2% of all the phytomass removed from the affected land during the 17 years of a fallow/crop cycle. With annual food requirements at 10 GJ per capita, this shifting cultivation would need nearly 3.5 ha of fallow and gardens per person and could support nearly 30 people/km².

extremely unproductive practice once the total phytomass contributions are taken into account. And the studies encountered a variety of location-specific practices, which does not make for any easy quantification of typical impacts.

Most recently, a global project on alternatives to slash-and-burn cultivation quantified a wide range of possible carbon stock changes (Palm et al. 2004). While undisturbed tropical forests averaged 300 t C/ha and managed and logged forests stored mostly between 150 and 250 t C/ha, short fallow (4–5 years), shifting cultivation replaced them with vegetation storing no more than 5–10 t C/ha—but long fallow (more than 20 years) and more complex alternatives (including agroforestry and intensive tree crops) could store an order of magnitude more of phytomass, between 50 and 100 t C/ha. Interestingly, the latest studies of shifting agricultures in Southeast Asia, the region with the highest remaining concentration of such farmers, have been reappraising the practice at a time when it is rapidly receding (Padoch et al. 2008).

Besides finding a surprising lack of agreement on the actual extent of the practice throughout the region, most of these studies document the replacement of shifting cultivation either by other forms of agriculture or silviculture (rubber or fruit trees, oil palms, plantation timber), as well as decreases in fallow length (Schmidt-Vogt et al. 2009). They stress its rationality in many tropical mountain environments, its suitability for smallholder operations, and, if properly practiced, its relatively high productivity. Moreover, case studies in Malaysia and Indonesia show that fallow length is a very weak predictor of subsequent yields and other factors (above all droughts, floods, and pests) are much more important (Mertz et al. 2008). Studies of vegetation recovery in northwestern Vietnam have demonstrated that the effect on biodiversity may not be as damaging as has been usually thought: after 26 years a secondary growth had half the species of the old-growth forest, and it was estimated that it would eventually reach a similar plant species diversity and about 80% of the original biomass (Do, Osawa, and Thang 2010).

Generalizations concerning pastoralism—which entails the metabolic conversion of grassy phytomass into meat, blood, and milk (and also wool and hides) through animal grazing—and fairly reliable summations of its worldwide claim on biomass are no less elusive. Pastoralism is perhaps best defined as a form of prey conservation whereby deferred harvests are profitably repaid by growing stocks of zoomass (Alvard and Kuznar 2001). This trade-off is more profitable for larger animals, but it is not surprising that the smaller species, sheep and goats, were domesticated first, because they have higher growth rates and are less risky to manage. The earliest domesticated animals—leaving dogs aside (Pang et al. 2009)—were sheep: their domestication process began about 11,000 years ago in Southwest Asia, and it was followed by goats and cattle (86,000 BCE), and later horses and water buffaloes (4000 BCE); domesticated pigs have been around nearly as long as sheep, chicken only about half that time (Ryder 1983; Harris 1996; Chessa et al. 2009).

Agropastoralism eventually became the dominant way of food production throughout Europe, North Africa, and West and Central Asia. Traditional pastoralist practices were strictly exploitative: there was no improvement of natural pastures, and labor (a significant share of it provided by children) was confined to herding (with a single person often responsible for hundreds of cattle and even larger herds of sheep or goats), guarding (sometime including the construction of temporary enclosures), and watering the animals. As a result, the ecological efficiency of pastoral subsistence—the quotient of the food energy of animal tissues consumed by humans and the forage energy intake of animals—was mostly less than 1% in seasonally arid African environments, similar to the efficiency of food obtained from hunting large native ungulates (Coughenour et al. 1985).

Milk and blood rather than meat were the primary food products (only old and ill animals were slaughtered). The variety of environmental settings (from poor semideserts to tall tropical grasses) and of domesticated herbivores precludes any easy generalizations about typical population densities of pastoral societies, whose actual practices have always ranged from the easily sustainable grazing of highly productive grasslands to the obvious overgrazing of only seasonally green semidesert grasses of low productivity. Moreover, in many cases the overall rate of phytomass consumption may not have changed all that much, as domesticated animals simply displaced wild grazers.

In most arid environments five to six head of cattle, or the equivalent zoomass in other species, are needed to support an average per capita food demand. The equivalence is commonly expressed in livestock units, whose size varies with breeds and feed supply: in East Africa, the region of some of the largest remaining pastoral populations, it is equal to 1.4 head of cattle, one camel, or 10 sheep or goats. When accurate head counts of grazers are known it is possible to quantify fairly accurately the effects on the stores and productivity of phytomass for a specific location or for a small and relatively homogeneous region, but such an account will miss some important qualitative differences, above all the impacts of different modes of grazing by cattle, camels, sheep, and goats. An East African example illustrates the challenge (box 8.2).

Because of significant differences in body weights and pasture qualities, any global account of grazing impacts can aim only at the right order of magnitude—providing that good approximations of total numbers of all major grazers are available. The earliest available global estimate is for 1890, when the totals were just over 400 million head of cattle, nearly 100 million horses, and about 750 million goats and sheep (HYDE 2011). The zoomass (live weight) of these domesticated grazers was about 150 Mt, ten times greater than a likely mass of some five million African elephants that were living in the wild by the end of the nineteenth century.

There is also no simple generalized formula to predict a degree of competition between domestic and wild herbivores: the outcome depends on the species involved and their densities, on seasonal stresses, and on other forms of interference (fences, setting fires to prevent bush and tree growth). Wild grazers on savannas coexist because of resource partitioning, that is, choosing different kinds, sward heights, and ages of available forage, and the introduction of cattle may hardly change that. Although cattle are usually considered to be nonselective roughage grazers, in Africa

Box 8.2
An Example of Grazing Impacts: East Africa

In East Africa, where a large-scale expansion of grazing began a millennium ago, the minimum number of cattle needed to produce enough food has been five to six head per capita; as a result, the most commonly encountered population densities were between just 1–2 people/km^2 (Helland 1980; Evangelou 1984; Coughenour et al. 1985). Similar population densities could be supported with three camels and about 30 sheep per capita. Standard metabolic equations can be used to calculate approximate annual feed requirements, assuming that the average weight of cattle is 200–250 kg, which is not much greater than for wildebeests and somewhat less massive than adult zebras (Voeten and Prins 1999).

The bovine feed requirement would be then about 40–45 MJ/day, and with an average grass phytomass power density of 11 MJ/kg, this would translate to annual intakes of up to about 1,500 kg of dry-matter feed per head. But this does not mean that cattle herds have been reducing the NEP of East African grasslands by 1.5 t per head every year. With lower-intensity grazing (in the absence of any cattle) the accumulation of high-grass phytomass may lead to larger fires rather than providing more nutrition to wild herbivores (Holdo et al. 2007). More important, exclosure studies have shown that in high-rainfall years, chronic grazing by a mixture of cattle and wild ungulates can actually increase annual NPP on nutrient-rich glades (due to deposition of nitrogen in feces and urine) while suppressing it on nutrient-poor bushland sites (Augustine and McNaughton 2006).

they may participate in a three-way partitioning with wildebeests and zebras, and their presence may overlap with zebras' resource use only the early wet season, and with wildebeest grazing only in the early dry season (Voeten and Prins 1999). And while in Kenya's arid regions wild grazers completely avoid areas that are heavily grazed by livestock, they mix with livestock in semiarid rangelands, and may actually graze close to settlements for protection from predators (Suttie, Reynolds, and Batello 2005).

Historical sources make it clear that populations of pure pastoralists were relatively common during antiquity as well as during the Middle Ages. Repeated encounters of settled societies with roaming pastoralists—be it due to long-distance invasions, protracted armed conflicts, or, particularly in China's case, attempts to manage the ever-present nomadic threat with walls, treaties, and trade—were one of the key unpredictable and disruptive factors of their premodern history (Adas 2001). Modern appraisals see a properly practiced pastoralism not as an archaic way of food production that is conducive to land degradation and can provides only negligible economic benefits but as a rational, environmentally appropriate, and potentially rewarding option to exploit the limited primary productivity of arid ecosystems (Rodriguez 2008).

But pastoral societies have been in retreat for centuries owing to a combination of natural and human factors. Unpredictable production in arid climates can be easily aggravated by overgrazing; once other options become available (such as moving to nearby farm settlements or migrating to distant cities), prolonged droughts pushed many pastoralists permanently off their ancestral lands; rangeland is reduced through the encroachment of neighboring agricultural societies coveting the best grazing lands; and recurrent attempts by centralized governments either restrict nomadic roaming or force the populations to resettle.

For these reasons, a gradual, and since the beginning of the modern era a surprisingly rapid, expansion of land devoted to animal grazing is not an indicator of a relative increase in the world's pastoral populations but a reflection of two separate processes. First, the higher population growth that first became evident during the early modern era (average annual growth rates were less than 0.1% before 1700, and during the eighteenth century they averaged about 0.5%) pushed increasing numbers of pastoralists into less inviting areas, whose poor grazing translated into lower productivity. Second, the extension of large-scale commercial cattle production primarily onto the vast grasslands of North America and Australia during the nineteenth century (particularly its latter half) and onto the South American and African grasslands throughout most the twentieth century added nearly twice as much pastureland in 200 years as the total arable land accumulated after millennia of settled cultivation—without any expansion of true pastoralist populations.

Approximate estimates of the world's grazing land show a slow growth during the first 1,500 years of the Common Era, from about 100 Mha at its beginning to about 220 Mha in the early sixteenth century (HYDE 2011). By 1800 that total had more than doubled, to just over 500 Mha, and then it expanded 2.5 times during the nineteenth century, mainly as a result of the westward population advances in the United States, Canada, and Australia (for a gain of some 250 Mha of pastures) and the large expansion of grazing throughout sub-Saharan Africa (a gain of some 200 Mha) and Central Asia. And the expansion rate was even slightly higher during the twentieth century (a 2.6-fold gain), when the greatest gains came in Latin America (quadrupling the 1900 total to about 560 Mha, mainly thanks to the conversion of Brazilian *cerrado*), followed by sub-Saharan Africa (more than doubling, to about 900 Mha) and several regions of Asia.

At the beginning of the twenty-first century pastures occupied about 34 million km^2, or more than twice the area devoted to the cultivation of annual and perennial crops (15.3 million km^2)—and Europe is now the only continent where less than 50% of all land used for food production is in pastures (the rate is just short of 40%): in North America the share is just over 50%, in Asia over 60%, in Africa over 80%, and in Australia close to 90% (Rodriguez 2008). But even if we define traditional pastoral societies as those deriving at least 50% of their livelihood from grazing animals, their numbers have been declining for many generations. Sandford (1983) put the total number of true pastoralists at fewer than 23 million people during the 1970s (with some three-quarters in Africa and the rest in the Middle East and South and Central Asia), and since that time the numbers have declined substantially.

The three countries with the largest pastoralist populations, Sudan, Somalia, and Ethiopia, have suffered from years of war, while the pastoralists in both the eastern and western parts of the continent have seen their pastures degraded by drought and overgrazing and reduced by the expansion of cropping and nature reserves. Pastoralism now accounts for significant shares of agricultural output only in the poorest arid countries of Africa (Sudan, Mauritania, Niger, Ethiopia), and even there it now contributes less than 10% of the total GDP. That share is less than 2% in the Andean countries (alpacas) and a small fraction of 1% in Australia and Spain (cattle and sheep).

Traditional Agricultures

Net energy returns in early farming were no or only marginally higher than in foraging, and peasant diets became even more plant-based than was the case with many foragers, but only permanent cropping could support higher population densities, by combining the cultivation of a large variety of annual and perennial plants with the breeding of several species of domesticated mammals and birds and, in some societies, also aquatic animals. These new food production systems arose gradually beginning about 11,000 years before the present and spreading to most of the Old World by 4000 BCE (Harris 1996; Zeder 2008).

The earliest domesticated plants—cereals and pulses (leguminous grains), einkorn and emmer wheat (*Triticum monococcum* and T. *turgidum*), barley (Hordeum vulgare), lentils (*Lens culinaris*), peas (*Pisum sativum*), and chickpeas (*Cicer arietinum*)—were cultivated as early as 8500 BCE, Chinese millet and rice between 7000 and 6000 BCE, New World squash by 8000 BCE, corn by 4500 BCE; and flax (*Linum usitatissimum*) was the first fiber crop (Zohary and Hopf 2000; Kirch 2005 Abboe et al. 2009). Animal husbandry was dominated by cattle, water buffaloes, camels, horses, sheep, and goats in the Old World but was limited to llamas, alpacas, and guinea pigs in South America. The most common domesticated birds were fowls of Southeast Asian origin, ducks, geese, and pigeons throughout the Old World; turkeys were the only birds domesticated in America.

Grains—primarily wheat, rice, barley, rye, and oats in Eurasia, millets in Africa, and corn in the Americas—eventually became the most important staples, and legumes (beans, peas, lentils, and soybeans in the Old World, beans in the Americas) supplied most of the dietary protein. Tubers—mainly sweet potatoes, taro, and cassava—were the staples in tropical societies, though after 1600 Andean potatoes diffused worldwide to become a major source of carbohydrates. These staples were supplemented by a wide variety of vegetables and fruits and in most societies also by the cultivation of oilseeds (two leguminous species, soybeans and peanuts, as well as sunflower, rapeseed, sesame, and olive).

Low-yield cropping remained the norm even in the economically most advanced regions until the modern era: in much of Atlantic Europe until the eighteenth century, on the Great Plains of North America until the late nineteenth century. Gradual intensification of inputs aimed at securing higher and more stable yields relied on leveling and terracing fields, building often elaborate irrigation systems, and recycling organic matter, but it resulted in only slowly rising yields rather than in fundamental productivity breakthroughs, and higher outputs had to be secured primarily by expanding the cultivated land.

In China this process entailed centuries of cropland expansion both in the arid northern regions and in the rainy south suited for rice cultivation, where it had to be usually preceded by deforestation; in Europe, cropland expansion led to the gradual colonization of less productive land and large-scale deforestation, first in the Mediterranean and later in Atlantic, Central, and Eastern Europe. In North America it led first to extensive deforestation in New England and in the mid-Atlantic states and later (after 1850) to a rapid conversion of the continent's interior grasslands to crop fields; a similar process took place in Ukraine and parts of European Russia, southern Siberia, and Kazakhstan.

Modification of this extensive farming began first by the introduction of more intensive practices in the most densely settled parts of Eurasia, but low-yield agriculture dominated the Americas even during the nineteenth century. Land claims of extensive cropping can be easily appraised by reviewing the best evidence of crop yields in ancient societies and in their medieval successors. Because of the Nile's minuscule gradient (1:12,000), there was no perennial canal irrigation in ancient dynastic Egypt, just levees, drainage channels, and storage basins used to manage

Box 8.3
Predynastic Egypt: Population Density and Grain Farming

Excavations indicate that Egypt's predynastic settlements were about 2 km apart and that they used only a quarter of the outer part of the 1.5-km-wide floodplain for cultivating grains, whose yield was about 600 kg/ha (Hassan 1984). This would prorate to about 75 ha of farmland per settlement and an annual harvest of 45 t of grain. With 25% of the harvest reserved for seed and at least another 10% lost in storage, there would be about 30 t of food grain (food energy density of 15 MJ/kg). Assuming (rather liberally) average daily food requirements of 10 MJ per capita and 75%–85% of that total supplied by staple cereals, that would be enough to feed about 150–170 people (and up to 200, with slightly lower daily per capita needs).

But because the Nile's annual fluctuations could cut the long-term average yields by as much as one-half, the minimum supportable population was perhaps no higher than 75–85 people, or just 1–1.1 people/ha of cultivated land (or 100/km^2). Egypt's population density rose from 1.3 people/ha in 2500 BCE to 1.8 in 1250 BCE, to 2.4 by 150 BCE (Butzer 1976), and 2.7 Mha cultivated during the Roman rule produced food for at least seven million people (one-third of them outside Egypt), with an implied a population density of about 2.5 people/ha of arable land. After more than 1,500 years of decline, the average density rose to about 5 people/ha (500/km^2) only after 1800.

annual flood flow; moreover, the absence of efficient water-lifting devices restricted irrigation to low-lying areas in the river's valley. As a result, there were no summer crops, and the population density supportable by annual flood soon reached its limit (box 8.3).

Fragmentary Chinese information offers similar rates. Prior to the third century BCE there was very little irrigation, double cropping, or crop rotation, and population densities ranged from just 1 person/ha in the arid north to 2 people/ha in the rice-growing south; they rose to about 2.8 people/ha during the early Ming dynasty (1400 CE) and to 4.8 people/ha two centuries later during the early Qing dynasty (Perkins 1969). By 1900 the mean was about 5 people/ha, Buck's (1937) detailed rural surveys resulted in a highly accurate mean of 5.5 people/ha of cultivated land around 1930, when many irrigated southern regions could support 7 people/ha (700/km^2).

Because corn (a C$_4$ species) is an inherently higher-yielding crop than wheat and rice (both C$_3$), the most fertile area of ancient Mesoamerica could support higher population densities than in the Old World. The highest rates were achieved thanks to the Aztec *chinampas*, rectangular fields raised by using mud, crop residues, and grasses and 1.5–1.8 m above the shallow waters of Lakes Texcoco, Xalco, and

Xochimilco. By the time of the *conquista* (1519) there were about 12,000 ha of *chinampas* in the Basin of Mexico able to sustain population densities ranging from fewer than 3 people/ha to as many as 13–16 people/ha, and the mean for the entire basin was about 4 people/ha (Sanders, Parsons, and Santley 1979). Similar densities could be supported by raised-field cultivation of potatoes around Lake Titicaca, the core area of the medieval Inca Empire (Denevan 1982). Indigenous North American cropping has received less attention, but McLauchlan's (2003) pollen analysis showed that even in southwestern Ohio the cultivation of starchy or oily seeds began as early as 4,000–3,500 years ago.

Fairly reliable historical statistics show that the U.S. grain yields remained very low even during the first quarter of the twentieth century: in 1866 (at the end of the Civil War) the wheat harvests averaged just 750 kg/ha, during the next five decades they rose only to a fluctuating plateau between 800 and 900 kg/ha, and they permanently surpassed 1 t/ha only during the 1940s (USBC 1975); similarly low rates had prevailed in Canada. But because of the late nineteenth-century opening up of the U.S. Great Plains and Canada's prairies, these low yields satisfied not only the rising domestic food demand but also produced plenty of high-quality feed for record numbers of draft animals and a surplus for export: in 1900 the land planted to supply the latter two demands accounted for, respectively, about 25% and 10% of all cultivated U.S. cropland (Smil 2008).

Well-documented history of traditional European farming shows that a long period of stagnating and fluctuating medieval yields ended only during the late eighteenth and the early nineteenth century. During the first two centuries of the Common Era (the apogee of Roman imperial power), cropping in Italy or Gaul (with typical wheat yields at just 400 kg/ha) could support no more than 1 person/ha, a millennium later typical wheat yields in Atlantic Europe were only marginally higher. Reconstructions of English wheat yields indicate average population densities of no more than 1.5 person/ha (100/km^2) at the beginning of the thirteenth century (box 8.4).

There are two simple explanations why food production in traditional agricultural societies—despite its relatively high need for claiming new arable land—had a limited impact on natural ecosystems: very low population growth rates and very slow improvements in prevailing diets. Population growth rates averaged no more than 0.05% during the antiquity and they reached maxima of just 0.07% in medieval Eurasia—resulting in very slow expansion of premodern societies: it took Europe nearly 1,500 years to double the population it had when Rome became an

Box 8.4
English Wheat Yields, 1200–1800

The earliest estimates of European wheat yields are derived from information on relative returns of the planted seed: during the early medieval period these ratios were only twofold and during the years of inclement weather harvested grain could be less than the planted amount. By 1200 the ratios ranged mostly between 3 and 4, and could surpass 5 (Bennett 1935; Stanhill 1976). Conversions of these (usually volumetric) ratios to mass averages can be done only approximately; they indicate yields of just over 0.5 t/ha by 1200 and an irreversible doubling to more than 1 t/ha only half a millennium later.

After reducing the average yield of 0.5 t/ha by at least 10% to account for post-harvest (storage) losses to fungi, insects, and rodents, and then by a third to be set aside for the next crop's seed, only about 300 kg of whole grain (or 4.5 GJ) are available for food consumption (the production of white flour would reduce the mass by another 15%); a yield of 1 t/ha, a storage loss of just 5%, and a quarter of the harvest set aside for seed would leave about 700 kg (10.5 GJ) for food. Because in Europe's traditional agricultural societies as much as 85% of all food energy (at least 8 MJ/day, or nearly 3 GJ/year) came from staple grains, medieval harvests of 0.5 t/ha could support only about 1.5 person/ha (150/km²), while those of around 1 t/ha (reached by 1700) could sustain between 3 and 4 persons/ha (300–400/km²) planted to cereals.

empire, and Asian doubling was only a bit faster, from the time of China's Han dynasty to the late Ming period.

And no traditional agriculture was able to produce enough food to assure more than basic (and overwhelmingly vegetarian) diets even during the best years and to prevent recurrent periods of hunger and famine that in some societies persisted even into the nineteenth century (Ireland 1845–1852, China and India 1876–1879 and 1896–1902). Low crop yields also limited the use of draft animals (particularly horses, whose heavy draft required concentrate feed; in contrast, the much less demanding cattle could be fed only grass and crop residues) and the production of animal foods. As a result, the global area of cropland grew only marginally during the first millennium of the Common Era (from 130 to 150 Mha), and it then took another seven centuries to double, to 300 Mha (HYDE 2011).

If we assume average food yields equivalent to 300 kg of grain/ha in the year 1000 and 400 kg/ha in 1700, the global harvests amounted, respectively, to roughly 50 Mt and 120 Mt of grain equivalent, enough to feed (assuming that plant foods had supplied 90% of all food energy needs averaging 3.5 GJ per capita per year) about 200 and 500 million people; this compares to global population estimates of about 300 million people in the year 1000 and 600 million people in the year 1700, leaving the remainder (respectively, about a third and roughly 15%) to be fed from foraging, shifting cultivation, and animal grazing. Similar approximations show that the eighteenth century brought still relatively small but clearly noticeable gains in average yields and population densities. The global area of cultivated land rose from 300 Mha in 1700 to about 420 Mha in1800, a 40% expansion, while the population increased from 600 million to nearly one billion, close to a 65% gain, and hence a 20% decline in the per capita availability of arable land, which implies a similar increase in average productivity.

Expanded cultivation also produced more crop residues. Annual seed crops had evolved to be tall (Donald and Hamblin 1984): traditional rice varieties grew to be up to 1 m tall, wheats as much as 1.5–1.8 m, corn up to 3 m, and sorghum to around 4 m. This tallness was valued because their stems and leaves provided residual phytomass that could be used as animal feed and bedding, as fuel, and as raw material, and hence they were usually harvested as carefully as the crop itself. Crop residues make inferior fuels: their bulk and their relatively low energy content mean that fires require a large volume of the material and frequent stoking, but they were the only readily available phytomass in deforested regions.

Traditional cultivars had invariably low harvest index (HI), the ratio of crop yield and the total aboveground phytomass, including harvested parts and residual stalks, stems, twigs, leaves, or pods. The HI for traditional wheat cultivars was as low as 0.2–0.3, for rice cultivars no higher than 0.36 (Donald and Hamblin 1984). Those residues that were not removed to be used elsewhere were left on the ground to decay naturally (sometime after they were grazed by animals), but a more common choice was either to plow them in to replenish soil organic matter or burn them. Although crop residues are low in nitrogen (usually no more than 0.5% of dry weight), some straws are relatively high in potassium, and in any case, their recycling is a rich source of organic carbon.

Farming intensification continued in most European countries after the recovery from an overproduction-induced depression in the early nineteenth century. Two German examples illustrate these changes (Abel 1962). In 1800 about a quarter of German fields were fallowed, but the share was less than 10% by 1883. The average annual per capita meat consumption was less than 20 kg before 1820, but it was almost 50 kg by the end of the century. Three-crop rotations were replaced by a variety of four-crop sequences: in a popular Norfolk cycle, wheat was followed by turnips, barley, and clover, and the practice of six-crop rotation was also spreading.

Applications of calcium sulfate, and of marl or lime to correct excessive soil acidity became common in better-off areas.

A slowly growing use of better-designed implements also accelerated during the nineteenth century. By the middle of the century yields were rising in every important farming region as the rapidly intensifying agriculture was able to supply food for growing urban populations. After centuries of fluctuations, population densities were rising steadily. In the most intensively cultivated Atlantic regions of the continent they reached 7–10 people/ha of arable land by the year 1900. But these levels already reflected considerable energy subsidies received indirectly as machinery and fertilizers produced with coal and electricity. European farming of the late nineteenth century became a hybrid: still critically dependent on animate prime movers but increasingly benefiting from many inputs requiring fossil energy.

Modern Cropping

The gradual intensification of traditional farming involved innovations with immediate effects on typical yields (and hence on the required land), as well as measures that had only a marginal impact on harvests. Increasingly elaborate irrigation techniques and regular recycling of organic wastes belonged to the first category: they improved average harvests, or at least reduced the annual fluctuation in yields. On the other hand, the substitution of animal draft for human labor or the adoption of better field machinery brought much higher labor productivities but often had only a minimal effect on average yields: by 1900, American field work was highly mechanized (with large horse-drawn steel plows, grain reapers, binders, combines, and steam-powered threshers)—but average grain yields remained low.

Major gains in terms of average land productivity came only with the assiduous combination of crop rotations (particularly with the regular cultivation of pulses or cover crops used as green manures), multicropping (reducing or eliminating fallow periods), and high rates of organic (animal and human waste, composts) recycling. Some of the most densely populated regions of Eurasia reached the highest levels of crop productivity that could be supported without synthetic fertilizers and pesticides and without improved cultivars; as already noted, by the 1920s China's traditional farming could feed 5–7 people/ha of cultivated land, and similar levels were reached in the most fertile parts of Japan, on Java, and in parts of Atlantic Europe.

While these innovations reduced the area of agricultural land needed to secure adequate diets, three trends had the opposite effect. The first one was higher population growth, the second one was a much more common reliance on more powerful draft animals, and the third one was a gradual improvement in typical diets, particularly manifesting in higher intakes of animal foodstuffs. During the eighteenth century, annual population growth rates in Europe were on the order of 0.4%, and during the nineteenth century they rose to around 0.7%, the latter rate being an order of magnitude higher than typical medieval rates. North American rates, thanks to immigration, were even higher, well above 2% after 1850.

The earliest use of draft animals called for only a limited expansion of phytomass harvests. Oxen used for field work in ancient Egypt or in Roman and medieval farming subsisted on readily available roughages (cereal straws, any accessible grazing) and hence did not require any additional land for feed crop cultivation and did not lower the potential grain supply of peasant families. This changed only when better-off farmers in Atlantic and Central Europe began using more powerful but also more demanding horses. By the early nineteenth century a pair of average European horses would need nearly 2 t of feed grain a year, a mass about nine times that of the total food grain consumed annually by their master. And the cultivation of North American grasslands led to an even greater demand for animal feed.

At the beginning of the twentieth century a team of dozen powerful horses used for field work in the grain fields of Dakotas or Manitoba needed about 18 t of oats and corn per year, about 80 times the total amount of food grain consumed by their owner. Obviously, only land-rich countries could afford this burden: growing concentrate feed for those 12 horses would claim about 15 ha of farmland. This would have been an excessive burden even in the land-rich United States, where by 1900 an average farm had almost 60 ha, but only a third of it was cropland. As a result, only large grain growers on the Great Plains could feed a dozen or more horses, and the countrywide ownership averaged only three horses or mules per farm (USBC 1975). Expansion of the cultivated area and the mechanization of field work (including the use of horse-drawn combines, some pulled by more than 30 animals) brought an unprecedented demand for feed crops, which eventually added up to a quarter of America's harvested land (box 8.5).

Outside North America, agricultural mechanization powered by draft animals proceeded at a much slower pace, and in some of the world's poorer regions it relied only on less demanding water buffaloes (in most of Asia's rice-producing countries) and oxen. The third important reason for the expansion of farmland was a gradual improvement in average diets: after centuries of basic subsistence typical food intakes began to include more animal foodstuffs, whose production required a

Box 8.5
Farmland Needed for America's Draft Animals in 1918

The total number of draft animals working on America's farms peaked during the second half of the second decade of the twentieth century at nearly 27 million, including more than 21 million horses and more than 5 million mules, with working stock two years and older making up about 85% of that total, or nearly 23 million animals (USBC 1975). With an average daily need of 4 kg of grain for working animals and 2 kg of concentrate feed for the rest (Bailey 1908), the annual concentrate feeding requirement would have been at least 35 Mt of corn and oats. With prevailing yields at about 1.5 t/ha for corn and 1.2 t/ha for oats and with three-quarters of all grain supplied by corn, this would have required the planting of about 25 Mha of feed grains.

Working horses also needed at least 4 kg of hay a day, while 2.5 kg of hay a day were enough for nonworking animals. Hay requirements would have thus added up to no less than 35 Mt/year. With average hay yields of between 2.5 and 3 t/ha, some 13 Mha of hay had to be harvested annually. The total land required to feed America's draft animals was thus around 38 Mha, or 26% of all area harvested in those years and significantly more than the area planted to exported crops (equal to about 17% of the total). For comparison, the USDA calculated the total feed requirements for 1918 to be just short of 37 Mha (or 25% of the total), an excellent confirmation of my independently derived total.

higher output of concentrate feed (cereal and leguminous grains) besides the roughages (grasses, straw) needed by ruminants.

Traditional agriculture of the early modern era (1500–1750) and of the great industrialization period (1750–1900) was still powered only by animate energies (human and animal muscles); it renewed soil fertility only by fallowing, the recycling of organic wastes, and rotation with leguminous crops; and it relied on a diversity of local cultivars. But cumulative improvements in agronomic management raised labor productivity and brought the best yields very close to the peaks attainable with premodern inputs. An hour of medieval field labor produced just 3–4 kg of grain; by 1800 the European mean was around 10 kg, and by 1900 it was well over 40 kg. And net energy gains in grain production—the quotients of available food energy and overall animate energy inputs—rose even more impressively, from less than 25 for typical medieval harvests to around 100 by 1800 and to 400–500 a century later (Smil 1994). These gains allowed masses of rural labor to shift into urban manufacturing, construction, and transportation sectors. In the United States, more than 80% of the labor force was in agriculture in 1800; a century later the share was down to 42% (USBC 1975).

Improved record keeping makes it much easier to trace the rising yields during the eighteenth and nineteenth centuries. Better implements (allowing timely completion of critical field tasks), a progressive reduction of fallow, increased manuring, regular crop rotations, and the diffusion of new cultivars were the key ingredients of the process. As expected, yields began to rise first in some of the most densely populated regions, including the Low Countries, England, and Japan. Farmers in densely settled Flanders were as much as two centuries ahead of their German or French neighbors in reclaiming the region's wetlands, gradually abandoning extensive fallowing, adopting several standard crop rotations, and applying increasing amounts of organic fertilizers.

In the eighteenth century, Flanders' organic recycling (manure, night soil, oil cakes, and wood ash) averaged 10 t/ha (Slicher van Bath 1963), and some fields received every year as much 30 t/ha. By 1800 less than 10% of arable land on Dutch farms was fallowed, and there was a close integration of cropping with livestock production (Baars 1973). In England, rotations that included legume cover crops— such as the four-year succession of wheat, turnips, barley, and clover—had doubled or even tripled the rate of symbiotic nitrogen fixation (Campbell and Overton 1993), and Chorley (1981) considered their regular adoption to be as significant for the society as the contemporaneous diffusion of steam power.

English wheat yields began a steady, slow increase already after 1600, but a real surge in productivity came only between 1820 and 1860, a result of extensive land drainage, regular crop rotations, and higher rates of manuring: by 1850 many counties had harvests of 2 t/ha (Stanhill 1976). Dutch wheat yields doubled during the nineteenth century to 2 t/ha. But neither in the Netherlands nor in England could these yield increases satisfy the combined effect of population growth and a rising demand for better diets, and both countries saw a substantial expansion of their cultivated area: in the already highly intensively cultivated Netherlands it grew by a quarter between 1750 and 1900, and in the UK the increase was nearly threefold (HYDE 2011). The same forces acted in Japan, where the average rice yield surpassed 1 t/ha already during the early seventeenth century, 100 years later it was about 1.5 t/ha, during the first full decade of the Meiji era (1870) it reached 2 t/ha, and in 1900 the average yield of husked rice was about 2.25 t/ha, or 25% higher if expressed as unmilled paddy rice (Miyamoto 2004; Bassino 2006).

Even when compared with the best mid-nineteen-century performance, modern plant breeding eventually tripled the Japanese rice yields and more than quadrupled the Dutch rate. Its most important goal has been to boost HI, that is, to increase the proportion of photosynthate channeled into harvested tissues (Donald and Hamblin 1976; Gifford and Evans 1981; Hay 1995).

These efforts have succeeded in pushing modern harvest ratio close to or even above 0.5 (and for rice close to 0.6). Obviously, with an HI of 0.5, phytomass is equally divided between crop and residual mass, while in traditional cultivars, whose HIs were often as low as 0.25 and rarely higher than 0.35, residual phytomass weighed 1.8–3 times as much as the harvested tissues.

Modern short-stalked wheat cultivars incorporated a semidwarf Norin-10 gene and were introduced by the International Maize and Wheat Improvement Center during the 1960s (Dalrymple 1986). During the same time IRRI released its IR8 variety (HI 0.53), which yielded as much as 10 t/ha. Corn yield began to improve already before World War II with the introduction of hybrids seeds in the United States starting during the 1930s, and by the early 1960s all U.S. corn was hybrid (Smil, Nachman, and Long 1983). Roughly 30 years after the introduction of high-yielding varieties came the products of the latest great breeding wave aimed at guaranteed higher yields, genetically engineered (transgenic crops), starting with corn incorporating a bacterial toxin, then both an herbicide-tolerant corn and soybeans (Fernandez-Cornejo and Caswell 2006). Because this innovation met with considerable consumer and regulatory resistance, it has not been rapidly extended to the two staple food crops, wheat and rice.

All of the far-reaching changes that affected cropping during the twentieth century can best be demonstrated using the statistics of major Western nations and Japan and combining J. L. Buck's pre-1930 studies with nationwide statistics for China, which begin only in 1949. Between 1950 and 2000, the average yields of America's three most important crops (when ranked by their value in the year 2000), corn, soybeans, and wheat, rose respectively 3.5, 1.7, and 2.5 times, to about 8.5 t/ha, 2.5 t/ha, and 2.8 t/ha (USDA 2000). During the same period China's rice, wheat, and corn yields increased respectively 3, 5.4, and 1.7 times, while Japanese rice yields doubled between 1950 and 2000, rising from 3.2 t/ha in 1950 to 6.4 t/ha (SB 2010). Without these improvements the area needed to grow staple food crops for expanding and better-fed populations would have to have been substantially larger.

Dietary Transitions and Meat Production

The transition from foraging to cropping, from a mobile to a sedentary existence, increased the average population densities (eventually by as many as four orders of magnitude), but in doing so it had to transform typical diets. The most obvious component of this change was the consumption of meat: not all foraging societies were highly carnivorous (and in the case of maritime cultures, animal protein came overwhelmingly from fish and shellfish), but without a doubt, the diet of megafaunal hunters of the late Paleolithic contained much more meat than the diet of the first Neolithic cultivators who planted legumes and cereals and supplemented their diet with regular hunting. In turn, those diets contained more meat than did the typical diets of peasants in medieval or early modern societies. As a result, the average quality of nutrition was generally lower among traditional cultivators than among the foragers with access to plenty of animals because meat is not only a convenient source of high-quality, easily digestible protein, it is also an excellent source of vitamins A, B_{12}, D, and iron and zinc.

The consequences of this shift are well confirmed by anthropometric studies that show diminished statures of sedentary populations when compared with their foraging predecessors (Kiple 2000). While there are no reliable data that would allow us to calculate representative regional or national averages of actual meat consumption during antiquity or the centuries of premodern agricultural intensification, the best generalizations would be as follows: even in the relatively well-off societies, average per capita meat intakes were mostly just 5–10 kg/year, and in most subsistence peasant societies meat consumption was minimal, as small quantities were eaten only once a week, with more consumed only on a few festive occasions.

And these low intakes did not rise even during the early modern period because more pastures were converted into arable land in order to feed larger populations. In addition, intensification of farming made bovines more valuable as prime movers (for both field work and transport) and sources of manure, and as a result, milk and cheese rather than meat were the most commonly consumed animal foods. During the early 1790s the meat consumption of poor English and Welsh rural laborers was only about 8 kg/year (Clark, Huberman, and Lindert 1995). In France, large numbers of peasants ate meat only at Easter and at weddings; Toutain (1971) calculated that meat intake supplied less than 3% of all food energy in 1800.

In the most densely populated regions of premodern Asia a very similar situation persisted well into the twentieth century. Japan represented the most extreme case, with its Buddhism-inspired imperial bans on meat eating going back originally to the seventh century and renewed repeatedly on many occasions afterward. Meat intakes averaged still less than 2 kg/year during the seventeenth century (Ishige 2001), and Buck's (1930) data for the early 1920s show the range of Chinese annual intakes between less than 0.3 kg/year in Hebei and 5 kg/capita in Jiangsu, with a nationwide mean of less than 3 kg/capita (Buck 1937). Even by the 1960s the average per capita meat intake was below 10 kg/year in all of Asia's most populous nations, China, India, and Indonesia.

Not surprisingly, intakes were higher among richer urbanites, rations were even higher for some armies, and premodern meat consumption was relatively high in North America as well as in Argentina and, later, in Australia and New Zealand. The dietary transition began first in the growing industrial cities of Europe and North America. Germany's average annual per capita meat consumption was less than 20 kg before 1820, but it was almost 50 kg by the end of the century (Abel 1962; French meat consumption remained low until 1850, then doubled to more than 50 kg by the 1930s (Dupin, Hercberg, and Lagrange 1984); and the British per capita consumption had roughly tripled during the nineteenth century to almost 60 kg by the year 1900 (Perren 1985). This pre-1900 dietary shift had a fairly limited impact on the biosphere, and not just because it affected only a small share of the world's population. The key explanation of this modest impact is in the prevailing modes of animal food production.

The traditional raising of domestic animals did not compete for arable land with food crops. Free-ranging poultry received only small amounts of grain as an occasional supplement, free-roaming pigs rooted for any edible biomass and were fed any available crop and kitchen waste, and cattle and water buffaloes, as well as goats and sheep, were raised by grazing (on grasslands as well as on harvested fields) and by feeding crop residues and hay. These animals made only a marginal additional claim on arable land, but they were also relatively inefficient users of feed and took much longer to reach slaughter weight than modern breeds. Chicken were commonly killed only at 3–6 months of age (compared to about six weeks for today's mass-produced, and also significantly heavier, broilers), while pigs attained their slaughter weight only after 12–15 months (compared to just short of six months from weaning to slaughter for modern pigs).

In 1900 there were about 1.3 billion large domestic animals, including 500 million bovines, about 200 million pigs, 750 million sheep and goats, and 100 million horses. If we were to assume that 10% of all animal food production in 1900 was energized by concentrate feeding (a generous allowance), and if we were to use high take-off rates (80% of all pigs and 15% of all bovines slaughtered annually) and a feed/live weight conversion efficiency of 6:1 for pigs and 10:1 for cattle, we would end up with a total feeding requirement of less than 40 Mt of grain equivalent. This could be produced, even if we assume average low yields of just 1 t/ha, on some 40 Mha of arable land, or not even 5% of about 850 Mha of the total cultivated area in 1900. Even if the actual total were twice as large, it would still be less than half the difference between the lowest (710 Mha) and the highest (935 Mha) estimate of arable land for 1900 (HYDE 2011).

In contrast, the modern production of animal foods depends on concentrates, that is, grains (mainly corn, but also wheat and barley) and tubers (cassava, potatoes) as source of carbohydrates, and soybeans as the high-protein ingredient (supplemented by oilseed cakes and by various crop and food-processing residues). Most of the cropland in affluent countries is now devoted to growing feed rather than food: during the first decade of the twenty-first century nearly 60% of U.S. farmland was used to grow corn, soybeans, hay, and sorghum, the four leading feed crops (USDA 2009). Increasing animal food intakes in land-poor countries could be done only by expanding imports.

In Japan, the world's largest food importer, feed corn became the country's most important imported grain during the late 1960s, when it surpassed imports of wheat for food consumption. Recent annual corn imports have been around 16 Mt/year, or more than 6% of America's corn crop, and their production required about 1.8 Mha, an equivalent of nearly 40% of Japan's arable land. Less than 10% of the imported wheat is used as animal feed, but barley imports are about equally split between food and feed uses, and sorghum, the fourth largest imported grain, is used only for feeding. In total, by the year 2000 Japan's feed grain imports had reached 27 Mt, by far the largest in the world and equal to almost exactly 10% of the global cereal trade.

And more land was claimed by feed grown to be produced Japan's meat imports. After decades of protectionist measures the country began to opens its meat market during the late 1980s, and the imports rose from about 0.7 Mt in 1985 to almost 2.8 Mt by the year 2000, making Japan the world's largest buyer of meat. In addition, Japan has been an importer of dairy products and eggs, amounting to altogether more than 600,000 t of animal protein imported in the year 2000. The average yield of all animal protein was about 200 kg/ha, implying a total land claim of about 3 Mha to produce feed for the export of animal foods to Japan (Smil and Kobayashi 2012). Alternative calculation methods yield similar totals, between 2.5 and 3.5 Mha for the year 2000 (Galloway et al. 2007).

Calculations of global feed requirements for the year 2000 show an enormous increase in phytomass claimed by the production of animal foodstuffs. In that year the total number of large animals approached 1.6 billion, including 1.3 billion heads of cattle, 164 million water buffaloes, 70 million horses and mules, more than 20 million camels, 900 million pigs, and 1.75 billion sheep and goats (FAO 2011c).

There were also more than 14 billion chickens and about 1.7 billion other birds (ducks, turkeys, geese). Even if no more than about half of all feedstuffs came from concentrates and average feeding efficiencies were fairly high—feed/product conversion ratios of 1:1 for milk, 4:1 for poultry meat and eggs, 8:1 for pigs, and 10:1 for cattle—the global requirement would be no less than 1.3 Gt of grain equivalent.

For comparison, the FAO's (2011b) balance sheets show about 38% of all cereals, 24% of all tubers and pulses, and 10% of all oilseeds were fed directly to livestock; adding to this cereal milling residues and oil cakes results in a total grain equivalent of nearly 1.3 Gt, an excellent confirmation of the just presented estimate. Producing this feed would have required, even with a relatively high average yield of 2.5 t/ha, some 520 Mha of arable land, or no less than a third of the total cultivated area in the year 2000 (see also Steinfeld et al. 2006). Although higher intakes of animal foods were a key component in the early stages of the modern dietary transition they had, at least until the 1940s, resulted in only limited demand for new arable land, and other key ingredients of the changing dietary composition were more important.

These included a higher consumption of sugar and plant oil, as well as a greater variety of vegetables and fruits. Per capita sugar consumption was minuscule in all traditional societies. It began to rise with production from cane plantations (first in the Caribbean, later also in Africa and Asia as well as in Hawaii and a few southern U.S. states), and in temperate latitudes it took off only with a large-scale commercial extraction from beets during the nineteen century. Plant oils began to compete with animal fats, and rising incomes created a demand for better-quality vegetables and fruits. At the same time, staple grain and tuber intakes declined, and legumes saw an even greater reduction. All of these trends got under way in parts of Europe and North America before the end of the nineteenth century, and after World War II they spread to Latin America and Asia.

9

Biomass Fuels and Raw Materials

Wood supplied virtually all the energy needed for space and water heating, cooking, and a growing range of artisanal manufactures and industrial processes for far longer than most people think. Fossil fuels were known, both in Asia and in Europe, since antiquity, but the first society in which coal became more important than wood was England of the early seventeenth century—and that remained a great exception. Reasonably reliable estimates and actual output statistics show that in France, coal began to supply more than half of all fuel needs only sometime during the mid-1870s, and in the United States the tipping point came in the mid-1880s. Fossil fuels gained more than half the Japanese energy supply in 1901, in Russia only during the early 1930s (Smil 2010b). Nearly smokeless and sulfur-free, charcoal was a preferred fuel for indoor heating and cooking; in China it was also the raw material for making an indispensable material of that ancient civilization, its black writing ink.

But charcoal's most important use was in smelting metals: it was the only metallurgical fuel used to reduce copper, tin, lead, and iron ores. The Bronze Age (bronze is an alloy of copper and tin) began nearly 6,000 years ago, while iron items first appeared appear more than 3,000 years ago. Several ancient societies had even solved the difficult challenge of making high-quality steel by reducing cast iron's carbon content, and medieval artisans could produce high-quality swords in Kyōtō or Damascus (Verhoeven, Pendray, and Dauksch 1998). But until large blast furnaces made it possible to mass produce iron utensils, tools, implements, and machines, all societies lived through millennia of a wooden age: steel became inexpensive only during the last generation before World War I (Smil 2005).

Charcoal's inefficient production was a leading contributor to extensive deforestation, first in the arid Mediterranean and in northern China, and during the early modern era also in Atlantic and Central Europe. And traditional biomass fuels remain the dominant or only source of heat energy for hundreds of millions of poor

Charcoal was the most valuable phytomass fuel in all advanced traditional societies. These two engravings from Diderot and D'Alembert's famous *L'Encyclopédie, ou Dictionnaire raisonné des sciences, des artes and des métiers* (1751–1780) show the construction of a charcoal kiln and its firing and breaking up.

peasants in sub-Saharan Africa and the least prosperous parts of Asia and Latin America. Wood and charcoal are still widely used in Brazil, the country that also leads the world in the use of ethanol; unfortunately, other national programs promoting liquid biofuels (most notably the U.S. corn-based ethanol and European biodiesel) make little sense as they do not offer any environmental, energetic, or economic advantages.

Wood was also the quintessential raw material of all preindustrial civilizations: metal tools, implements, machines, and weapons were much less common than their wooden counterparts. As for building materials, stone and bricks have been common choices since prehistoric times, but in many parts of the world wood retained its primacy in construction into the twentieth century. Japan's pre–World War II residential buildings were overwhelmingly wooden, and sawn wood remains

(continued)

the structural foundation of North American housing. And without sawn wood there would have been no preindustrial transportation, either on land (wood was used to construct vehicles from simple wheelbarrows to heavy wagons and fancy carriages) or on water (the evolution of wooden vessels spans more than five millennia)—and no affordable and leak-proof storage of liquids (in casks and barrels). There would also have been no railroads: wood was always only a marginal locomotive fuel compared to coal, but wooden ties (sleepers) were, and still are, essential to distribute the load and maintain the gauge.

Higher disposable incomes mean that the ownership of wooden furniture is more widespread than ever, with a typical Western family owing more than a score (or two) of wooden chairs, tables, desks, beds, dressers, and stands (leaving aside all those still very common wooden floors). The only major difference between traditional artisanal furniture-making and modern industrial manufacturing is that veneers and plywood have largely displaced solid wood. And the advent of modern papermaking during the latter half of the nineteenth century created another large market for wood as the source of pulp.

Fuelwood in Traditional Societies

The preindustrial combustion of wood had an undesirable common characteristic: poor efficiency in converting the fuel's chemical energy into heat, which led to high consumption rates even for the simplest heat-demanding tasks. Moving the open fires indoors did little to improve their efficiency but did result in chronic exposure to high levels of indoor air pollution, a problem that has persisted in many of the world's poorest societies (especially in Africa and Asia) into the twenty-first century (McGranahan and Murray 2003). Typical wood combustion efficiencies rose a bit once the richer traditional societies moved their fires into stone or brick fireplaces. Some food (meat on spits, stews in cauldrons) could be cooked a bit less wastefully than in the open (thanks to the controlled ascent of heat through a flue and its reverberation from fireplace walls), but fireplaces did a poor job heating large rooms: their radiated heat reached only the immediate vicinity of the hearth, and on cold days the room's warm air was drawn outside to support combustion; moreover, combustion was dangerous (fire hazard, CO emissions) and difficult to regulate.

Some early space-heating arrangements—the Roman *hypocaust,* whose oldest remains date to the third century BCE (Ginouvès 1962), and the Korean *ondol* led hot gases through brick or stone flues under floors, while the Chinese *kang* was a heated platform that was also used as a bed at night—were both more efficient and more comfortable. Chimneys became common in Europe only during the late Middle Ages (elsewhere only centuries later), and iron-plate stoves followed only during the seventeenth century. These stoves raised typical combustion efficiencies to at least 15%–20%, compared to less than 5% for open fires as well as most fireplaces. After 1860, household heating and cooking in European cities rapidly converted to coal or coal gas, but the continent's rural areas and both urban and rural populations outside Europe continued to rely on wood or, in forest-poor lowlands, on crop residues.

The third major category of wood and charcoal consumption has been in the production of goods: in small-scale artisanal manufactures, in brick, pottery, and china kilns, in forging metals, shoeing horses, and brewing beer. And starting in the nineteenth century a fourth category could be added, wood combustion to generate

Box 9.1
Wood Consumption in the Roman Empire

Average per capita rate is constructed by adding conservative estimates (based on the first principles) for major wood consumption categories. I assume that bread baking (mostly in urban *pistrinae*; only rich households baked their own bread) and cooking (mostly in numerous *tabernae*) required 1 kg of wood/day per capita. At least 500 kg of wood/year were needed for space heating that was required for the roughly one-third of the empire's population that lived beyond the warm Mediterranean in the temperate climates of Europe and Asia Minor. To this must be added an average annual per capita consumption of 2 kg of metals (much of it for military uses), which required about 60 kg of wood per 1kg of metal.

This adds to annual per capita demand of 650 kg (roughly 10 GJ) or about 1.8 kg/day—but as the Roman combustion efficiencies were uniformly low (no higher than 15%), the useful energy derived from burning that wood was only on the order of 1.5 GJ/year, a modern equivalent of just 35 kg of crude oil (or nearly 50 liters, or one tankful, of gasoline). For comparison, when Allen (2007) constructed his two Roman household consumption baskets he assumed average consumption of nearly 1 kg wood/capita a day for what he called a respectable alternative and just 0.4 kg/capita for a bare-bones budget (none of these rates included fuel for metallurgy and manufacturing).

steam to power locomotives and boats. This use was almost everywhere rather rapidly superseded by coal, but wood continued to supply small workshops and local industries in many low-income countries throughout the twentieth century. Quantifying preindustrial wood combustion is highly uncertain: there is no reliable information either for antiquity or for the Middle Ages. My reconstruction of fuel demand in the early Roman Empire (that is, during the first two centuries of the Common Era) ended up with at least 10 GJ per capita (box 9.1), while Galloway, Keene, and Murphy (1996) put the average annual wood consumption in London around 1300 at 1.5 t of air-dried wood (20–25 GJ) per capita.

Recent studies of those Asian and African societies where phytomass remains the principal source of heat show similarly low rates, with an annual per capita consumption of less than 10 GJ among the poorest inhabitants in the warmest regions. But typical wood consumption rates were always higher in heavily wooded central and northern Europe. Some traditional manufacturing tasks were highly wood-intensive. German medieval glassmaking required as much as 2.4 t of wood—nearly all of it burned to obtain potassium rather than to generate heat—to produce 1 kg of glass (Sieferle 2001). Only about 3% of all wood heated the glassmaking process, but that still translated to as much as 1 GJ/kg of glass; in contrast, today's large-scale industrial glass production needs only about 15 MJ/kg, mostly in the form of natural gas. Salt production in interior locations—evaporating brines in large wood-heated iron pans—was another inefficient process consuming large volumes of wood: a kilogram of salt required as little as 5 kg and as much as 40 kg of wood.

Residential consumption could reach very high levels in colder forested areas (Sieferle 2001). In 1572, the Weimar court needed annually 1,300 cords of wood, or some 3,000 m³, a volume that would require about 600 ha of coppice growth. In the early eighteenth century the University of Königsberg allocated 70 m³ of wood for every employee, that is, at least 500 GJ per household, or as much as 100–125 GJ per capita. That was exceptionally high: during the eighteenth century, annual wood consumption in Germany was typically on the order of 50–60 GJ per capita (Sieferle 2001). The Austrian mean in 1830 was 73 GJ per capita (Krausmann and Haberl 2002), and the residential demand was similar in the nineteenth-century United States: household heating consumed up to 50 GJ per capita, and rising industrial needs added as much or more. Schurr and Netschert (1960) estimated that all final uses prorated to an average U.S. per capita consumption of nearly 100 GJ during the 1850s. In comparison, the contemporaneous Parisian rates were minuscule as annual per capita fuelwood consumption fell from less than 15 GJ (1.8 m³) in 1815 to less than 5 GJ by 1850 (Clout 1983).

Charcoal

Wood's bulk and often high moisture content made charcoal a preferred choice for indoor combustion and a perfect reducing agent for smelting metals. Charcoal is a slightly impure carbon with a high energy density of nearly 30 MJ/kg, but its mass density is low, ranging from 115 kg/m³ for charcoal made from pines to 189 kg/m³ for fuel derived from tropical hardwoods. Charcoal burns without any particulate or sulfurous emissions. In Japan, charcoal was burned in *kotatsu*, braziers placed in pits (about 40 cm deep) cut into the floor and covered with a table and a blanket: this may have warmed legs and even an upper body clad in traditional Japanese clothes, but left most of the room uncomfortably cold. In England, heating pots and braziers remained the norm well into the early modern era when the large-scale use of coal was common: the British House of Commons was heated by large charcoal fire pots until 1791.

But household consumption of charcoal was a small fraction of the demand for charcoal as metallurgical fuel, a reducing agent in smelting first the ores of color metals (copper, tin, and lead) and later, and more important, iron: charcoal was the only source of carbon used to reduce ores in preindustrial societies (Biringuccio 1540). Metallurgical charcoal is made from wood (reeds and cotton stalks were also used) by pyrolysis, that is, heating in the absence of oxygen; its energy density is twice that of air-dried wood or dry straw and 20%–35% higher than that of common bituminous coals. Traditional charcoal production in simple earthen kilns (wood piles covered with mud or turf) was extremely wasteful. Typical yields were only between 15% and 25% of the dry wood charge by weight, and in volume terms, as much as 24 m^3, and rarely less than 10 m^3, of wood were required to make a ton of charcoal.

The 1:4 ratio is perhaps the best conservative approximation of large-scale charcoal:wood yields, and with 15 MJ/kg of air-dried wood, traditional charcoal-making thus entailed an energy loss of about 60%. But the burning charcoal could reach 900^{0}C, and the use of a forced-air supply (using hollow bamboo or wooden tubes, later bellows) could raise the temperature close to 2,000°C, well above the melting point of all common metals. Copper was the first metal that was produced in relatively large amounts; it melts at 1,083°C and lead melts at 232°C, while iron melts at 1,535°C; consequently, the substitution of bronze by iron brought a major increase in charcoal demand.

Forbes (1966) estimated that some 90 kg of wood were needed to smelt 1 kg of copper; assuming 900 kg of wood per tree in 40 years and 300 trees/ha (or about 6.75 t/ha a year), a smelter producing 1 t Cu/day would have required nearly 5,000 ha a year, and if such operations persisted in a region for a century, the cumulative demand would have reached 500,000 ha of forest. An estimate for Cyprus (Constantinou 1982) showed forest demand equal to 16 times the island total area, a calculation that shows a "tendency toward exaggeration" (Williams 2006, 78). But even if we consider that the demand extended over a period of many centuries, or even two millennia (and hence a large part of it would have come from regenerated growth), there is no doubt that copper smelting caused extensive deforestation around the Mediterranean.

Iron smelting eventually became much more efficient, but the metal was produced in much larger quantities. The smelting was done first in only partially enclosed hearths and yielded small (around 50 kg) masses (blooms) or slag-contaminated metal that could be forged but not cast. Iron production in bloomery hearths and forges needed as much as 20 kg of charcoal (up to 80 kg of wood) per kilogram of hot metal (Johannsen 1953). These specific needs declined with the adoption of simple shaft furnaces that were operated in Europe, India, and China (Chakrabarti

1992; Needham 1964), and then decreased even more with the introduction of the first blast furnaces. These taller structures appeared first during the fourteenth century in the Rhine-Meuse region and could produce cast iron (called pig iron, as the metal was cast into shapes resembling chunky pigs). Their lower specific demand was far outweighed by the expanded production. which led to acute wood shortages in some regions.

Sussex is now an affluent county, but by the mid-sixteenth century, when a single furnace claimed annually a circle of forest with a radius of about 4 km, its inhabitants petitioned the king to shut down many furnaces, wondering "what number of towns are like to decay if the iron mills and furnaces be suffered to continue" and where the wood for their houses, wheels, and barrels would come from (Straker 1969, 115). During the first decade of the eighteenth century an English blast furnace typically worked for eight months a year (between October and May) and smelted about 300 t of pig iron (Hyde 1977). With 8 kg of charcoal per 1 kg of iron and 5 kg of wood per 1 kg of charcoal, a single furnace would have needed some 12,000 t of wood during its eight-month campaign, and the overall English demand for metallurgical charcoal would have claimed more than 1,000 km^2 a year; a century later the U.S. demand added up to an equivalent of nearly 2,500 km^2 of mature forest cut every year (box 9.2).

In Europe, Sweden continued to be a heavy charcoal user throughout the nineteenth century. In 1850, 25% of the country's wood harvest was used to make charcoal (Arpi 1953). By 1900 the typical charging ratio was less than 1.5 kg of charcoal per 1 kg of metal, and the best Swedish furnaces needed as little as 0.77 kg (Greenwood 1907). Brazil is now the world's only consumer of metallurgical charcoal, made from fast-growing eucalyptus trees, with large plantations mostly in the state of Minas Gerais (Peláez-Samaniegoa 2008). Charcoaling is done in hemispheric or cylindrical brick kilns, which exposes workers to a number of possible injuries and to hazardous smoke (Kato et al. 2005). No by-products are recovered, and the conversion efficiency remains no better than 25%. Less than 10% of it is used by households and in cement production, close to 10% goes for making ferroalloys, and 75% is charcoal for blast furnaces. A typical requirement has been about 2.9 m^3 (or 725 kg) of eucalyptus-derived fuel to smelt 1 t of hot metal (Ferreira 2000).

Biofuels Today

At the beginning of the twenty-first century, biomass fuels were still more important than either hydro- or nuclear electricity—at least when their contribution is expressed

Box 9.2
Charcoal Requirements for Iron Smelting

In 1810, the earliest year for which the nationwide data are available, the United States smelted about 49,000 t of pig iron. With an average rate of 5 kg of charcoal (at least 20 kg of wood) per kilogram of hot metal, this production required approximately 1 Mt of wood. Even if all that wood were harvested in old-growth eastern hardwood forests, which contained around 250 t of aboveground phytomass per hectare (Brown, Schroder, and Birdsey 1997), and even if all stem, bark, and branch phytomass could be used in charcoaling (both being excessively generous assumptions), an area of about 4,000 km² (a square with a side of about 63 km) had to be cleared in that year. By 1840 all U.S. iron was still smelted with charcoal, but by 1880 nearly 90% of it was produced with coke.

Without the switch from charcoal to coal-based coke, even the forest-rich United States could not carry on for another century: in 1910 the country smelted 25 Mt of iron, and even with just 1.2 kg charcoal (or about 5 kg of wood) per 1 kg of hot metal it would have needed 125 Mt of wood. Harvesting this wood from a mixture of first- and second-growth forests averaging 125 t of aboveground phytomass per hectare would have claimed 10,000 km²/year; in just ten years an unchanging rate of iron smelting would have cleared 100,000 km² of forest, an area only slightly smaller than Pennsylvania—but by 1920 the U.S. output was more than 35 Mt of iron, and annual wood harvests would have to be (even with efficiency improvements) about a third higher.

These realities should be kept in mind by all advocates of green energy futures. Blast furnaces produced about 50 Mt of iron in 1900, 580 Mt in 2000, and nearly 1,000 Mt (1 Gt) in 2010. By 1910 the average worldwide ratio of dry coke to hot metal had declined to just 0.45:1, which means that the global need for pig iron smelting was about 450 Mt of coke. Replacing all of this with charcoal would require—even when assuming just 0.7 t of charcoal per 1 t of hot metal and a 3.5:1 wood: charcoal ratio— about 2.5 Gt of wood, the total roughly equal (assuming average 650 kg/m³) to the world's harvest of about 3.9 Gm³ of roundwood. And even if that wood were to come from high-yielding tropical eucalyptus (producing annually 10 t/ha), its growth would occupy some 250 Mha, an equivalent of more than half of Brazil's Amazon forest.

in terms of gross available energy. But comparisons with other sources of primary energy cannot be done with a high degree of accuracy because there is still a fairly large uncertainty concerning the global annual rate of phytomass combustion; there are also major disparities in average conversion efficiencies of phytomass and fossil fuels. Biofuels contributed about 9% of the world's total fuel supply in the year 2000. But as the overall conversion efficiency of fossil fuels is now about 40%, the combustion of 340 EJ of fossil fuels produced nearly 140 EJ of useful energy. Even if we assume a conversion efficiency of 20%, the world's biofuels yielded less than 10 EJ of useful energy, or less than 6% of the useful energy contributed by fossil fuels.

The demand for wood and crop residues for household cooking and heating would be even higher if it were not for improved stoves. Schumacher (1973) was an early proponent of their diffusion, but most of the designs introduced during the 1970s proved impractical: usually too expensive, inconvenient to use, difficult to repair, and short-lived (Kammen 1995). The most important change came with China's National Improved Stove Programme, launched in 1982 and originally intended to provide 25 million units within five years. The program far surpassed its initial expectations. By the end of 1997 about 180 million stoves had been adopted by nearly 75% of China's rural households (Wang and Ding 1998). Because these stoves are typically 25%–30% efficient, they were credited with saving close to 2 EJ a year (Luo 1998).

While the traditional biofuels retain their importance in rural areas of many low-income countries, their use is now marginal in all but a few affluent nations. After their displacement by fossil fuels they began to make a slight comeback following the rise in oil prices during the early 1970s, but their share for all OECD countries was only 4.3% in 2009, compared to 3% in 1973 (IEA 2010). The highest national shares of combustible renewables in the total primary energy supply are now in Sweden and Finland (in both countries about 20%, mostly wood-processing and logging wastes used by industry) while the contributions are below 7% in Germany, 5% in France, less than 4% in the United States, and less than 1.5% in Japan.

Given the high rates of urbanization and a nearly universal urban access to natural gas and electricity, the household use of biofuels in affluent countries will remain minuscule. But many proponents of renewable energy have argued that biofuels could make a major difference is in supplying liquids for road transport. This is an old proposition: Henry Ford favored ethanol a century ago, and his famous Model T could, like the Brazilian flex vehicles today, run on gasoline, ethanol, or a mix of the two fuels (Solomon, Barnes, and Halvorsen 2007). Brazil pioneered the large-scale production of sugar cane-based ethanol with its Programa Nacional do Álcool, which was launched in 1975 and eventually led to more than half of Brazilian cars using anhydrous ethanol, whose lower heating value of ethanol is only 21 MJ/L, compared to 32 MJ/L for gasoline (Smil 2010a).

The U.S. production of corn-based ethanol began in 1980, but rapid expansion of output began only in the year 2002: by 2005 the shipped volume had nearly doubled, to 15 GL, and by 2009 it had surpassed 40 GL, with the maximum volume mandated at 136 GL of ethanol by the year 2022. That goal implies a 17-fold expansion of ethanol output in 20 years, but even if gasoline demand remained stable at the 2010 level of about 520 GL, biofuel would supply only about 17% of the 2022 demand. Low power densities of corn-derived ethanol (averaging only about 0.25 W/m^2) limit its eventual contribution. Even if the entire U.S. corn harvest (averaging 280 Mt/year between 2005 and 2010) were converted to ethanol, the country would produce some 2.4 EJ of biofuel, or just 13% of its total gasoline consumption in 2010. Moreover, corn-based ethanol has many negative consequences.

Studies of energy costs of corn-derived ethanol have shown values ranging from a significant energy loss (Pimentel 2003) to a substantial energy gain after taking credits for by-products of fermentation used for animal feeding (Kim and Dale 2002). But these accounts ignore the environmental degradation inherent in any intensified and expanded corn cultivation: the perpetuation of a massive crop monoculture, an increased demand for irrigation water, the leaching of nitrates and the ensuing eutrophication of streams, and the extension of the anoxic dead zone in the coastal waters of the Gulf of Mexico (Smil 2010a). And the post-2002 expansion has taken place only thanks to the substantial government subsidies needed to make the fuel competitive with gasoline (Steenblik 2007).

Sugarcane is a much better choice as a feedstock for ethanol fermentation: the entire process has a power density of nearly 0.5 W/m^2, and according to Macedo, Leal, and da Silva (2004), typical cultivation and fermentation practices in the state of São Paulo have an energy return of 8.3, and the best operations may have rates just in excess of 10. But de Oliveira, Vaughan, and Rykiel (2005) found those calculations underestimate the energy cost of sugarcane cultivation, and they put the net energy return at less than 4. But the return is a secondary matter compared to the availability of suitable land: no other large populous country is as land-rich as Brazil, and this puts limits on the mass-scale production of cane ethanol (box 9.3).

And even the United States and Brazil would have to deal with higher food prices if they embarked on the large-scale cultivation of biofuel crops, a policy that the

Box 9.3
Limits on Ethanol Production from Sugarcane

Sugarcane (*Saccharum officinalis*) is a perennial tropical grass whose average global yield is now about 65 t/ha (FAO 2011d). In the wet tropics it does not need any irrigation; the best Brazilian cultivars do not need any nitrogen fertilizer (or only minimal supplementation), thanks to endophytic N-fixing bacteria in stems and leaves; and the ethanol production does not require any external fuel as it can be entirely fueled by the combustion of bagasse, the fibrous residue that remains after expressing the juice from cane stalks. Sucrose is about 12% of the harvested crop (or 7 t/ha), and even if it were converted to ethanol as efficiently as it is now in Brazil—producing 82 L/t (Bressan and Contini 2007)—1 ha of cane would yield less than 5,500 L of ethanol.

Some 23 Mha were planted to sugarcane in all tropical and subtropical countries in 2010, and even if all that land produced cane for ethanol, the annual fuel yield would be about 2.6 EJ, equal to about 7% of the world's 2005 gasoline consumption of roughly 37 EJ. In order to supply the global gasoline demand, the entire demand cane would have to be planted on some 320 Mha, or 20% of the world's arable land. But high-yielding sugar cane can be grown only in the tropics, and this means that about 60% of all the farmland that is now under cultivation in that region would have to be devoted to cane for ethanol, clearly an impossibly high claim.

arable land shortage makes impossible (or quite irrational) in China, India, or Indonesia. There is no need for academic debates regarding that choice: as Jean Ziegler, the UN Special Rapporteur on the Right to Food, put it, "It is a crime against humanity to convert agriculturally productive soil into soil which produces foodstuffs that will be burned as biofuel" (Ferrett 2007). Other crop-based options, ranging from biodiesel made from rapeseed to the exploitation of tropical species, including oil palm and jatropha, offer no overall advantages compared to corn or sugarcane, and that is why the future of liquid biofuels is now seen to depend on the success of cellulosic ethanol, on enzymatic conversion of cellulose in crop residues and wood.

Availability, performance, and cost of the requisite enzymes and problems with scaling up bench experiments to large continuous or batch operation producing billions of liters a year are the most obvious challenges. But is should be also remembered that feedstock supplies are neither as abundant nor as easily harvested as is often assumed. Corn stover, America's most abundant crop residue, illustrates some of these challenges. Different assumptions regarding stover:grain ratios, high average moisture content, and recycling requirements have resulted in estimates of annual U.S. stover harvests as low as 64 Mt and as high as 153 Mt of dry matter (Kadam and McMillan 2003).

Less than 5% of all stover is regularly harvested as cattle feed, and the rest is recycled. Stover availability fluctuates with yields by as much as 20%, and Pordesimo, Edens, and Sokhansanj (2004) argued that its typical fresh-weight supply is 20% less than the total resulting from a commonly assumed 1:1 stover:grain ratio. Another common assumption is that 3–4 t of stover could be harvested per hectare on a sustainable basis, but Blanco-Canqui et al. (2007) found that stover harvesting at rates above 1.25 t/ha changes the hydraulic properties of soil and has a pronounced effect on earthworm activity (complete removal cut the number of earthworm middens by 95%).

A fairly conservative approach is to assume that with conventional tilling, about 35% of stover could be removed without adverse effects, and that the rate rises to about 70% for no-till farming. This would give a weighted national mean of about 40%, and it would mean that about 80 Mt (in dry weight) of stover could be removed annually, a theoretical equivalent of roughly 20–25 GL of ethanol—worth no more than 3% of the U.S. gasoline consumption in 2010. And using phytomass-based liquid biofuel is even more questionable given the utter lack of system appropriateness: the negative consequences would be easier to accept if the fuels were used in highly efficient vehicles (in excess of 50 mpg), not by a fleet that now averages less than 30 mpg.

A combination of environmental, agronomic, social, and economic impacts thus makes modern liquid biofuels a highly undesirable choice. Resource constraints (the availability of farmland, competition with food crops) mean that even their large-scale production would have a marginal effect on the overall supply. The production of cellulosic ethanol would use relatively abundant crop residues, but they are often difficult to collect and expensive to transport and store, and, once gathered, their structural cellulose and lignin are not easy to break down to produce fermentable sugars. Moreover, crop residues are not useless wastes whose best use is an enzymatic conversion to ethanol; they are an extremely valuable resource that provides a number of indispensable and irreplaceable agroecosystem services that range from the recycling crop of the three macronutrients as well as many micronutrients and the replenishment of soil organic matter to the retention of moisture (through their spongelike action), and prevention of both wind and water erosion. Consequently, any excessive harvesting of this only apparently waste phytomass is inadvisable.

Wood as a Raw Material

Crop residues incorporate more than half of the world's crop phytomass, and their traditional off-field uses for feed, bedding, and household fuel remain widespread.

But who is now thatching roofs in Europe or North America, who is wearing straw coats to protect against rain or weaving *waraji* (straw sandals) in Japan? True, Japanese still produce *tatami*, woven rice straw mats whose dimensions (variable, but mostly between 85 and 95 cm × 180–190 cm) used to define the size of rooms—but that is a clear exception: in affluent countries crop residues are no longer used either in construction or in manufacturing. Wood is a different matter.

Many of wood's past (and until fairly recently large-scale) uses have also disappeared (nobody is building massive wooden ships, still dominant in the mid-nineteenth century) and others have been reduced (new wooden bridges are relatively uncommon)—but wood continues to have many obvious, universal, and lasting uses (USDA 2010). Most people would list sawn wood (timber) as the principal structural component; more finely cut, trimmed, veneered, often bent and treated wood is used for furniture; simply shaped wood is found in hand tools and gadgets ranging from kitchen utensils (mallets, pins, skewers) to garden implements (stakes, rakes, trellises); and pulped wood makes paper. But most people would not think immediately of an enormous variety of past wooden uses, including weapons, tools, and farming implements (spinning wheels, looms, rakes, flails) and pipes.

And even some of today's less obvious but still very common wood uses might escape those lists. Concrete and steel are now the dominant structural materials in construction and tubular steel is the preferred material for scaffolding, but wood is still used for concrete-forming molds, and in large parts of Asia bamboo scaffolds still dominate. Worldwide, most railroad ties are still wooden, as are the underground mine props or crates used to package heavy machinery. Other persistent uses include the finishing components of houses (stairs, doors, door and window frames, floors), barrels and casks for wine and spirits, specialty woods for musical instruments and for sport implements, and many applications of plywood (layered wood veneers whose commercial use began during the 1850s).

Even the affluent countries do not have any disaggregated statistics for key wood consumption categories before the late nineteenth century (at best there are estimates of aggregate timber use for some earlier dates), and even basic order-of-magnitude approximations of aggregate wood consumption are very difficult to construct because of an enormous variety of uses. Buildings have ranged from flimsy shacks to massive log structures, vessels from small river boats with displacement of a small fraction of 1 t to large naval sailing ships whose construction required several thousand tons of sawn timber; and life spans of wooden structures extend from years to centuries, or even to more than thousand years: wood for Japan's Hōryūji pagoda in Nara was cut 1,400 years ago.

Moreover, we simply do not know how many wooden houses, palisades, bridges, ships, or heavy weapons were built by the traditional settled societies on four continents (Australian aborigines did no build any permanent wooden structures) during some five millennia of their existence. As a result, we will never know how much wood ended up in the buildings of ancient China's large rectilinear cities or how much of it sank to the bottom of the ocean with the naval and merchant fleets of great voyaging Atlantic powers. The best I can do is to offer some revealing partial quantifications. Matters improve greatly once we reach the nineteenth century.

Wood requirements for traditional housing spanned a large range of volumes and weights even when the comparison is restricted to relatively simple but fairly comfortable structures whose ownership could be commonly afforded. At one end of the spectrum would be *minka*, traditional Japanese rural and urban house built of wood, bamboo, straw (for often heavy thatching), mud, and paper, whose ownership became fairly widespread during the Tokugawa era (Morse 1886). At the other end of the range would be a log-house (be it the old Scandinavian *stock hus*, the traditional Russian *izba*, or the log cabins of early American settlers), a relatively massive structure with sizes ranging from a single room to large, multilevel designs.

Minka (literally, people's houses)—with their simple post-and-beam construction, with walls often made of clay or bamboo, with sliding doors and interior partitions frequently made of paper, and in rural areas often with only earthen floors—were lightweight. A 100 m² *minka* might need as little as 8 m³ of pine, cedar, or cypress wood for its posts, beams, roof, and floors (Kawashima 1986). In contrast, even a small log cabin made with logs 15 cm in diameter needed about 15 m³ for its walls, floor, ceiling, and roof, and a single-story 100 m² log house (with 25-cm-diameter logs for outside walls and with heavy wooden doors and partitions) would need easily 100 m³ of heavy logs. And large farmhouses in Germany or Switzerland required commonly 1,000 m³ of timber (Mitscherlich 1963). In some climates well-built wooden houses were inhabited for centuries, in other regions they needed frequent repairs and rebuilding, making any attempts to estimate cumulative wood consumption in traditional housing even more difficult.

But what can be done with a high degree of confidence is to compare the demand for construction wood with the demand for fuelwood. Even if we take the highly wood-intensive and rather large traditional Swedish log house, whose construction (including outbuildings) would need 100 m³, that mass would be quickly dwarfed by the wood needed for its heating. Typical annual consumption of fuelwood in Sweden of the early 1800s was nearly 5 m³ per capita (Gales et al. 2005), so five people inhabiting a house would need 20–25 m³ a year, and in just 50 years heating would require 1,000–1,250 m³ of fuel, or at least 12 times as much as went into construction.

In the much warmer Piedmont, Italy's relatively cold northern province, the annual preindustrial fuelwood consumption was close to 850 kg per capita or (assuming 550 kg/m³) about 1.5 m³ (Gales et al. 2005). A family of five thus needed 7–8 m³/year and would have burned more wood for heating than it used to build its house in only about 10–15 years. Of course, these comparisons ignore important qualitative differences (straight logs and better-quality woods are needed for construction, while any woody phytomass, including fallen branches, bark, stumps, and waste wood, made suitable fuel), but they make a critical point: fueling wooden villages and even more so large wooden cities put a much greater strain on nearby forests than did the initial construction. Moreover, it was one thing to import wood from longer distances for what often amounted to a once-in-a-lifetime construction project and another to rely on long-distance imports of fuelwood.

During the late nineteenth century and throughout the twentieth century, new construction relied less on wood and more on structural steel and concrete, and eventually also on aluminum and plastic materials. The United States and Canada have been the only two Western countries where the building of single-family houses continued to be dominated by wood (studs for frames, plywood for outer walls, wooden roof trusses, joists, floors, and staircases), but changing construction methods and a greater reliance on other materials have reduced the specific demand for lumber, a trend that was partially countered by a steadily increasing size of an average U.S. house (doubling from 110 m² in the early 1950s to 220 m² by 2005).

But wood for buildings has been far from the only market claiming large amounts of wood as a raw material. After 1500, most of Europe's maritime societies engaged in building large oceangoing fleets and navies, and this quest had translated into high demand for quality timber. England was the best example of these needs, but other maritime powers (France, Spain, Portugal, Holland) were not far behind. These needs were notable because of greatly increased ship sizes. Typical seagoing vessels remained small from antiquity throughthe Middle Ages, with hardly any growth of typical displacements between the centuries of the early Roman Empire and the time of the first, late medieval long-distance European voyages, made by the Portuguese.

We have detailed information about wood requirements for different kinds of vessels. The extremes are a single tree (for a dugout canoe or, after planking, for a small boat) and a small forest of more than 6,000 oak trees needed to build the largest (more than 100 guns) triple-deck warships of the first half of the nineteenth century. For the wood requirements of some of the earliest seafaring vessels we have Homer's description of Odysseus building his boat (the shipbuilding passage in the *Odyssey* 5:23–253): "Twenty trees in all did he fell, and trimmed them with the axe; then he cunningly smoothed them all and made them straight to the line."

Assuming, rather generously, 1.5 m³ of wood per tree puts the upper limit of timber required to build that mythical Bronze Age ship at about 30 m³; with wood waste at no less than 30% and an air-dried weight of no more than 600 kg/m³, the actual volume of the ship timber would be just over 20 m³ and its mass would be about 12 t. Much larger ships (such as those for transporting Egyptian obelisks to Rome) were built in antiquity, but their voyages were limited to coastal and cross-Mediterranean traffic. The first long-distance voyages by Europeans were done by the Vikings, who used their elegant long ships to roam the North Atlantic (all the way to North America), the Mediterranean, and the Black Sea.

A well-preserved Gokstad ship (built around 890 CE and discovered in Norway in 1880) displaced 20 t, and its construction (including the mast and 16 pairs of oars) required the wood of 74 oaks. The ships that made the first long-distance voyages from Europe half a millennium later were still fairly small: Vasco da Gama's *São Gabriel* (it reached India in 1498) displaced only about 80 t, Magellan's *Victoria* was only marginally larger, at 85 t, and Columbus's *Santa María* displaced about 110 t. With hulls, masts, and spars accounting for 65%–70% of the total displacement (the rest being sails, ballast, supplies, armaments, and crew), those ships contained between 50 and 75 t of sawn timber (Fernández-González 2006). Wooden ships reached record sizes, and several European navies built them in record numbers, during the last century before the introduction of steam propulsion (the first oceangoing steamer came during the 1830s), and the volumes of timber delivered to shipyards substantially surpassed the timber incorporated into new vessels (box 9.4).

Shipbuilding claims expressed in terms of actual forest harvests depended on the species and the density and maturity of the exploited stands. The maximum density for the best shipbuilding timber—fully mature (after some 100 years of growth) British oaks—was about 100 trees/ha, and with every such tree yielding 2.8 m³ (two loads) of wood, the cut amounted to 280 m³, or about 200 t/ha. A slightly denser but immature stand yielded only one load per tree, or about 120 t/ha. Building a ship containing 4,000 t of wood (requiring about 6,000 t of delivered timber) thus claimed anywhere between 30 and 50 ha. That much wood would have sufficed to provide an annual supply of fuel for as many as 1,500 families—but the comparison is revealing only in strictly quantitative terms, as any woody phytomass can

Box 9.4
Timber Required to Build Large Naval Sailships

An original French design of large (about 54 m long at the gun deck), two-decked battleships carrying 74 guns and crewed by up to 750 men became the dominant class of naval sailing vessels during the late eighteenth and early nineteenth century: the British Royal Navy had eventually commissioned nearly 150 of them (Watts 1905; Curtis 1919). A typical 74-gun ship required about 3,700 loads of timber (a load being equal to 1.4 m^3), including nearly 2,000 loads of compass, or curved grain, wood. With *Quercus robur* density (air-dried, seasoned wood) at 650 kg/m^3, the entire ship consumed about 3,400 t of wood. The total mass of timber varied due to differences in the thickness of planking (the heaviest one could be in excess of 10 cm) and in the construction of keels, decks and masts (the latter usually of lighter straight pine with average air-dry density of less than 400 kg/m^3), on the vessel's size (gun deck lengths of 74s varied between about 50 and 56 m) and on the mixture of wood species used.

The total mass of wood used to build a 74-gun ship ranged from less than 3,000 t to nearly 4,000 t. Larger ships needed less material when measured in wood loads per gun: for the largest 120-gun vessels the rate was as low as 40 loads per gun, for 74-gun vessels it was around 50, and for frigates (22 guns) it was as much as 100 loads per gun. In absolute terms the highest totals (such as for the 121-gun *Victoria*) were in excess of 6,000 loads, or as much as about 5,500 t of oak wood. But more timber had to be ordered because 30%–40% of the material delivered to a shipyard ended up as waste. Linebaugh (1993) showed that the gap was often much larger: chips, pieces of wood no more than 90 cm long, could be taken offsite by workers, to be used as fuel or timber for their houses and furniture or sold for profit. As a result, up to 60% of all timber delivered to a shipyard to build a 74-gun ship ended up elsewhere.

be burned, but only the mature trees of a few oak species (often also from specific areas) were considered the best choice for ship timber.

We also have a comprehensive estimate of the Royal Navy's wood requirements during the last decades of the wooden ship era. In 1810, the fleet's tonnage was some 800,000 tons, and with roughly 1.5 tons of timber needed per ton, it took approximately 1.2 million loads, or 1.1 Mt if all of it were oak and less than 1 Mt if it were a mixture of hard- and softwood species dominated by oak. The average lifetime of naval ships was about 30 years, but substantial repairs were needed after 15 years. When these realities are taken into account, the annual rate of timber consumption for new construction and repairs prorated to about 110,000 loads, or at least 90,000 t of timber a year.

After the age of large wooden ships ended (steel became the top choice by the 1880s), wood continued to be used for building smaller vessels, boats, and yachts (but is now limited largely to their interior finishes, as is the case with cruise liners)—and casks. Amphoras, ceramic jars used to transport wine, oil, and processed foodstuffs (Twede 2002), were the most common containers of ancient commerce, but liquid-tight wooden barrels whose bilge was made from hardwood staves and whose tightness was assured by iron hoops came into use during the Roman time, and until the nineteenth century they were the mainstay for transporting liquids and processed foods (wine, oil) on ships (Twede 2005). Most wooden barrels needed between 50 and 60 kg of wood to build, but it is impossible to make a good estimate of their cumulative production (wood decay leaves few very old barrels, precluding any informed guesses about the rate of past cask use).

Coal's ascendancy eventually eliminated the use of wood as the leading fuel, but it had created two new major markets for wood, first for props in underground coal mining and then as an indispensable enabler of the first form of mechanically powered and coal-fueled long-distance land transportation. Wood requirements in Western coal mining were typically between 0.02 and 0.03 m^3/t of coal, but in deforested China specific consumption has been always much lower. My best estimate is that the global demand for mining timber surpassed 20 Mm^3 (about 15 Mt) by 1900 and 40 Mm^3 by 1950, reaching about 2% of that year's total roundwood harvest. Two post–World War II trends reduced the need for mining timber: increasing share of surface mining (in the United States it rose from 25% in 1950 to 66% by the year 2000) and the more frequent use of movable steel roof supports in highly efficient long-wall mining. A massive post-2000 expansion of Chinese mining generated new demand, but with low specific use (now averaging only 0.005 m^3/t of extracted coal), mining timber accounts for less than 10% of China's wood consumption (SFA 2009).

The rapid expansion of railways needed ties (sleepers), and for more than a century (before concrete ties began to claim increasing shares of the market), wood was virtually the only material, and in most countries it still is the best choice, to use: by the end of the twentieth century 94% of the U.S. ties were wooden. and worldwide only about 15% of all ties were made of other materials (Sommath et al. 1995). Ties transfer the loads to the crushed stone ballast and foundations and maintain the desired gauge; hardwoods (ash, beech, maple, and oak) have been the best choice. The only advantage of softwoods (pine, Douglas fir) is that they absorb creosote (applied to limit insect and fungal damage) more readily.

The average expected service life of treated ties increased from 35 years before World War II to 40–50 years (James 2001). The shortest spans (around 30 years) are in the hot and humid climates, the longest in hot and arid regions, and the mean

Box 9.5
Sawn Wood for Railroad Ties, 1830–1900

The standard rate of tie installation is about 1,900 (1,888) ties per kilometer of track in the United States (where the rails are joined by spikes) and 1,500 in Europe (where iron or steel chairs are used for joining). A single tie ($255 \times 22.5 \times 17.5$ cm) weighs between 71 kg for pine and 99 kg for oak (Webb 2011), and hence 1 km of tracks needs as little as 107 t of sawn wood in Europe and as much as 188 t in the United States. The railroad age began in 1830 with the Liverpool-Manchester link (56 km). By 1860 the worldwide length of railroads had surpassed 100,000 km, and by the end of the nineteenth century the UK had about 30,000 km of track, the European total was about 250,000 km, Russia had 53,000 km, the U.S. total had surpassed 190,000 km, and the worldwide length of railroad tracks had reached 775,000 km (Williams 2006). This means that between 1830 and 1900, the expansion of railroads consumed at least 100 Mt of sawn wood for the initially emplaced ties and, with an average annual replacement rate around 3%, at least another 60 Mt for track renewal.

---

in the eastern United States is 46 years: consequently, annual rates of 3% for the decades before 1950 and 2% afterward would be a good approximation of typical replacement demand. Fairly good historical statistics for the post-1830 expansion of railroads make it possible to calculate the highest annual demand for new ties, as well as to estimate the cumulative sawn wood demand for ties initially emplaced and used in subsequent track maintenance during the nineteenth century (box 9.5).

Finally, wood is harvested to make paper. The production of mechanical pulp became common during the 1870s, but this inferior product (with high lignin content, prone to yellowing) is now used only for newsprint, toilet paper, cardboard, and building board. Chemical pulp was made first by the alkaline (soda) process and later (starting during the 1870s) by the acid (sulfite) method, which, unfortunately, produces notoriously brittle and crumbling paper (Smil 2005). This drawback was solved by the sulfate process that became commonly used after 1900 and that now accounts for about 65% of the world's wood pulp production. These innovations converted tree cellulose into the cheapest mass-produced commodity of modern civilization and led to a still continuing increase of printed matter: during the 1850s, European and American paper output was less than 5 kg per capita; by 1900 the U.S. rate had surpassed 30 kg/capita, and a century later it was more than 200 kg per capita; and by 2010 the global mean was above 50 kg per capita (FAO 2011e).

III

Adding Up the Claims: Harvests, Losses, and Trends

Terraced rice fields covering formerly forested hills of South China's Yunnan province exemplify the extent of anthropogenic land use changes and the intensity of harvests. A high-resolution image can be downloaded at http://upload.wikimedia.org/wikipedia/commons/1/16/Terrace_field_yunnan_china.jpg.

Three fundamental limitations make any large-scale, long-term accounting of human claims on the biosphere's production challenging—and uncertain. First, these claims belong to three different categories that might simply be labeled extraction, management, and destruction, but a closer look shows that there are overlaps and blurred boundaries. Second, no single measure can adequately express the increasing extent and the overall magnitude of human claims on the biosphere. Third, although a combination of several revealing variables provides a better assessment of these claims, it does not bring a fully satisfactory appraisal because of the many uncertainties in quantifying the natural baselines and assessing the true extent of human interventions. Every one of these summary descriptions requires a closer examination.

Fishing and whaling, the killing of elephants for ivory, the capture of wild animals for the pet market, and the collecting of wild plants and mushrooms are common examples of extractive activities. Agriculture is the dominant activity in the management category ("constructive transformation" could be another term): natural ecosystems are replaced by managed agroecosystems that produce annual crops or are composed of perennial species, and although either the original vegetation has been completely replaced or its makeup has been changed beyond recognition, the site continues to produce phytomass and to support nonhuman heterotrophs. In contrast, the destructive transformation of natural ecosystems alters or entirely eliminates a site's primary production capacity as it destroys natural vegetation either to extract minerals or to expand settlements: industrial facilities and residential and commercial buildings also need associated space for requisite transportation networks, the storage of raw materials, and waste disposal.

But complications and qualifications are obvious. Extractive activities target particular species, and as long as they are well managed, they can continue with minimal adverse effects. But even in such cases there may be a great deal of associated damage, for the harvesting of a target species may kill or injure other species (even those belonging to a different class of animals, as exemplified by dolphins caught in tuna purse seines). And once the target species gets overexploited or harvested to extinction, such actions will have changed not only the composition of affected communities and ecosystems but also their long-term dynamics and overall productivity: an entire ecosystem can be affected by such degradation, sometime irreversibly. Quantifying the human intervention only in terms of the specifically extracted biomass captures only part of the actual ecosystem effect. At the same time, even a complete removal of a key heterotroph may have no effect on the overall level of primary productivity or may actually enhance it: the removal of elephants, whose heavy browsing of woody phytomass and toppling of trees help keep savannas open, is a key reason for the regrowth of shrubs and trees.

Permanent cropping replaces natural vegetation with domesticated species and creates new agroecosystems that are nearly always much less diverse in comparison with the original communities. The contrast is obviously greatest where the dominant cultivation takes the form of monocultures planted over vast areas. Crop rotations offer a more rational management approach, but even they will underperform compared with their natural predecessors. Agriculture's effects on the standing phytomass and on primary productivity fit a clear pattern of overall loss in all instances where mature and rich natural ecosystems were replaced with a low-yielding annual crop in a climate whose short vegetation period makes only one harvest possible.

Converting a patch of a boreal forest to a wheat field, a transformation common in medieval Europe or nineteenth-century Canada, replaced an ecosystem storing several hundred tons of aboveground phytomass per hectare with an agroecosystem whose peak preharvest aboveground phytomass was at best 3 t/ha. Bondeau et al. (2007) modeled this effect globally and found that global vegetation has been reduced by about 24% as a result of agriculture, and that the annual phytomass harvest was reduced by 6–9 Gt C during the 1990s. They also displayed their findings in a map grading the phytomass storage and net primary productivity (NPP) difference between the existing farmland and previous natural vegetation.

For most of the world's cultivated land the difference between carbon stocks in cultivated plants and in the vegetation of natural ecosystems that used to occupy their place is more than 2 kg/m^2 (200 kg C/ha). Moreover, and somewhat counterintuitively, the difference is greatest (more than 1 t C/ha) in some of the world's most productive agricultural regions, including the Corn Belt in the United States and parts of Europe, Russia, and China, as even high wheat, corn, and rice yields remain far below the levels of carbon stored in mature forests or rich grasslands that were converted to fields. The only area where agriculture accumulates more carbon than natural vegetation would have done in the same places is where irrigation boosts productivity in arid regions, such as Egypt or Pakistan and northern India: there the crops will store more than the short-grass communities they replaced.

On a small scale, such gains are possible even in areas where the losses are dominant. Converting temperate grassland into a crop field producing an annual harvest of corn with alfalfa as a winter crop can result in an overall phytomass and productivity gain. The NPP of a good European meadow may be 10–15 t/ha, while a modern corn crop will yield 8–10 t/ha of grain and an identical mass in stover, and a single cut of nonirrigated alfalfa can add more than 5 t/ha, and as many as five cuts are possible (Ludwick 2000). Bradford, Lauenroth, and Burke (2005) quantified such a productivity gain for the Great Plains of the United States, the country's premiere agricultural region, and concluded that compared with native vegetation, cultivation is increasing the region's NPP by about 10%, or nearly 100 Mt/year. But while the primary productivity (and even a temporary phytomass storage) have been enhanced, the specific composition has been simplified, and the biodiversity of the entire ecosystem (as well as some of its natural services: in the case of row crops such as corn, the key concern would be the compromised protection against soil erosion) has been reduced.

Matters are no easier when looking at the transformations of forests. Shifting agriculture may remove all natural vegetation within a cultivated patch, but it lets the original plant communities reassert themselves after a period of cropping, and if the regeneration period is relatively long and the temporary plantings are surrounded by still intact growth, the regenerated forest may regain most of its original biodiversity. A great deal of historical evidence shows that even some large-scale deforestation followed by decades or even centuries of permanent cropping is substantially reversible.

One of the best-documented examples of this large-scale recovery is the case of Massachusetts's forests. In 1700, some 85% of the state's area was covered by forests, but 150 years later Henry David Thoreau noted in his *Journals* that "our woods are now so reduced that the chopping this winter has been a cutting to the quick." By 1870, when the clearing reaching its greatest extent, only about 30% of the state was covered by trees, but by the end of the twentieth century the cover was back up to about 70% (Foster and Aber 2004). The history of Massachusetts forests also illustrates how the interplay of natural and human factors determines long-term outcomes. The 1938 hurricane, which damaged more than 70% of all wood volume in the Harvard Forest, was a perfect reminder of the fact that only a minority of trees in the U.S. Northeast can live out their maximum natural life span. as the region is (infrequently but assuredly) subject to major hurricanes. And the massive death of hemlocks, originally one of three dominant species in the state's natural forests, illustrates the impact of pests against whose attack there is no known defense.

Contrasts of single measures, even if they are such critical variables as the total standing phytomass or its annual productivity, cannot capture the true impact of human transformations of natural ecosystems—but neither can a list of species that have been endangered or eliminated by such changes, or a calculation of complex indices designed to quantify the overall loss of biodiversity: after all, even simplified anthropogenic ecosystems can be aesthetically pleasing, highly productive, and, when properly managed, also fairly resilient. The best way to judge the human impact is to look at a number of indicators. Chapter 10 concentrates on the evolution of postglacial global phytomass storage and on the land-cover changes brought about by food production, deforestation, industrialization, and urbanization. Chapter 11 takes apart the concept of NPP appropriation by humans, and chapter 12 concludes the book with some reflections on the evolution, extent, impact, and future of the human presence in the biosphere.

10

Changing Land Cover and Land Use

The most obvious, and conceptually the simplest, indicator of the human impact on the biosphere's productivity and phytomass storage is the total area of the natural ecosystems that have been transformed by human action. To put these changes into an evolutionary context, I will begin with a brief review of phytomass storage during the past 20 millennia, since the last glacial maximum (LGM), when North America north of 50°N and much of Europe beginning at only a slightly higher latitude were covered by massive continental glaciers. Primary productivity and carbon strongly rebounded during the next 15 millennia, reaching maxima by the mid-Holocene, some 5,000–6,000 years ago at the time of the first complex civilizations. This was followed by millennia of locally severe and regionally substantial transformations whose global impact remained still relatively minor.

The most common kind of conversion was to create new fields by converting forests, grasslands, and wetlands (and in some regions also deserts) to new croplands and pastures. Next in overall importance have been the claims made on forest phytomass to remove timber and firewood and, starting in the nineteenth century, wood for making paper. These human transformations of ecosystems began to accelerate during the early modern era (after 1600) and reached an unprecedented pace and extent thanks to the post-1850 combination of rapid population increases and economic growth, marked by extensive urbanization, industrialization, and the construction of transportation networks.

Tracing these changes can be revealing even when most of the conclusions must be based on estimates and approximations, and a closer look is rewarded by a more nuanced understanding. At one extreme are those areas whose plant cover has been entirely lost by conversions to constructed impervious surfaces (such as pavement) in urban and industrial areas and transportation corridors: the primary production of these areas has been completely eliminated. At the other extreme are natural

Deforestation in the tropics has been responsible for the most of the global phytomass loss during the twentieth century. This satellite image from 2010 (acquired by the Moderate Resolution Imaging Spectroradiometer on NASA's *Terra* satellite) shows the extent of forest clearing in the state of Rondônia in western Brazil. The image can be downloaded at http://earthobservatory.nasa.gov/Features/WorldOfChange/images/amazon/amazon_deforestation_2010214_lrg.jpg.

grasslands that are lightly grazed by well-managed numbers of domestic animals (alpine meadows with cattle and sheep) or natural forests where harvests remove only annually renewable fruits, seeds, or nuts (such as the collection of Brazil nuts in an old-growth tropical rain forest) or are limited to a few valuable animals (such as the trapping of small mammals for their pelts in Canada's boreal forest): the primary production of these areas remains virtually (or largely) intact.

In between these two extremes is a wide continuum of human interventions. Many urban and most suburban residential areas have retained some of their site's potential primary productivity thanks to lawns, parks, and street trees: these new anthropogenic ecosystems are highly fragmented and have a low biodiversity and low productivity when compared with their natural predecessors, but on a small scale they might approach or even surpass the performance of the ecosystems they replaced. Selective logging (including its extreme mode, which uses helicopters to remove large tree trunks from steep slopes) takes only some targeted trees, preserves much of the site's productivity, and does not change its regenerative potential, while harvesting timber by forest clear-cutting is analogous in its destructive impact to clearing land for field crops.

This most extensive of all anthropogenic land conversions is a unique hybrid of destruction and high productivity. The cultivation of annual or perennial crops is usually predicated on a near total elimination of a climax natural ecosystem, and most of the fields have lower primary productivities than the plants they replaced. But in many cases the difference in productivity is not that large, and good agronomic practices (multicropping with rotations that include high-yielding leguminous cover crops) may actually result in higher yields. The conversion of a short-grass Canadian prairie to an alfalfa field may have a minimal impact on overall primary productivity, and it will maintain such important ecosystem services as protecting soil against erosion, retaining moisture, and adding bacterially fixed nitrogen.

Similarly, conversions of tropical forests to rubber or cocoa plantations maintain a semblance of an arboreal ecosystem (indeed, the FAO classifies these plantations in a forest category), support a relatively rich complement of heterotrophs, and continue to provide protection against excessive soil erosion. But in both cases, a crucial difference remains: before conversion, the primary productivity of a natural ecosystem was entirely available for consumption by a variety of wild heterotrophs, and carbon and nutrients in the unconsumed organic matter were recycled (largely in situ) through bacterial and fungal metabolism.

The different consequences of these interventions and a continuum of invasive practices that makes it impossible to define clear intervention categories are the best arguments against any simplistic aggregation of areas "affected," "modified," "trans-formed," or "impacted" by human actions. And these categorical complications are not the only challenge in assessing the aggregate consequences of anthropogenic changes: whereas quantifying their large-scale extent has become easier thanks to modern remote sensing techniques, major uncertainties remain. The global monitor-ing of these changes became possible only with the launching of the Earth observa-tion satellites and with the gradually improving resolution and multispectral images produced by their sensors (the sequence was discussed in the first chapter). At the same time, those who produce global maps and data sets of changing land use do not do enough to stress many of the inherent limitations of the data, and hence the untutored users of these products have unrealistic opinions about their accuracy, reliability, and comparability.

What these limitations mean is perhaps best illustrated by taking a closer look at the global satellite monitoring of cropland. Most of today's gridded global land-use databases have a resolution of 5 arc-minutes, encompassing an area of about 9.2 × 9.2 km, or roughly 8,500 ha of farmland. Even on Canadian prairies with their large holdings averaging 400–500 ha, that resolution would aggregate all crops planted on about 20 different farms into a single data point. In most parts of Asia, such a resolution would homogenize hundreds of different farms planted to scores of different crops—and unlike on the Canadian prairies, with their single crop per year, nearly all of those farms would harvest at least two crops a year, many fields would be triple-cropped, and suburban vegetable fields would produce four to six crops every year. Obviously, a 5-minute resolution is good enough to identify large-scale patterns of land use but not to provide accurate assessments of specific plant composition or productivity, especially in multicropped areas.

Data series for the more recent presatellite eras must be assembled from the best available national statistics or periodic land-use mappings. These sources use non-uniform definitions of land-cover and land-use categories, their reliability varies from excellent to dubious, and while they may offer enough quantitative pegs to make reasonably good continent-wide estimates for nineteenth-century Europe or North America, they have little or no information for pre-1900 Africa and large parts of Asia. But even such questionable, fragmentary. and widely spaced numbers are largely absent for the Middle Ages and entirely unavailable for antiquity: land-cover reconstructions for those eras must rely heavily on assumptions based on anecdotal written evidence and, where available, on painstaking paleoecological reconstructions.

Twenty Millennia of Phytomass Change

Given the lasting uncertainties in quantifying the aboveground phytomass in today's ecosystems (leaving aside even greater errors in estimating the belowground biomass and its productivity), it may seem audacious to model global phytomass stores going back 20,000 years, to the last glacial maximum (LGM), or even "just" to about 10,000 years ago, to the preagricultural era of the early Holocene, or 5,000 years ago, to the mid-Holocene, the time of the first complex civilizations. The inherent uncertainties of such attempts are obvious, but the uncertain totals should be able to convey the magnitude of change between the LGM, when only small bands of dispersed hunters (totaling no more than few hundred thousand people) roamed the Earth, whose northern latitudes were covered by massive continental glaciers, and the era of maximum phytomass storage during the preagricultural period, when forests occupied large areas of Eurasia and North America.

Different studies have used global climate models and paleoecological, palynological, pedological, and sedimentological evidence to reconstruct past vegetation cover and carbon storage, usually expressing the latter as a combination of vegetation and soil carbon, with some studies including and others excluding the post-LGM peatland growth. Not surprisingly, there have been some extreme findings. After reconstructing past ecosystem distributions, Adams and Faure (1998) concluded that the total terrestrial carbon storage (plants, soils, and peat) 18,000 years before the present was only 931 Gt, compared to a total of 2,130 Gt C for the late twentieth century—the total arrived at by adding 560 Gt C in plants (Olson, Watts, and Allison 1983), 1,115 Gt C in nonpeat soils (Post et al. 1982), and 461 Gt C in peatlands (Gorham 1991)—a nearly 2.3-fold increase, for a net gain of about 1,200 Gt.

In contrast, Prentice and Fung (1990) claimed that (excluding peat) there was no significant increase in overall terrestrial carbon storage, putting the overall LGM-to-present carbon gain at 0 ± 50 Gt C: this would mean that the terrestrial biosphere was no (or not a major) carbon sink during the period of rapid deglaciation. But Prentice et al. (1993) presented a new model that showed the late twentieth-century terrestrial carbon storage to be at least 300 Gt C and as much as 700 Gt C higher than during the LGM. Moreover, seven other studies published by 1998 converged to a very similar range, indicating a gain of about 30% in the overall terrestrial carbon storage (in absolute terms, mostly 300–700 Gt C) from the time of the LGM to the late twentieth century (Peng, Guiot, and Van Campo 1998).

Two more studies came out in 1999, the first one echoing the just noted consensus and concluding that the total terrestrial carbon addition since the LGM amounted to 550–680 Gt (Beerling 1999), the other one ending up with LGM totals ranging from 710 Gt C less to 70 Gt C more than the present level (François et al. 1999). And two studies published in 2002 (for some reason, these reconstructions have not been pursued after that date) came up with a very similar result: total terrestrial carbon storage during the LGM was 821 Gt lower (Kaplan et al. 2002) or at least 828 Gt C (and as much 1,106 Gt C) lower (Otto et al. 2002) than at the beginning of the twenty-first century. After reviewing most of these studies (whose simple mean is about 650 Gt C), Maslin and Thomas (2003) concluded that the difference between estimates centering on 500 Gt C and those averaging around 1,000 Gt C can be explained by taking into account isotopically light emissions of CH_4 from gas hydrates.

For the mid-Holocene (6,000 years before the present) we have two sets of calculations: François et al. (1999) offered an inconclusive range of as much 132 Gt C less and 92 Gt C more than at present, while Beerling's (1999) two models ended up just 103 Gt C apart, with the lower total at 750 Gt of plant carbon and with 1,363 Gt C in vegetation and soils. As already noted in chapter 1, a reconstruction of natural ecosystems based on 106 major plant formations found that the maximum preagricultural standing phytomass was about 2,400 Gt, or around 1,200 Gt C (Bazilevich, Rodin, and Rozov 1971), while a reconstruction by Adams et al. (1990) found that the Earth's potential vegetation could store 924 Gt C.

By 2010 there were are at least eight explicit values for the preindustrial phytomass carbon (usually assumed to apply to the year 1700), ranging from 610 to 1,090 Gt C, with the most realistic range between approximately 620 and 910 Gt C (Köhler and Fischer 2004). The latest reconstruction of preindustrial terrestrial phytomass storage used an unprecedented approach as it tested a global carbon model against nearly 1,500 surface pollen spectra from sample sites in Africa and Eurasia. That approach yielded a total plant storage of 907 Gt C, with the largest share (about 20%) in tropical rain forest, fitting into the previously established most likely range (Wu et al. 2009).

Pongratz et al. (2009) combined a global climate model and a closed carbon cycle model with previous reconstructions of anthropogenic land-use changes (Pongratz et al. 2008) to estimate that between 800 and 1850 carbon emissions (direct from destroyed vegetation and indirect due to lowered NPP) amounted to 53 Gt, and that another 108 Gt C were added between 1850 and 2000. The net effect of these land-use changes (after accounting for the biosphere's carbon sinks) was

96 Gt C, or roughly a net loss of 200 Gt of phytomass during the second millennium of the Common Era. Other estimates of more recent losses attributable to anthropogenic land use range from 136 Gt C for the years 1850–1998 (Bolin et al. 2001) to 200 Gt C for the years 1800–2000 (House, Prentice, and Le Quéré 2002). Subtracting these estimates from about 900 Gt C of preindustrial terrestrial phytomass would leave us with anywhere between 600 and 800 Gt C by the late twentieth century.

In contrast to these values, the estimate of terrestrial phytomass stores for the year 1950 offered by Whittaker and Likens (1975) seems to be too high at 1,837 Gt or about 920 Gt C, while Post, King, and Wullschleger (1997) modeled the change in terrestrial carbon storage during the twentieth century and concluded that it increased from 750 Gt C in 1910 to 780 Gt C in 1990. But the first assessment report from the Intergovernmental Panel on Climate Change chose 550 Gt C as the most likely value for the 1980s (IPCC 1990), and the third assessment listed two global totals, a lower one of just 466 Gt C, from the German Advisory Council on Global Change (WBGU 1998), and a higher one of 654 Gt C, from Saugier, Roy and Mooney (2001), for the twentieth century's end. Of course, the compared values for the LGM, the mid-Holocene, the preindustrial era, 1950, and the late twentieth century come from studies that used a variety of approaches and assumptions and whose only commonality is a significant margin of error.

This makes it impossible to make any clear quantitative conclusions, but the trend appears to be indisputable and the relative magnitudes seem plausible. If we assume that during the LGM the total continental carbon stores were about 500 Gt C lower than at present and that at least 35%–40% of the LGM terrestrial carbon total was in plants, the peak glacial vegetation stored 175–200 Gt C less than did the terrestrial biosphere around the year 2000, or close to 500 Gt C; Köhler and Fischer (2004) put the most realistic range at 350–640 Gt C. By the mid-Holocene that total could have more than doubled to more than 1,000 Gt C, and human activities subsequently reduced it to no more than 900 Gt C by the onset of the industrial era and, most likely, to less than 700 Gt C by the year 2000.

The general sequence is undoubtedly correct: a reduced phytomass during the LGM, with a substantial gain during the Holocene (doubling does not seem excessive, as the total area of tropical rain forest had roughly tripled between 18,000 and 5,000 years before present and that of cool temperature forests expanded more than 30-fold [Adams and Faure 1998]), followed by millennia of gradual decline as a result of the extension of cropland and wood harvests, and then by accelerated deforestation losses since the mid-twentieth century. What lies ahead is uncertain because our understanding of NPP in a warmer world with higher atmospheric CO_2 levels is a mixture of confident conclusions and unsatisfactory speculations.

Thanks to our knowledge of the biophysical and biochemical processes that govern photosynthesis, we can make some useful predictions regarding the short-term plant response. Indisputably, the current partial pressure of the atmospheric CO_2 (at about 390 ppm, or nearly 0.04% by volume) is considerably below the concentration needed to saturate the photosynthesis of C_3 plants, the dominant species in both natural and managed ecosystems: species-specific responses show saturation at levels that are twice to three times the current tropospheric concentration. Moreover, C_4 species, whose optimum productivity is above 30°C, would benefit from warmer temperatures. This means not only enhanced but also more efficient photosynthesis operating (thanks to reduced stomatal conductance) with higher water-use efficiency.

Models of global NPP based on satellite observations indicated that climate change has already eased the CO_2 and temperature constraints on plant productivity and that between 1982 and 1999, the annual rate rose by 6%, or about 3.4 Gt C (Nemani et al. 2003). But during that time global photosynthesis was also boosted owing to volcanic aerosols that were emitted all the way to the stratosphere by the Mount Pinatubo eruption in 1991: that effect resulted from the fact that plant canopies use diffuse radiation more efficiently than they use direct-beam radiation (Gu et al. 2003). On the other hand, a pronounced European heat wave during the summer of 2003 depressed GPP by nearly a third and turned the continent temporarily into a anomalously significant (about 500 Mt C) net source of carbon (Ciais et al. 2005). And the best estimates for the decade between 2000 and 2009 indicate that the record-breaking average temperatures accompanied by extensive droughts had a global effect on terrestrial NPP, reducing it by as much as 2 Gt C/year (in 2005) and depressing it, on the average, by 0.55 Gt C/year (Zhao and Running 2010).

Should extreme heat events become more common as the average global temperatures rise, then both trends would have a significant cumulative impact on biospheric carbon storage. Long-term responses that would require acclimation and shifts of vegetation boundaries are extremely difficult to quantify because of dynamic links among NPP, radiation, temperature, and precipitation (both in terms of averages and in terms of seasonal and monthly fluctuations), atmospheric CO_2 levels, and nutrient availability. A single modeling exercise suffices to illustrate these uncertainties. Schaphof et al. (2006) applied different climate change scenarios based on five global circulation models—with the global temperature averages in

2071–2100 ranging from 3.7°C to 6.2°C above the 1971–2000 mean, and with the increases in annual continental precipitation between 6.5% and 13.8%—to the Lund-Potsdam-Jena global vegetation model and found the following large uncertainties.

Even the overall response of the biosphere was unclear, with changes ranging from the loss of 106 Gt C to the additional storage of 201 Gt C, with three out of five climate models indicating that the biosphere will be a net source of carbon (particularly thanks to boreal forests), one yielding a neutral outcome, and one implying a carbon sink, and with the response of the tropical rain forests of South America and Central Africa being most uncertain. Moreover, the relative agreement between different models was less for individual seasons than for the annual mean. This implies not only a large relative error—assuming 650 Gt C of plant carbon in the year 2000, the range would be +30% to –16%—but an absolute uncertainty (about 300 Gt C) nearly as large as the most likely total of plant carbon decline that the biosphere has experienced since the mid-Holocene.

Land for Food Production

No other human activity has been responsible for such a large-scale transformation of terrestrial ecosystems as food production. All kinds of natural ecosystems have been converted to permanent fields; this category of land use is still commonly labeled arable land, although in some countries substantial shares of it are now managed with reduced tillage or no-till cultivation. This land claimed no less than 1.380 Gha in 2010, or nearly 11% of all continental surfaces. The addition of permanent plantations (growing fruits, tea, coffee, cacao, palms) extends the total of about 1.53 Gha of agricultural land, or about 12% of the earth's ice-free surface (FAO 2011f). For comparison, the total based on combining various agricultural inventory data satellite-derived land coverage (at 5-minute or roughly 10 km resolution) came up with a nearly identical median value of 1.5 Gha and a 90% confidence range of 1.22–1.71 Gha in the year 2000 (Ramankutty et al. 2008).

These totals refer to the existing stock of farmland, not to the area of all crops that are actually planted in a calendar- or a crop-year; in national terms that aggregate can be both a bit lower and substantially higher. This is because in many countries some arable land is always fallowed, while variable shares of farmland are planted to more than one crop a year. Consequently, multicropping ratios—expressing the number of crops harvested regionwide or nationwide per unit of arable land every calendar year—range from less than 1 (in Canada or on the U.S. northern Great Plains, where some fields are fallowed and the rest are planted only once a year) to about 1.2 (China's current national average, reflecting northern single-cropping and southern multicropping) to more than 2 (in the most productive coastal and interior provinces of South China, where triple-cropping is not uncommon).

The total numbers of domestic animals that relied solely or largely on grazing eventually became too large to be supported by seasonal or year-round productivity of natural grasslands. As a result, the conversion of natural ecosystems to heavily grazed pastures has affected not only many grassy ecosystems but has been a major reason for tropical deforestation (particularly in many countries of Central America and in Brazil) and for the destruction of wetlands. According to the FAO, in the year 2000 the global area of permanent meadows and pastures was, at about 3.35 Gha, more than twice as large as the total area of agricultural land (FAO 2011f). Europe was the only continent where the reverse was true: its pastures occupy an area equal to only about 60% of its fields and orchards. In contrast, Asia's pastures cover twice the area of the continent's fields and plantations.

But a recent global account that was also based on inventory data came up with only 3.15 Gha of pastures, and its adjustment on the basis of satellite imagery resulted in an even smaller area of just 2.8 Gha (Ramankutty et al. 2008), while the History Database of the Global Environment lists 3.43 Gha (HYDE 2011). The FAO/HYDE total means that permanent pastures are the largest area of natural ecosystems that has been modified in various degrees by human actions. Croplands are next, and tree plantations and actively managed forests come third, with the total about 650 Mha (Noble and Dirzo 1997), followed by impervious surface areas. The FAO's land-use and land-cover statistics have been generally accepted as the most consistent and most reliable global compilation for the post–World War II period. Their main categories related to food production now list the following areas: total agricultural land, arable land, land devoted to permanent crops, irrigated land, fallow land, and permanent meadows and pastures.

Data are available online for all countries and as various regional groupings and global totals since 1961 and in print since 1947, with many data gaps in the earliest years. The information in the FAO's database relies on the official submissions of member countries and, when these are not available, on the best estimates prepared in the organization's headquarters in Rome. Either sourcing can be problematic, as some countries have supplied deliberate misinformation and as the internal estimates are often based on the best available but unreliable and dated information.

Perhaps most notably, China's official figures kept underestimating the country's arable land, and the FAO kept accepting this well-known misinformation, which put the country's arable land at no more than 95 Mha in 1995, about 30% less than the real figure.

During the 1990s, China's *Statistical Yearbook* cautioned that "figures for the cultivated land are under-estimated and must be further verified"—many years after it was known that the official total was wrong. The earliest remote sensing studies, based on imagery with inadequate resolution (Landsat, with a resolution of 80 m), indicated farmland area as large as 150 Mha, detailed sample surveys of the late 1980s came up with the range of 133–140 Mha (Smil 1993), and a correction using land survey data for 1985 ended up with 137.1 Mha for the year 1995 (Heilig 1997). The best remote sensing evaluation, using classified imagery from the *Keyhole* series of intelligence satellites, yielded the total of 133–147 Mha for the year 1997 (MEDEA 1997; Smil 1999a).

Only in the year 2000 did China's State Statistical Bureau finally relent and correct the total to 130 Mha (SSB 2000), a figure that was promptly adopted by the FAO. How many other underestimates of this kind are hidden behind the official figures submitted to the FAO by many Asian or African countries? And how uncertain are all those estimates made in the FAO's headquarters on the basis of old or fragmentary information for countries beset by endless violent conflicts, such as Afghanistan, Sudan, or Somalia? Of course, we now have satellite sensors whose resolution could distinguish the smallest vegetable fields in Asia, but their use would be prohibitively expensive, and hence the global mapping of cropland typically proceeds at much coarser resolution and, inevitably, its results carry substantial aggregation or omission errors.

The most accurate data on the totals of arable land as well as for actually planted and harvested areas are available for all affluent countries of North America, the EU, East Asia (Japan, South Korea, Taiwan), and Australia and New Zealand, and some of these data series are available for the entire twentieth century and part of the nineteenth century; they were collated in the annual statistics published by the International Institute of Agriculture, the FAO's Rome-based pre–World War II precursor, since 1910. The first global-scale effort to estimate the history of farmland expansion going back more than 150 years was offered by Richards (1990), whose series began with 265 Mha in 1700 and progressed to 537 Mha by 1850, 913 Mha by 1920, and 1.17 Gha by 1950. A new approach to the reconstruction of past farmland expansion came during the 1990s, when backcasting models provided results for a gridded ($0.5° \times 0.5°$) distribution of farmland.

Ramankutty and Foley (1999) ended up with 400 Mha for 1700 and 820 Mha for 1850, while Klein Goldewijk (2001) used a Boolean approach (based on available statistics and maps of natural potential land cover) to generate distributions of croplands, grasslands, and natural ecosystems and came up with the same values as Richards (1990). Pongratz et al. (2008) used historical maps of agricultural areas and population data as a proxy for farming activities to estimate that about 500 Mha of natural vegetation were transformed to cropland (slightly more than half of that total mainly by conversion of forests) and pastures (mainly by conversions of natural grasslands) between 800 CE and 1700. And Kaplan, Krumhardt, and Zimmermann (2009) developed a historical model of European deforestation according to which between 70% and 95% of land available for clearing for agriculture in Central and Western Europe had forest cover in the year 1000 (ranging from 95% in Poland to 69% in Italy), but by 1850 that share was generally below 10% (6% in France, only 3% in Germany).

The latest and the most systematic of these reconstruction efforts—HYDE 3, by Klein Goldewijk and van Drecht (2006)—modeled the changes in land use and land cover by using the distributions of population density, land suitability, distance to major rivers, and natural vegetation cover. Its global totals for cropland begin in 5,000 BCE with about 5 Mha, rise to about 130 Mha at the beginning of the Common Era, reach only a bit over 150 Mha a millennium later, and double to 300 Mha by the year 1700. These data address the problem of the inherently large errors of all pre-1800 estimates by offering not only the most likely totals but also a low and a high estimate: for the year 1700 the range was ±25% around the preferred base value, in 1800 it was about ±18%, by 1900 the differences were −10% and +22%, and the ranges changed little even by 1950, to −8% and +16% (HYDE 2011).

The productivity gains of the early modern era could not satisfy the additional food and feed requirements resulting from higher population growth rates, continuing urbanization, slowly improving diets, and a larger number of draft animals. As a result, the expansion of farmland began to accelerate during the eighteenth century (in some countries already after 1600), and the nineteenth century turned out to be the time of unprecedented (and also never to be repeated) conversions of grasslands and forests to arable land. This trend is clear even for countries with limited opportunities for farmland expansion: a steady increase in rice yields during Japan's Tokugawa era (about 55% between 1600 and the 1860s) was far below the rate of population growth, which itself experienced a nearly threefold rise, from 12 million to 33 million people during the same period (Smil and Kobayashi 2012). Major reclamation efforts were needed to put more land under cultivation, and its total rose from just over 2 Mha in 1600 to 3.2 Mha by the late 1860s.

Among the populous countries with the oldest cultivation, the greatest farmland expansion took place in China: the total increased by little more than 10% during the seventeenth century but rose by nearly by 25% during the eighteenth century (to about 85 Mha, largely thanks to forest conversions in the southern provinces) and a further 10 Mha were added by 1900, mostly in the three northeastern provinces suitable for large-scale plantings of wheat, corn, and soybeans. Between 1750 and 1900, Europe's cultivated area expanded by nearly 50%, that in South Asia doubled, and the conversion of temperate grasslands to grain fields became one of the formative developments in the history of Canada, the United States, Australia, and to a lesser extent Russia.

Russia's cultivated land nearly tripled between 1750 and 1900, with new cropland added both in the European part of the country (mainly in eastern Ukraine and in Russia's southern provinces between the Dniepr and Volga and north of the Caucasus), as well as in southern Siberia (Koval'chenko 2004). During the same period the North American total expanded from a relatively small area cropped in Britain's coastal colonies along the Atlantic and in Quebec to about 170 million ha, an increase of two orders of magnitude. During the nineteenth century the U.S. total expanded roughly 20-fold, to nearly 145 Mha, with most of that gain commencing only after 1850 with the great westward expansion of the American state and with mass-scale conversion of the Great Plains to grain fields (Schlebecker 1975).

The U.S. historical statistics list total land in farms (including cropland, pastures, wood groves, and fallow land) going back to 1850, when it amounted to about 117 Mha; by 1900 it was nearly three times as much at 336 Mha. In 1850 close to 60% of all land in farms was in the southern Atlantic states, and that region still led in 1900 (with 43% of the total), but the north-central states (essentially the Corn Belt from the Dakotas to Kansas and from Nebraska to Missouri) were a close second (with 38%), and two decades later they moved into first place. Land conversions in the western states were also significant, with the region's land in farms rising from less than 2 Mha in 1850 to nearly 40 Mha by 1900, with California being the leading contributor. Data on actually harvested U.S. cropland begin in 1880, with about 66 Mha: that total had nearly doubled by 1900, to 113 Mha, and by 1920 it had reached 139 Mha.

In 1750, the cultivated land in territories that later (in 1867) formed Canada was limited to less than 30,000 ha in the eastern maritime region, now the provinces of Nova Scotia, Prince Edward Island, and New Brunswick. By 1900 the total, at more than 30 Mha, was three orders of magnitude higher: the great nineteen-century expansion began first with the settling of Ontario (Reaman 1970) and then with the conversion of Canadian prairies to large rectangular crop fields (Spector 1983). This wave proceeded westward from Manitoba (it became a province in 1870) through Saskatchewan to Alberta (both gaining provincial status only in 1905), and it continued during the early twentieth century

Australian, and also Argentinian, contributions were comparatively much smaller. As there was no cultivated land in mid-eighteenth century Australia (James Cook's landing at Botany Bay took place in 1770, and substantial numbers of British settlers began to arrive only during the 1840s and 1840s), it is impossible to offer a comparative multiplier. By 1900 Australia's farmland had reached about 4.5 Mha, more than 30 times the 1850 total. Argentina's farmland expanded about 18-fold during the nineteenth century (to about 3.5 Mha), unlike Brazil's cultivated land, which grew "only" sixfold (its great twentieth-century extension is described in the next chapter). Globally, the 1750–1900 increase of cultivated land was more than twofold, from about 360 to 850 Mha, with American and Asian grasslands representing the largest category of natural ecosystems converted to cropland. At the same time, that period also set new records for loss of arable land because of an unprecedented rate of urban expansion and spreading industrialization.

During the twentieth century, arable land was added in all the countries that had undergone rapid nineteenth-century expansion: by the year 2000 Australia's total had grown more than 10-fold (by about 43 Mha) and Argentinian farmland had nearly sextupled (gaining about 22 Mha), while Canada added about 75% (also about 22 Mha) and Russia about 45% (130 Mha by 2000). In all these cases most of the new farmland was planted to grains, above all to wheat. The most ambitious program to expand wheat production was the Soviet cultivation of virgin lands (*tselinny*, east of the lower Volga and extending to western Siberia), pushed by N. S. Khrushchev to solve the country's chronic agricultural underperformance (Durgin 1962). Between 1953 and 1964 more than 42 Mha were plowed, more than 70% of it in northeastern Kazakhstan (*Tselinnyi krai*), and more than 70% of the added land was devoted to overcoming the chronic deficits in the Soviet grain production—but in the end, the program did not meet its goals, and the USSR had to turn to large-scale imports of U.S. grain during the 1970s (Gustafson 1979).

The U.S. farmland expanded by more than 40% during the first half of the twentieth century (adding about 67 Mha) before higher yields and the abandonment of cultivation in the eastern United States stopped any further growth. Significant areas of arable land were also added in the world's two most populous countries.

India's farmland expanded by about 40% during the twentieth century, with major gains in the northwest (Punjab) and the northeast (Assam). The best reconstruction of China's land-use changes shows a gain of more than 25% (30 Mha) between 1900 and 1980, from 113 to 143 Mha, mostly through grassland and wetland conversions in the northeast (Huang et al. 2010) and forest conversions in the southwest—followed by a loss of more than 5% between 1980 and 2005 (Liu and Tian 2010).

Unmistakable signs of saturation and stagnation, in some cases followed by substantial retreats, were most noticeable during the second half of the twentieth century in Europe. Between 1950 and 2000 the total farmland area hardly changed in the three largest agricultural producers, France, Germany, and Spain, and declined by 20% in the UK, by 23% in Italy, and by 25% in Sweden. Abandonment of cropland following the demise of the USSR was on a scale large enough (more than 200,000 km^2) to make a measurable difference to the global carbon balance: Vuichard et al. (2008) calculated that the land recovered by herbaceous plants increased the annual storage of carbon by nearly 500 kg C/ha, or by a total on the order of 10 Mt C.

In contrast, the twentieth century's greatest expansion of farmland took place in Brazil, where the farmland increased tenfold between 1900 and 1950, and that total then more than tripled by the year 2000. Most of these gains took place after the mid-1960s in *cerrado*, vast grassland with poor soils (made productive by appropriate fertilization) but that with adequate rainfall has become the country's largest beef- and soybean-producing area, and conversions to cropland and pastures have also encroached on the Amazonian rain forest, with major losses in the states of Rondônia, Mato Grosso, and Tocantins (Tollefson 2010).

In absolute terms, Brazil's post-1960 expansion of pastures has no parallel worldwide, but the land devoted to grazing has also seen major global gains since the beginning of the early modern era, for two main reasons. First, faster population growth that began during the early modern era has been pushing increasing numbers of pastoralists into less inviting areas where grazing requires more land for a unit of annual meat or milk productivity. Second, the extension of large-scale commercial cattle production onto the vast grasslands of North America and Australia during the second half of the nineteenth century and onto the South American and African grasslands throughout most the twentieth century added nearly twice as much pastureland in 200 years as the total arable land accumulated after some ten millennia of settled cultivation, but without any expansion of true pastoralist populations. Approximate estimates of the world's grazing land show a slow growth during the first 1,500 years of the Common Era, from about 100 Mha at its beginning to about 220 Mha at the onset of the early modern era (HYDE 2011).

By 1800 that total had more than doubled, to just over 500 Mha, and then it expanded 2.5 times during the nineteenth century (to nearly 1.3 Gha), mainly as a result of the westward advance of permanently settled territories in the United States, Canada, and Australia (for a gain of some 250 Mha of pastures) and a large expansion of grazing throughout sub-Saharan Africa (a gain of some 200 Mha) and in Central Asia. The expansion rate was even slightly higher during the twentieth century (2.6-fold gain), when the greatest gains came in Latin America (where the grazing area had quadrupled since 1900 to about 560 Mha), followed by sub-Saharan Africa (more than doubling to about 900 Mha) and several regions of Asia.

As already noted, at the beginning of the twenty-first century permanent meadows and pastures occupied about 34 million km^2, or more than twice the area devoted to the cultivation of annual and perennial crops (15.3 million km^2)—and Europe is now the only continent where less than 50% of all land used for food production is in pastures (the rate is just short of 40%). In North America the share is just over 50%, in Asia it is over 60%, in Africa over 80%, and in Australia close to 90% (Rodriguez 2008; FAO 2011f). Countries with the largest areas of grazing land are China (400 Mha), Kazakhstan (185 Mha), and Mongolia (115 Mha) in Asia, the United States (nearly 240 Mha) and Brazil (almost 200 Mha) in the Americas, and Sudan (close to 120 Mha) in Africa.

Deforestation and Reforestation

While there is no major uncertainty about what constitutes annually cropped land or permanent tree plantations, the definition of forest depends heavily on arbitrary limits set on the spatial extent and density of growth. By the year 2000 more than 650 definitions of forests had been collected in the course of the FAO's Global Forest Resources Assessment (FAO 2001). The most commonly cited national, continental, and global totals regularly published by the FAO (and available in the organization's electronic database) are now based on fairly relaxed criteria: they include all areas larger than 0.5 ha (while the 1990 assessment required a minimum of 100 ha) with trees higher than 5 m and a canopy cover of more than 10%, or trees able to reach these thresholds in situ. It does not include land that is predominantly under agricultural or urban use (FAO 2005).

The explanatory notes make even more liberal allowances: any areas under reforestation that are expected to reach the above-listed criteria can be included, and even any temporarily unstocked land that is expected to regenerate! As for the specific composition, bamboo and palms are included if their groves meet the basic criteria, as are those plantations whose trees are used primarily for forestry or protection: as a result, both rubber and cork oak "forests" are in. Moreover, the total also subsumes forest roads, firebreaks, and "other small open areas." Obviously, increasing the canopy cover requirement to more forestlike rates of 25%–35% makes a large difference, and the closed-canopy forest (100% coverage when seen from above) becomes an even smaller fraction of that total. In addition, some historical reconstructions and modern statistics combine a wide range of ecosystems with trees under a sweeping category of "forest and woodland."

The FAO's latest Global Forest Resources Assessment put the total forested are at just over 4 G ha, or 31% of all continental surfaces (FAO 2011g). National coverage varies widely, ranging from ten countries with no forest and 54 with forested areas smaller than 10% of their territory to forest-rich Brazil, Canada, and Russia. Primary forests (composed of native species with no obvious signs of human activities) make up 36% of the global total, 57% are other naturally regenerated forests, and 7% are tree plantations; the total phytomass storage was estimated at 289 Gt C. Comparisons with the end-of-the-century assessment indicate a relatively small but absolutely appreciable gain of nearly 150 Mha (FAO 2001, 2011g).

In contrast, the 2000 total of 3.96 Gha (95% natural growth and 5% forest plantations) was more than 500 Mha higher than the aggregate of 3.442 Gha reported in 1990 (FAO 2001). An increase of just over 15% in ten years is unthinkable, and indeed, it did not take place: the higher total is mostly the result of a new uniform forest definition. The FAO had previously used canopy thresholds of 10% for forests in low-income nations and 20% for those in industrialized nations, while the 2000 assessment used only a single 10% threshold. This change made the greatest difference in Kenya, where the increase was from 1.305 Mha to 18.027 Mha (a nearly 14-fold jump) and Australia: the continent's forested area nearly quadrupled, from about 40 Mha in 1990 to more than 157 Mha in the year 2000. The two earlier FAO assessments had totals only slightly different from the 1990 value of 3.442 Gha: 3.779 Gha in 1963 and 3.650 Gha in 1947, while the first attempt at assessing global forest resources, prepared by the U.S. Forest Service, ended up with 3.031 Gha for the early 1920s (Zon and Sparhawk 1923).

These comparisons would indicate that the global extent of forests was roughly a third larger in 2010 than in 1920, a counterintuitive finding given the intervening decades of large-scale deforestation. But such a conclusion would be wrong. As Mather (2005) stresses, it remains impossible to compile any reliable historical series for the global forest area even for the twentieth century. All of the estimates that try to reconstruct even longer periods of the Earth's changing forest cover have even greater error margins: they must be seen merely as useful indicators of likely trends, not as markers suitable for solid quantitative conclusions.

Differences among the chosen definitions of forest account for most of the uncertainties in reconstructing the preagricultural extent of forest biomes and calculating the degree of their subsequent reduction. Matthews (1983) set the global area of preagricultural closed forests and woodlands at 6.15 Gha and estimated that by 1980 it was cut by less than 1 Gha, to 5.2 Gha, a reduction of only about 15%. In contrast, Richards (1990) concluded that 1.39 Gha of forests were cleared in less than 300 years between 1700 and 1995, with at least 950 Mha (nearly 70%) in the tropics and less than a third in temperate ecosystems. Williams (1990) estimated the pre-1650 global forest clearings at just 99–152 Mha, followed by 60–75 Mha between 1650 and 1749, 159–168 Mha between 1750 and 1849, and 418 Mha between 1850 and 1978, for a cumulative total of 740–810 Mha. Ramankutty and Foley (1999) assumed that the potential forest and woodland vegetation covered 55.27 Mkm^2, that it was reduced by less than 5% by 1700 (to 52.77 Mkm^2), and that in 1992 it covered 43.92 Mkm^2 (4.39 Gha), roughly a 20% reduction of the potential cover.

Discrepancies in estimates of forest areas do not end with historical reconstructions: contemporary figures also have a surprisingly large range of uncertainty. Whittaker and Likens (1975) put the forested area at 50 Mkm^2, while Matthews (1983) settled on 33.51 Mkm^2 of forests and 15.23 Mkm^2 of woodland, and later chose a more restrictive definitions of forest, limiting it to areas of closed or partially closed canopies, resulting in a total of only 29–34 Mkm^2 (Matthews et al. 2000). Olson, Watts, and Allison (1983) listed 30.8 Mkm^2 of tree formations (woods) and 27.38 Mkm^2 of interrupted woods; Ajtay et al. (1979) had only 33.3 Mkm^2 in tropical, temperate, and boreal forests and another 2.5 Mkm^2 in Mediterranean shrublands, while Saugier, Roy, and Mooney (2001) put the totals in the same two categories at 41.6 and 2.8 Mkm^2. None of these totals agrees with the FAO's recent global assessments.

Estimates of historical rates of deforestation are even more questionable. There are no reliable global estimates of deforestation during antiquity and the Middle Ages. This absence holds even for China, a country with an exceptionally long record of local and county gazetteers that were used to record all kinds of events and changes. Such fragmentary and discontinuous information makes it impossible to reconstruct the advancing deforestation on a nationwide scale. Wang (1998 estimates that China lost a total of about 290 Mha of forests, with the loss roughly split between conversions to farmlands, settlements, and transport networks and excessive cutting for fuel and timber, most of which could have been prevented with proper management. The most comprehensive reconstruction based on archival sources goes back only to 1700 (He et al. 2007), when China had nearly 250 Mha of forests. Extensive deforestation reduced the total to about 160 Mha by 1900 and to 109 Mha (about 11% of the country's area) in 1949, the year of the establishment of the Communist state.

Quantifications of preindustrial deforestation are difficult even for Europe, with its centuries of often detailed land-use records. The greatest period of the medieval European forest clearing—what the French historians have called *l'âge des grands défrichements* (Bloch 1931)—took place between 950 and 1350, after which conversions slowed down for nearly two centuries (a result of epidemic population loss and slow population growth) before a faster wave of deforestation began during the early modern age, helped by a rising productivity associated with water-powered saws that could process cut timber first 10 and eventually 30 times faster than manual cutting with pit saws (Williams 2006).

One of the best-known attempts to capture the spatial dimension of this process is Otto Schlüter's (1952) reconstruction of forest cover in Central Europe (in this instance an area extending from Alsace to eastern Poland and from the shores of the Baltic and the North Sea to the Alps) around the year 900, contrasted with the well-mapped extent of forests in the same area in 1900. This study, based on a variety of linguistic evidence (place-names denoting deforestations), archaeological findings, and soil and vegetation history, concluded that about 70% of this large region was forested in the early medieval period, compared to no more than 25% by 1900. Richards (1990) estimated that 180 Mha of temperate forest were cleared in Eurasia and North America between 1700 and 1850 and another 130 Mha between 1850 and 1920, with Russia accounting for nearly half of the 1700–1920 total. Nearly 120 Mha of forests were cleared in the United States between 1630 (when the total stood at 423 Mha, or 46% of all land, with about 38% in the West) and 1907, when 307 Mha remained, with just over half in the West (USDA 2001).

During the twentieth century the total barely changed, dipping to 299 Mha by 1987 before moving slightly up to 302 Mha by 1997, which means that some 70% of the area that existed in 1630 is still (or again) covered by forests. As coal replaced wood as household fuel, the rates of deforestation declined in both Europe and North America. In contrast, the exploitation of tropical forests accelerated, reaching unprecedented rates during the last quarter of the twentieth century in West Africa as well as Southeast Asia, but the process reached its greatest and most worrisome extent in Amazonia, the world's most extensive tropical forest, shared by Brazil, Venezuela, Colombia, Ecuador, Peru, and Bolivia, where large-scale conversions were rare before 1970.

The first decade of accelerated Amazonian deforestation (enabled by new road construction and encouraged by the government) was concentrated most heavily in Rondônia, Mato Grosso, and Pará and coincided with the launching of the first Earth-observation satellites: their imagery showed a herringbone pattern of deforestation progressively thickening along newly cut roads. Pre-deforestation Amazonian forest covered about 4.1 Gha, and the additional nearly 850 Mha of *cerrado* and 92 Mha of inland water surfaces made up roughly 5 Gha of what is considered in Brazil the "legal Amazon." The total deforestation before 1970 amounted to no more than 35 Mha; in contrast, by 1978 nearly 80 Mha of forest were cleared and an additional 130 Mha included trees isolated by progressing cutting and exposed along its edges; a decade later these two totals rose, respectively, to 230 and nearly 360 Mha, their total prorated to an annual loss of 1.5 Mha (15,000 km^2, an area smaller than Kuwait) and the lost and disturbed area amounted to nearly 12% of the legal Amazon (Skole and Tucker 1993).

Deforestation rates increased during the 1990s and they remained high during the first five years after 2000: between 1996 and 2005 Amazonia lost 19.5 Mha, with cuttings moving deeper into the heart of the region in the states of Acre and Amazonas. That is why a changed attitude of the Brazilian government, reflected by recent statistics, is perhaps the most promising development in an overwhelmingly depressing realm of tropical tree loss: in 2008 it announced its target to cut deforestation rates to 20% of their 1996–2005 level, and between 2005 and 2009 they had already fallen by 36% (Nepstad et al. 2009). An important part of this effort has been to put 51% of the remaining forest into various parks and reserves.

But nothing will come easily. In 2009 a group of Brazilian and U.S. researchers thought that the end of deforestation in the Brazilian Amazon was a real possibility within a decade (Nepstad et al. 2009). Just two years later satellite monitoring in March and April 2011 recorded destruction of 593 km^2, or a 473% increase in cleared area compared to the same months of 2010 (INPE 2011). And in April 2012 the lower house of Brazil's National Congress passed a bill whose provision could greatly reduce forest protection and reaccelerate deforestation (Tollefson 2012).

The last decade of the twentieth century and the first decade of the twenty-first century have also seen very high deforestation rates in Southeast Asia (particularly in Indonesia's Kalimantan and Sumatra, in Thailand, and in Myanmar) and in parts of tropical Africa (above all in the basin of the Congo River), and slower but continuing losses of remaining old-growth forests in virtually every tropical and subtropical country. In contrast, the temperate and boreal forests in the affluent countries have continued to expand. This reforestation has multiple beginnings, going back to the seventeenth century and including changes in attitude (the cult of trees in early modern England with its spacious and meticulously designed estate parks and gardens), long-range planning (Jan Baptiste Colbert's pre-1680 planting of oak forests to supply timber for the French navy of the nineteenth century), and advances in forest management, particularly the German contributions to the development of highly productive forestry based on non-native but fast-growing coniferous species, mainly Norway spruce and Scotch pine (Bernhardt 1966).

In Europe, slow pre–World War II gains accelerated after 1950 as higher crop yields led to the abandonment of large areas of marginal cultivated land and as large-scale afforestation programs and stricter requirements for immediate replanting of harvested forests stopped the net losses from commercial tree cutting. Some of these gains have been small, particularly in highly forested Scandinavian nations, but other countries have recorded substantial expansion of the tree cover: between 1950 and 1998 by 22% in Italy, between 1947 and 1988 by 30% in Germany, and between 1947 and 1997 by 32% in France, and the relative gains have been even higher in terms of the growing stock, 2.7-fold in Italy and threefold in France (Gold 2003). The latest data (UNECE 2010) show continued strong gains. Between 1900 and 2005 Europe's forest cover increased by almost 13 Mha (roughly the equivalent of Greece), annual additions to the total growing stock averaged 358 Mm3 (an equivalent of the total growing stock of Slovenia), and the total carbon stock in aboveground phytomass increased by 2 Gt to 53 Gt C.

But no nation has planted more trees during the second half of the twentieth century than China. Large-scale reforestation campaigns claimed to restore tree cover on 28 Mha between 1949 and 1979, but most of those plantings failed, and the forested area (referring to fully stocked productive forests with a canopy cover of at least 30%) was put at 122 Mha, or 12.7% of China's territory, in 1979 (Smil 1993). More successful reforestation campaigns began during the early 1980s (mostly in the southeast and the southwest), and the satellite-based national land cover data set showed forested area at nearly 138 Mha in the year 2000 (Liu et al. 2005). Most of the Chinese gains have been monocultural plantings of fast-growing trees; so it has been in India where the latest nationwide forest survey found a 5% expansion during the preceding decade (MEF 2009) but, as noted by Puyravaud, Davidar, and Laurance (2010), this was misleading, as the automated analysis of remote sensing lumps natural forests with the plantations of introduced species (eucalyptus, acacia, teak, rubber trees, pines), whose area has been expanding since the early 1990s.

Too many countries are still clearing their forests at an unacceptable pace, but the worldwide numbers have been improving. Global rates of deforestation rose from 12 Mha/year between 1950 and 1980 to 15 Mha/year during the 1980s and to about 16 Mha/year during the 1990s before they declined to around 13 Mha/ year between 2000 and 2010 (FAO 2011g). During that decade primary forests decreased by 40 Mha, but thanks to the expanding reforestation the net annual global loss fell from 8.3 Mha/year during the 1990s to 5.2 Mha/year between 2000 and 2010 (FAO 2011g). An important factor contributing to these diminishing losses has been the growing protection of forests. Nearly 8% of all (FAO-defined) forests now falls within the more strictly protected areas (the International Union for Conservation of Nature management categories I–IV), and the total rises to 13.5% when the less strictly protected categories are included (Schmitt et al. 2009). Protected shares range from about 3% of all temperate freshwater swamp forests to 28% for temperate broadleaf evergreens, and 9%–18% for most tropical forests.

What these countervailing deforestation-reforestation trends mean in terms of forests as sources or sinks of carbon remains uncertain. Between 1990 and 2005 the U.S. forest sector sequestered annually about 160 Mt C, about half of it in living and dead trees and a quarter in wood products in landfills (Woodbury, Smith, and Heath 2007). Better management and expansion of European and Russian forests have made them net sinks on the order of 300–600 Mt C/year (Nabuurs et al. 2003; Potter et al. 2005; Beer et al. 2006). The latest value for the European forests as a sink is about 75 g C/m^2 a year, with about 30% of that total sequestered in forest soils and the rest added to phytomass increments (Luyssaert et al. 2009). Chinese and Japanese forests have been also net sinks (Pan et al. 2004; Fang et al. 2005). Piao et al. (2009) concluded that China's forests (above all in the south) store annually 190–260 Mt C.

Amazonia was seen as either a large net source of carbon (-3 Gt C/year) or a major sink sequestering as much as 1.7 Gt C/year (Ometto et al. 2005), but a study of vertical profiles of atmospheric CO_2 found that (after subtracting carbon emissions from land use changes) tropical ecosystems may currently be strong carbon sinks (Stephens et al. 2007). According to Lewis et al. (2009) during the recent decades this sink averaged 1.3 Gt C/year. And Harris et al. (2012) concluded that between 2000 and 2005 gross carbon emission from tropical deforestation averaged about 0.8 Gt C, only 25–50% of previous estimates.

Urban Areas, Impervious Surfaces and Energy Supply

Humans prefer to settle on fertile soils, and this means that urbanization and industrialization have led to disproportionately large net losses of primary productivity.
O'Neill and Abson (2009) quantified this reality by analyzing the average potential
NPP in urban areas and comparing it to the average potential NPP of regions and
biomes containing the cities. They concluded that on the global scale, the potential
NPP of urban areas is about 20% higher than the NPP of their regional surroundings, and they also found an opposite phenomenon, namely, that the areas set aside
for protection as parks had a lower potential NPP. Taking a different approach,
Imhoff et al. (2004b) calculated that urbanization in the United States has reduced
the country's overall carbon sink capacity by about 1.6% of the pre-urban rate, a
reduction large enough to nearly offset the overall gain of 1.8% created by the
conversion to intensively farmed agricultural land.

Although cities have a history of more than five millennia, an overwhelming
majority of agricultural populations had always lived in small settlements, and the
high habitation densities of such villages or farmsteads (including sheds for animals
and storehouses for harvested crops) made for very small spatial claims compared
to the land needed for food and fuel production. Ancient and medieval cities also
had very high residential densities within their stone or brick walls, and because
low yields required all but a small fraction of the population to be engaged in food
production, city sizes remained restricted. For example, Rome, the largest city in
European antiquity, occupied an area of just 15 km^2 within the Aurelian walls, and
with its exceptionally high population total of about one million people, it averaged
nearly 700 people/ha, a high density even by modern urban standard (Smil 2010c).

The overall situation changed little even at the beginning of the early modern era:
historical reconstructions of urban populations indicate that in 1600, only about 5%
of people lived in cities, compared to about 4% in 1500 and about 2.5% in the year
1000 (Klein Goldewijk, Beusen, and Janssen 2010). In the Western world cities began
to expand beyond their medieval walls only with the onset of large-scale industrialization. In North America they grew with the rising immigration, industrialization, and
the westward settling of the continent during the latter half of the nineteenth century,
and in Latin America, Asia, and Africa the period of rapid urban growth began only
during the twentieth century, mostly after World War II, and in China (where migration to cities was restricted during the Maoist decades) only during the 1980s.

By 1900 urban residents made up 15% of the global population, and 100 years
later the share had reached about 47%. According to Klein Goldewijk, Beusen, and

Janssen (2010), the total built-up area of the cities rose from about 47,000 km^2 (an area a bit larger than Denmark) to 538,000 km^2 (an area a bit smaller than France). The last total is just one of eight recent studies that calculated the total global urban area for the year 2000, with results ranging over an order of magnitude, from 270,000 km^2 to as much as 3.52 Mkm^2. Potere et al. (2009) assessed the accuracy of these studies by using a random sample of 10,000 high-resolution validation sites and 140 medium-resolution city maps and concluded that the assessment based on the new MODIS 500 m resolution global urban map is the most accurate: it put the total urban land at 657,000 km^2 in 2001 (Schneider, Friedl, and Potere 2009). This study classified as urban all areas where built structures occupied more than 50% of the total area, and it included all nonvegetated, human-constructed elements with a minimum area of more than 1 km^2.

This finding has several interesting implications. First, the total is only about 20% higher than the one cited in the last paragraph that was used to derive the historical trend of urban land claims, and hence it basically confirms the validity of the progression of land claimed by cities from a total on the order of 10,000 km^2 in 1500 to 50–60 times as much half a millennium later. Second, the claim of some 660,000 km^2 equals only about 0.44% of the Earth's continental area. and even when allowing a 50% error, the total would be no more than 1 Mkm^2 or less than 0.7% of ice-free land. This surprisingly small share of land needed to house half of humanity (the 50% mark was surpassed in 2007) is due to very high residential densities (Demographia 2010).

In the year 2000, citywide means of population densities were about 13,000/km^2 in Tokyo, 16,000/km^2 in Seoul, 20,000km^2 in Paris, and 40,000/km^2 in Manila. Hong Kong's most densely populated district (Kwun Tong) had 50,000/km^2, and Tokyo's four central wards had a daytime density of about 55,000/km^2, as high as the daytime population of Manhattan, where the Wall Street area packs in close to 250,000 people/km^2 during working hours. Residential densities on the order of 50,000 people/km^2 imply more than 2 kg/m^2 of anthropomass, the rate unmatched by any other vertebrate and three orders of magnitude higher than the peak zoomass recorded for large herbivorous ungulates in Africa's richest grassland ecosystems. What is even more stunning is that such human densities surpass even those of all microbial biomass that normally dominates the weight of heterotrophs in natural ecosystems.

Urban and industrial areas and transportation corridors are, of course, far from homogeneous: they include such impervious surface areas (ISA) as buildings, roofed and paved surfaces (roads, sidewalks, and parking lots), and aboveground storages of materials (fossil fuels, ores) that are devoid of any vegetation, as well as plenty of grassy surfaces around buildings, roads, and railways, and not only individual trees or small groves in parks and along transportation corridors but often substantial tree cover in low-density residential suburban and exurban developments. Perhaps the most accurate ISA total for the United States is found in a USGS study that ended up with just over 90,000 km^2 for the 48 states (USGS 2000).

Elvidge et al. (2004) used remote sensing imagery to estimate that in the United States, impervious surfaces covered about 113,000 ± 13,000 km^2, and Churkina, Brown, and Keoleian (2010) ended up with a somewhat higher total of 141,000 ± 40,000 km^2 for the year 2000. They also estimated that grass (with a carbon density of just 0.5 t/ha) covers about 40% of urban and 66% of all exurban areas, and that forest (with a carbon density of about 85 t/ha) accounts for just over 25% in each of these categories (defined, respectively, as one housing unit for every 4,000 m^2 and every 4,000–162,000 m^2). These assumptions mean that urban and exurban areas of the United States stored at least 3.5 (the most likely range being 2.6–4.6) Gt C in the year 2000. For comparison, they also calculated that carbon stored in the woody phytomass used in buildings and furniture was on the order of 1 Gt, and that landfills harbored about 2 Gt C.

The first global inventory of constructed ISA used a 1 km^2 grid of the brightness of satellite-observed nighttime lights and population counts calibrated using 30 m resolution ISA of the United States (Elvidge et al. 2007). This study found that the worldwide total of constructed ISA in the year 2000 was about 580,000 km^2, or 0.43% of land surface (equivalent in area to Kenya or Madagascar), and its total for the United States (about 84,000 km^2) was very close to the USGS study. The U.S. ISA was only a few percentage points smaller than the world's largest national total, in China (87,000 km^2); India, Brazil, and Russia completed the top five.

The ranking is very different in per capita terms. After the small desert countries of the Persian Gulf and even smaller-city states (whose ISA is exceptionally high a share of the national territory) are set aside, the highest rates are in the Northern Hemisphere's affluent countries, and particularly in the higher latitudes: Canada led, with about 350 m^2 per capita, followed by Finland (320 m^2), the United States (about 300 m^2), Norway (235 m^2), and Sweden (220 m^2), while the ISA rates for most other countries (both high- and low-income) were between 50 and 150 m^2 per capita. Perhaps the most remarkable (and counterintuitive, for such a highly urbanized society) is Japan's low rate of just 114 m^2 per capita—but it is perfectly explicable, given the exceptionally high density of housing and the constraints on construction arising from the country's lack of nonagricultural flat land (85% of Japan's surface is mountainous with often fairly steep slopes).

The great concentrations of humanity in urban areas (increasingly in megacities), the affluent lives of some 20% of the world's population, and the rising economic aspirations of billions of inhabitants of low-income countries have been made possible by unprecedented levels of energy consumption. Although the extraction and conversion of fossil fuels, the dominant sources of primary energy supply, proceeds with high power densities, those processes still make significant land claims. My detailed estimate for the United States in the early 2000s showed that the extraction of fossil fuels claimed less than 500 km², and so did the mostly coal-fired thermal electricity generation; fuel processing (including the refining of imported crude oil) needed about 2,000 km² (Smil 2008).

Areas entirely stripped of vegetation, built up, or paved are thus fairly small, and much larger claims are made by pipeline and transmission rights-of-way: the former added up to less than 12,000 km², the latter had reached about 9,000 km². Those rights-of-way preclude many land uses (including regrowth of natural vegetation) but allow for harvests ranging from crop fields to grazing land and Christmas tree plantations. By the year 2000 the United States also had an outstanding land debt of some 4,000 km² of unreclaimed coal mine land (some of it with natural regrowth). The aggregate of land claimed by the U.S. fossil fuel energy system was thus about 25,000 km² (no more than 0.25% of the country's territory, or an equivalent of Vermont), and net additions during the first decade of the twenty-first century were less than 1,000 km² a year. Estimating the worldwide extent of land claimed by modern energy infrastructures is much more uncertain.

My calculations for the beginning of the twenty-first century (generally based on fairly liberal assumptions) resulted in no more than 300,000 km², or an Italy-sized area occupying roughly 0.25% of the Earth's ice-free land (Smil 2008). But less than 10% of that total was claimed by extraction and processing of fossil fuels and by the generation of thermal electricity generation, and more than half of the total was due to water reservoirs for hydroelectricity generation. But attributing all of that water-covered area to energy infrastructure is arguable because many reservoirs have multiple uses, supplying water for irrigation, industries, and cities and providing opportunities for recreation. A rough proportional attribution of reservoirs' surfaces would cut the total claim of global energy infrastructures to less than 200,000 km², and if all the land occupied by reservoirs and transmission rights-of-way was left aside (the latter choice justified by the fact that the land can be used for crops or grazing), then the overall land claim would be more than 75,000 km², roughly the area of Ireland.

11

Harvesting the Biosphere

In an attempt to quantify what clearly appeared to be a disproportionate share of the Earth's photosynthetic production that is claimed (directly and indirectly) by its most sapient species, Vitousek et al. (1986) chose net primary productivity (NPP) as the baseline and expressed the overall effect of harvests and modifications as the fraction of NPP appropriated by humans. A very wide range of possible appropriations offered by that paper—from as little as 3% for their low calculation to as much as 40% for their high estimate—made its use for any heuristic purposes, and even more so for any policymaking purposes, rather questionable.

But this did not prevent the authors from concluding "that organic material equivalent to about 40% of the present net primary production in terrestrial ecosystem is being co-opted by human beings each year" (Vitousek et al. 1986, 372). In 1990 a new assessment concluded that humans divert for their own uses biomass energy flows amounting to 23.5% of the biosphere's potential NPP (Wright 1990). No new findings were published during the 1990s, and the next decade brought three new assessments of the human appropriation of the global net primary production (HANPP became yet another awkward acronym), the first by an American trio (Rojstaczer, Sterling, and Moore 2001), the second by a larger American team (Imhoff et al. 2004a), and the third by a group of Austrian researchers (Haberl et al. 2007).

I will first review the key findings of all these HANPP studies and show how different assumptions and different choices among many uncertain totals available for key stores and fluxes can result in very different outcomes. In order to judge the degree of human impacts on the global environment, it is highly desirable to quantify our biomass harvests and compare them with the best estimate of the biosphere's productivity. At the same time, we must be aware that this is very much a work in progress: while it would be an exaggeration to say that such studies can produce

American corn has been the country's most important crop. It is grown annually on 30–35 Mha of farmland (about 20% of all arable land), its recent yields have been around 10 t/ha, and its annual harvest has fluctuated around 300 Mt. The photograph from FREEBigPictures. com can be downloaded at http://freebigpictures.com/plants-pictures/corn-field.

almost any result, the range of outcomes that can emerge from relying on plausible combinations of defensible methods and published information is still simply too wide to offer a reliable guidance. Reducing these uncertainties and coming up with a more accurate appraisal of human claims on the biosphere should be a major goal of modern environmental science.

Most of this chapter tries to make a small contribution toward this goal. I present three rather straightforward but fairly exhaustive sets of quantifications that detail direct biomass harvests for human consumption and their collateral impacts. The first section quantifies the current extent of global crop harvests and their evolution during the twentieth century. In the second section I do an analogical examination of aquatic harvests that are now also included in the expanding aquacultural production. The chapter's closing section reviews the extent and recent history of harvesting woody phytomass for fuel, timber, and other uses.

Human Appropriation of the Net Primary Production

The first assessment of human claims on the NPP defined "appropriation" by resorting to three levels of estimates (Vitousek et al. 1986). The low estimate included only the share of the NPP that people use directly as food, fuel, fiber, or timber. Food consumption was calculated for the population of five billion people by using the FAO's average per capita food supply of 2,500 kcal/day. That rate reflects average per capita food availability according to the food balance sheets prepared by the FAO for all of the world's nations, but it is not, as the authors incorrectly assumed, "average caloric intake": the actual food intake is significantly lower (in some affluent countries by as much as 40%) than the mean per capita supply (Smil 2000, 2008).

This low calculation assumed that during the late 1970s people consumed annually 910 Mt of biomass, including 760 Mt of phytomass and 150 Mt of zoomass, and that it took about 2.9 Gt of phytomass to produce all animal foodstuffs. Vitousek et al. (2006) derived the total of 2 Gt of marine phytomass needed to produce the global fish harvest of 20 Mt of dry mass by assuming that the average fish caught was feeding at the second trophic level, and ended up with 2.2 Gt of woody phytomass (for timber, pulp, and fuel). This added up to about 7.2 Gt of phytomass, or roughly 3% of the annual NPP as estimated by Ajtay et al. (1979).

The intermediate calculation added all NPP that was "co-opted" by humans, that is, "material that human beings use directly or that is used in human-dominated ecosystems by communities of organisms different from those in corresponding natural ecosystems," a taxing definition that also includes biomass destroyed during land clearing or conversion. The count began with the NPP of all croplands (15 Gt/year) and all pastures that had been converted from other ecosystems (9.8 Gt/year); to this was added the phytomass of natural grasslands that was either consumed by grazing livestock (800 Mt) or destroyed in anthropogenic fires (1 Gt). The forest part of the total account included all phytomass cut and destroyed during wood harvesting and shifting cultivation and in the course of establishingf plantations (total of 13.6 Gt). The grand total of 40.6 Gt of "co-opted" terrestrial phytomass amounted to 30.7% of the global NPP estimated by Ajtay et al. (1979).

Finally, the high estimate includes all of the phytomass from the intermediate calculations, as well as all "productive capacity lost as a result of converting open land to cities and forests to pasture or because of desertification or overuse (overgrazing, excessive erosion)," and the authors thought this total "seems a reasonable statement of human impact on the biosphere" (Vitousek et al. 1986, 368). These additions brought the grand total to 58.1 Gt, an equivalent of 38.8% of the global NPP estimate of Ajtay et al. (1979). This finding led to the most quoted sentence of the entire report: "Thus, humans now appropriate nearly 40% of potential terrestrial productivity " (Vitousek et al. 1986, 372), and the authors added that the humans also affect much of the remaining 60%, "often heavily," but offered no quantifications of that claim.

The second attempt to quantify the human appropriation on the Earth's NPP was expressed in energy terms. Wright (1990) estimated this impact in order to find the likely effect on the survival of wild species whose existence depends on metabolizing the available primary production, the supply of which is reduced by food harvesting, wood gathering, and other human interferences. Wright sought to compare human appropriations not with the actual NPP but with the amount of primary production that would be available in natural ecosystems "in the absence of human impact by expressing them as a percentage of pre-impact global net primary production" (Wright 1990, 189). The entire exercise was thus predicated on a fairly accurate knowledge of potential primary production, and Wright based his estimate on the already cited work of Bazilevich, Rodin, and Rozov (1971) and Olson, Watts, and Allison (1983), and opted for the total of 160 Gt/year of phytomass, equivalent to about 2,800 EJ.

Wright's appropriation count focused on what he called production effects, that is, impacts that changed (lowered) the natural potential of the NPP; such impacts included forage consumption by livestock (64 EJ), habitat destruction (the conversion of natural ecosystems to cropland and urban areas, altogether worth 480 EJ), and habitat degradation (the conversion of forests to grazing land, desertification, and the production of secondary growth in forest and field patches, adding up to 112 EJ). The grand total of 656 EJ was equal to 23.5% of the Earth's annual potential production, and by using specific energy curves for groups of heterotrophs, Wright estimated that this human diversion of natural energy flows translated into endangerment of 2%–7% of terrestrial species and predicted (conservatively, as he noted) that by the year 2000, 3%–9% of the world's species could be extinct or endangered.

The third attempt to calculate the human appropriation of primary production (Rojstaczer, Sterling, and Moore 2001) was essentially an update of the 1986 assessment: the authors followed the original template for the intermediate calculation, arguing that the low estimate, which dealt only with direct consumption,

yielded an unrealistically low assessment, while the high estimate required speculation as to the NPP lost as a result of human actions. Their evaluation differed
from the 1986 account of Vitousek et al. in three important ways: it was limited
to the terrestrial NPP; they used the most recent data available at the time of
their analysis (many derived from continental- or global-scale measurements);
and they estimated the uncertainty ranges for all parameters with inadequate literature references and derived an estimate of NPP variability by using a stochastic
simulation.

Their mean result of HANPP—39 Gt of dry matter, or 20 Gt C—was equal to
32% of terrestrial NPP, matching almost precisely the intermediate value estimated
by Vitousek et al. (1986). This was a mere coincidence because most of the parameters used in this analysis had substantially different values. More important,
Rojstaczer, Sterling, and Moore (2001) concluded that the variance in their estimates
of parameters resulted in poorly constrained confidence intervals of ±27 Gt (14
Gt C) and hence in a more than fivefold range for HANPP of 12–66 Gt of dry
matter. Not surprisingly, they had to admit that "the error bounds are so wide that
mean estimates of HANPP like that obtained here and earlier have limited utility"
(Rojstaczer, Sterling, and Moore 2001, 2550). Translated into shares of NPP, their
results mean that at the end of the twentieth century, humans "appropriated" as
little as 10% and as much as 55% of all terrestrial photosynthesis.

The fourth attempt to quantify the global rate of human appropriation of photosynthetic products defined the measure as the amount of terrestrial NPP that is
required to produce foodstuffs and fibers consumed by humans, including the phytomass that is lost during harvests as well as during the processing of harvested
tissue into final products (Imhoff et al. 2004a). Unlike the previous two attempts,
which were based on biome and global averages, this account also presented a rough
spatial distribution of the appropriation by using the FAO's data for 230 countries
in seven key categories: plant food, meat, milk, eggs, and wood for construction and
fuel, paper, and fiber.

To acknowledge the many inherent uncertainties, Imhoff et al. (2004a) presented
the HANPP estimates in low, intermediate, and high versions (amounting, respectively, to 8, 11.54, and 14.81 Gt C), and assumed the annual NPP was 56.8 Gt C;
as a result, human appropriations in the different versions represented about 14%,
20%, and 26% of the biosphere's annual primary production. The global HANPP
mean of about 20% was derived from continental averages ranging (for intermediate values) from about 6% for South America to 80% for South Asia, with Western
Europe just above 70% and North America just below 25%. As would be expected,

the spatial variation was even greater at smaller scales, ranging from zero to more than 30,000% in major urban centers.

Haberl published his detailed account of the HANPP for Austria, first in German and then in English (Haberl 1997), and the method he chose for that assessment was eventually used by a group he led to prepare yet another global estimate of human claims on the biosphere (Haberl et al. 2007). That study followed Wright's (1990) suggestion and defined the HANPP as the difference between the net productivity of an ecosystem that would be in place in the absence of humans (potential NPP, labeled NPP_0) and the net productivity that actually remains in an existing ecosystem (labeled NPP_t). NPP_t is thus the difference between the NPP of the actual vegetation (NPP_{act}) and the NPP harvested by humans (NPP_h): $NPP_t = NPP_{act} - NPP_h$, and $HANPP = NPP_0 - NPP_t$.

NPP_0 was calculated using the Lund-Potsdam-Jena model of global vegetation (PIK 2012), and the FAO's country-level data were used to derive crop and forest harvests. Aggregate global HANPP added up to 15.6 Gt C, or nearly 24% of potential NPP, with 53% of the total attributable to phytomass harvests, 40% to land-use-induced changes in primary productivity, and 7% to anthropogenic fires.

A regional breakdown showed HANPP values ranging from 11% for Australia and Oceania to 63% for South Asia, with Western Europe averaging 40% and North America 22%. Haberl et al. (2007) also recalculated the global HANPP according to the method used by Vitousek et al. (1986) and, despite the substantial difference in definitions and accounting procedures, wound up with a virtually identical value of 37%. Table 11.1 summarizes the mean HANPP values of these five assessments and also lists their ranges: the mean of means is just over 25%, the extreme low values are just 3%–10%, and the maxima range from 26% to 55%. Except for Wright (1990), these rates were calculated using standard estimates of global NPP values close to 60 Gt C, but if Welp et al. (2011) are correct, those values may be as high as 75–88 Gt C. In that case, all the values presented in table 11.1 would have to be reduced by at least 20% and perhaps by as much as 32%, and the mean rate of the five HANPP studies would thus fall to between 17 and 20%.

National assessments of HANPP have been done for only a few countries, but in some of these cases the analysis has been extended into the past and has uncovered some clear historical trends. The longest span has been covered by the HANPP study of the UK between 1800 and 2000, performed in Austria (Musel 2009). During those two centuries—when the country's population nearly quadrupled, cropland first expanded and then declined, the forested area nearly doubled, and urban land increased by an order of magnitude—the HANPP showed first a steady

Table 11.1
Comparison of the Global Estimates of the Human Appropriation of the Net Primary Productivity (HANPP)

Study	Period of Estimate	HANPP Mean (%)	HANPP Range (%)
Vitousek et al. (1986)	Late 1970s	27	3–39
Wright (1990)	1980s	24	
Rojstaczer, Sterling, and Moore (2001)	1990s	32	10–55
Imhoff et al. (2004b)	1990s	20	14–26
Haberl et al. (2007)	2000	24	

Note: Except for Wright, the rates were calculated using standard (consensus) estimates of global NPP close to 60 Gt C/year.

decline during the nineteenth century (from 71% in 1800) followed by an increase until the later 1950s and then another decrease to 68% by the year 2000.

Krausmann (2001) presented only a slightly shorter trajectory for Austria between 1830 and 1995, with the HANPP declining from about 60% in 1830 to 48% by 1970 and then rising slightly to 51% by 1995. Because of the higher productivity of managed crop and forest phytomass, harvests in the mid-1990s were more than 70% above the 1840s mean, yet 23% more phytomass remained in the terrestrial ecosystem than in 1830. The calculation of HANPP values for the Philippines between 1910 and 2003 (the period of massive deforestation, a tenfold increase in population, and impressive gains in average crop yields) showed a rapid increase from claiming 35% of potential primary production in 1910 to slightly above 60% by 1970, with constant values thereafter (Kastner 2009).

Schwarzmüler (2009) examined the changes in Spain's HANPP between 1955 and 2003, a time when the country's cropland declined and its forest cover expanded and when, thanks largely to higher crop yields, its phytomass harvest grew by more than 50% to more than 100 Mt of dry matter a year. At the beginning of that period Spain's HANPP was about 67% of potential production; half a century later it had declined to 61%, still rather high when compared to other high-income countries' HANPP and, of course, well above the global mean. Finally, Prasad and Badarinth (2004) calculated that between 1961 and 1998, India's HANPP increased by about 28%, a much slower growth than the expansion of India's population, which more than doubled during the same period.

Vitousek et al. (1986) reduced a complex exercise based on numerous questionable assumptions and relying on rough estimates with undefined error margins to a simple claim that humans already "co-opt" about 40% of the planet's photosynthetic production. If true, this finding implies highly worrisome consequences for the future of humanity. If the combination of further population growth and higher per capita consumption were to double, the overall human demand for the products of primary production (a development that could easily be foreseen to occur before the middle of the twenty-first century), then we would have to harvest all but a small share of the products of terrestrial photosynthesis, transforming every cultivable area into cropland or a tree plantation and leaving virtually nothing to sustain natural ecosystems.

That would clearly be an impossible situation, a development incompatible with the perpetuation of a well-functioning biosphere. To my surprise, there was no immediate (or later) critical reexamination of that original claim; instead, the 40% rate of "appropriation" or "co-optation" attracted considerable attention, and it became a frequently cited indicator of the progressing human impact on the biosphere. A comprehensive analysis of the concept and, even more so, a detailed examination of all individual assumptions and approximations that went into constructing the first HANPP assessment and its successors, introduced in the previous section, could easily fill a long book chapter rather than a short subsection.

That is why I will focus on only a few key problems, some general and integral to any construction of additive indicators or indices, others specific to the HANPP exercises. Questions and reflections should start with the meaning and appropriateness of a highly problematic denominator chosen to calculate the appropriation ratio. As I explained in some detail in the first chapter, the NPP is a theoretical concept, not a variable that is subject to direct measurement or a physical entity that can be left alone or harvested—and hence it is incorrect to say that people can use it, directly or indirectly. And while Imhoff et al. (2004a) estimated all belowground production, most of the other published NPP rates refer only to aboveground production, but that exclusion is not always clarified either in the original publication or in the references made to HANPP studies.

Some HANPP studies make the distinction explicit, and some even use the acronym aHANPP. This restriction creates two obvious problems. First, grasslands store more phytomass underground rather than aboveground, and their belowground NPP is in most cases considerably higher than their shoot productivity: its share is roughly 50%–65% in tall grasses, 60%–80% in mixed-grass prairie, and 70%–80% in short-grass ecosystems (Stanton 1988). Large herbivores only rarely remove belowground phytomass, and hence calculating the HANPP of grasslands by considering only shoot NPP and shoot consumption by grazers misrepresents the true dynamics of primary productivity in grasslands. In contrast, harvests of aboveground tissues of annual crops leave behind dead roots, and Imhoff et al. (2004a) included them among human appropriation—but the Austrian group excludes them from HANPP calculations because that dead phytomass is fully available to decomposers and soil heterotrophs.

The second problem with limiting the NPP account to only the aboveground phytomass concerns the harvesting of belowground tubers, roots, and seeds. The first category represents a massive harvest of white and sweet potatoes, yams, cassava, and taro, totaling some 700 Mt worldwide; the second category includes sugar beets (now more than 200 Mt/year) and many vegetable root crops, ranging from carrots, onions, and garlic to celeriac and Japanese *gobo*; the most prominent component of the third category is peanuts, whose harvest is now close to 40 Mt/year. The grand total of these underground harvests is now thus on the order of 1 Gt of fresh phytomass, but their "appropriation" is charged against the aboveground NPP!

More important, even before humans begin any harvesting, the NPP of crops and forests is reduced, often substantially, by incessant heterotrophic attacks. Here the fundamental realities of phytomass harvests clash with both the choice of the analytical denominator (NPP) and the correct understanding of the key operative term ("appropriation"). On the most general semantic level we should ask what is meant by saying that humans "appropriate" (or "co-opt") a certain share of the Earth's annual photosynthetic production. "Appropriate" may be a felicitous choice of a verb intended to capture the entirety of human intervention: it is obviously superior to "consumed," as the connection of the latter verb to phytomass evokes first food, and on second thought also wood for construction and pulp.

The HANPP may be defined in ways that subsume not only direct harvests of phytomass for food, feed, and fuel but also many indirect claims humans make on the biosphere's photosynthetic production: the annual burning of grasslands to maintain open pastures for large herds of domesticated animals is spatially the most extensive example of such interventions. But the very first definition of that inclusive verb in *Merriam-Webster's Dictionary*, "to take exclusive possession of," also shows that the way it was used by Vitousek et al. (1986) is not accurate: the biosphere operates in ways that make it impossible for humans to take exclusive possession of any phytomass.

Viral, bacterial, and fungal infections affect every staple grain: infestations by fungi, including wheat rust (*Pucinia triticum*), rice blast (*Pyricularia oryzae*), and corn ear rot (*Aspergillus flavus*), are particularly common. Insect depredations can reduce the yield or claim virtually the entire productivity of tree stands on scales ranging from local to semicontinental: such long-standing pests as mountain pine beetle (*Dendroctonus ponderosae*) and spruce budworm (*Choristoneura fumiferana*) and Asian longhorned beetle (*Anoplophora glabripennis*), a relative newcomer to North America, are common examples of invasive invertebrates that inflict massive, large-scale, and chronic damages. To these losses must be added those induced by the periodically devastating locust (*Schistocerca gregaria*) swarms, which can destroy crop harvests on regional and even national scales.

Vertebrate depredations range from elephants eating and trampling African crops to deer and monkeys feeding on corn and birds picking off ripe grapes in vineyards around the world. And highly variable, but in aggregate not insignificant (often on the order of 5%), shares of crops are not harvested because of lodging of stalks (particularly common in grain and oil crops), shattering, and preharvest sprouting of grain. These incessant probings and sustained or episodic attacks by heterotrophs and physical damage by wind and rain can be only partially warded off by applying antibacterial agents, fungicides, and insecticides, by breeding (at least somewhat) resistant cultivars, or by frightening away the invaders, restricting their access by nets, or killing them.

Undeniably, even in modern, highly managed agroecosystems, where a great deal of energy and investment is spent on minimizing any heterotrophic depredations, as well as in tree plantations, the terms NPP and NEP are far from identical: the preharvest heterotrophic consumption remains substantial and will be never eradicated. Obviously, the correct denominator should be the net ecosystem production (NEP), not the NPP—but we would have to know its value at the time of actual harvests in a particular year because highly variable weather and the presence of pests will result in annual NEP averages that will easily fluctuate ±10% even around a short-term (say, five-year) mean.

Further, the harvested phytomass is subject to a second wave of heterotrophic attacks during its storage: bacteria, fungi, insects, and rodents assert their claim before the feed or food harvests can be consumed; improperly stored grain in low-income countries is particularly vulnerable (more than 5% of it may be lost before consumption), and tubers in the tropics sustain even higher preconsumption losses. A case can be made that these storage losses should be classed under human appropriation, but their obvious beneficiaries are commensal or wild heterotrophs, and this reality contradicts the claim of an "exclusive possession" of phytomass harvested by humans for their own use.

All of the difficulties arising from the choice of NPP as the principal analytical denominator are greatly magnified when the calculation calls (as in all analyses done by the Austrian authors publishing HANPP estimates) for the NPP of potential rather than actual vegetation. Although all large-scale estimates of actual NPP have been subject to substantial errors because of their reliance on theoretical models or on combinations of models and remote sensing, smaller-scale studies of HANPP could use rather accurate measures of NEP derived from actual field measurements. Such an option is, obviously, impossible when quantifying potential production, and the outcome will be only as good as the initial inputs of the key variables driving the calculation and the overall ability of a chosen vegetation model to reproduce the real world.

All Austrian-authored assessments calculate the potential NPP using the Lund-Potsdam-Jena global dynamic vegetation model, one of more than a dozen more or less complex models used to simulate photosynthesis on large scales (Sitch et al. 2003). An intercomparison of such models used to simulate actual annual production showed a substantial range of results, with the highest global NPP value twice the lowest rate; even after four extreme values were excluded, the extreme values of the remaining 12 assessment differed by 40% (Cramer et al. 1999). Choosing a different model or changing some of the input parameters of a preferred model may thus result in differences easily amounting to 10%–15%.

Sometimes we can safely identify the direction of the most likely error, but when faced with half a dozen or a dozen uncertainties we cannot be sure, even if each one of them is relatively small (no greater than 5%), what the aggregate effect will be. The fortuitous cancellation of opposite errors would make no difference, but simple arithmetic shows that if both the harvested phytomass total and the chosen (actual or potential) NPP value have minimum unavoidable errors of ±15%, then the extreme HANPP shares would be about 27% less and 33% more than the mean rate, bracketing a range of possible outcomes that is simply too large to assess real concerns and to inform effective policies.

Finally, there has been no uniform approach to calculating the HANPP, and the published values are often cited without making clear what the numbers represent. A *sensu stricto* definition of human appropriation of photosynthetic products is rather self-evident as it includes all crop harvests (be they directly for food or for animal feeding, raw materials, or medicinal or ornamental uses) and all harvests of woody phytomass (be it for fuel, construction timber, or roundwood to be made into plywood and furniture or pulped to make cardboard and paper for packaging, printing, and writing). That is the first, low-estimate choice by Vitousek et al. (1986).

A *sensu lato* definition is much more elastic: there is no clear natural cutoff for inclusion, while many impacts that should be obviously included are difficult to quantify in a satisfactory manner. Grazing by domestic animals should be included, as that kind of consumption would not take place without humans—and yet all grazing that is done in a truly sustainable manner (by keeping the animal densities well below the available annual phytomass production or by ensuring that they do not compete with the resident wild grazers) does not diminish the overall photosynthetic capability of a site: in fact, some grazing may promote growth. And if the phytomass were not consumed by domesticated herbivores it would not necessarily be "appropriated" by other vertebrates: the ungrazed grass would die during the winter or arid season and eventually decompose. Moreover, domesticated grazers also return much of the partially digested phytomass in their wastes, actually promoting patchy grasslands productivity.

This is not the only case in which strict logic would demand an adjustment of the actually harvested phytomass. Increasingly common conservation tillage and no-till practices either recycle most of the residual phytomass (straws, stalks) or do not remove any of it, leaving it to decomposers and other heterotophs. Significant shares of cereal straws removed for the bedding and feeding of ruminant animals are returned to fields (made available to heterotrophs) as manures. And most logging operations do not remove the tapering treetops and any branches from forests. All of these realities should be considered when calculating the difference between the NPP of potential vegetation and the NPP remaining in ecosystems after harvests: in some cases the difference will be marginal, in others it will be substantial. By a logical extension, the regular burning of grasslands that is done to prevent the reestablishment of woody phytomass should be also included in the HANPP, as should all phytomass burned by shifting cultivators and all forest fires caused by human negligence or arson.

A comprehensive global estimate of phytomass consumed in anthropogenic fires used the best available published estimates of the share of human-induced large-scale vegetation fires in different countries (mostly between 80% and 95% in the tropics, but only 15% in Canada) and a set of assumptions to calculate the biomass burned in small (shifting cultivation) fires (Lauk and Erb 2009). The exercise resulted in the annual burning of 3.5–2.9 Gt of dry matter, with one-third (1–1.4 Gt) attributed to shifting cultivation and with sub-Saharan Africa's grassland fires accounting for the largest share of the total (2.2 Gt/year).

Other studies of African burning show how uncertain that total is. The median burning interval is about four years, but some grasslands in the Sahel are not burned for up to 20 years, while annual fires are common in the Guinean zone. This makes for substantial interannual fluctuations, and different assumptions regarding the density of the burned phytomass result in annual aggregates whose extreme difference has been more than eightfold, 0.22 versus 1.85 Gt/year (Barbosa, Stroppiana, and Grégoire 1999). The latest published annual rate is for the years 2001–2005, when about 195 Mha of African grasslands were burned annually, releasing about 725 Mt C (Lehsten et al. 2009).

But adding this uncertain total to the HANPP is questionable because nearly all of the released carbon will be incorporated into new grass growth following the burning, and because many tropical and subtropical grasslands have been always subject to widespread natural seasonal fires: fire has been a recurrent event in tropical forests of Amazonia (Cochrane and Schulze 1999), as well as a dominant driver of carbon balance in Canadian boreal forests (Bond-Lamberty et al. 2007), and hence it would not be easy to quantify only the net increase in fire activity due to deliberate burning. Moreover, the productivity of many fire-adapted ecosystems actually benefits from regular burning, and so an accounting challenge would be even greater: quantifying only that part of deliberate burning that reduces overall productivity. Similarly, the productivity of fire-adapted forests may actually increase after a fire as fast-growing new trees have lower autotrophic respiration than does old-growth forest.

Moreover, an analysis of a global sedimentary charcoal data set shows that the recent rates of anthropogenic burning are actually much lower than in the past (Marlon et al. 2009a). The record shows a prolonged decline in biomass burning that lasted since the beginning of the Common Era to about 1750; then came a marked rise peaking around 1870, followed by a sharp downturn. Yet the post-1870 period has seen the most rapid land-use changes as well as rising temperatures, and hence the downturn cannot be explained by reduced human activity or a cooler climate: the fragmentation of vegetated areas, the emergence of generally less flammable landscapes, and active fire suppression are the most likely causes. These long-term trends in anthropogenic burning were confirmed for the last 650 years by an analysis of concentrations and isotopic ratios of atmospheric CO preserved in a South Pole ice core (Wang et al. 2010): it shows a pronounced decline in Southern Hemisphere burning between 1350 and 1650, followed by an undulating rise peaking during the late nineteenth century, followed by a decline to levels lower than at any time since 1350.

Yet another illustration of a comparatively large extent of premodern fires comes from comparing wildfires and set fires in North America. Kay (2007) argued that the potential for fires started by North American natives was most likely many hundreds to many thousands of times greater than the known lightning ignition rates; this led him to conclude that, compared to aboriginal burning, fires caused by lightning may have been largely irrelevant for at least 10,000 years. And Stephens, Martin, and Clinton (2007) estimated that fires ignited by lightning and Native Americans in California consumed annually about 1.8 Mha. That amounts to nearly 90% of the total area affected annually by wildfires in the entire United States during the years 1994–2004, a decade that was characterized as "extreme" as far as wildfire activity was concerned. Such a description is a perfect example of a common ahistorical approach to many natural and anthropogenic phenomena.

The flooding of vegetation by large reservoirs created by dams is obviously an anthropogenic act that curtails potential phytomass production and should be included in an appropriation account—but so should an often considerable (but also unwelcome, owing to its effects on evapotranspiration, the oxygen content of water, and the survival of fishes) production of phytoplankton and aquatic macrophyta created by water storage, particularly in the tropics. And while clear-cutting of a forest will remove a significant amount of phytomass and reduce a site's productive potential, such changes will be relatively short-lived in regions with efficient forest management, where a replanted forest may eventually reach a productivity almost as high as the original growth's.

Higher productivities of both field crops and well-managed forests may result in smaller areas devoted to these managed harvests, and as natural vegetation fills the vacated space, the national HANPP will decline. As noted in the previous section, this has indeed been the case in three of a few countries for which HANPP trends are available, Austria, Spain, and Britain. At the same time, the reduced areas devoted to intensive cropping and high-yield tree plantations experience greater environmental burdens (greater fertilizer, pesticide, and herbicide applications and increased nitrogen losses, including through greater leaching and the resulting eutrophication of waters) and may be subject to less desirable agronomic practices (increased monocropping, reduced crop rotation, soil compaction by heavier machinery). Obviously, such a decline in the HANPP cannot be seen as a purely desirable development.

In sum, human appropriation of the global NPP is not just a poorly defined measure whose quantification depends on an abstract modeled value and on a concatenation of variables that cannot be known without considerable error margins.

More fundamentally, it is a concept whose consensual unambiguous formulation is hardly possible, whose practical applications are made questionable by a number of logical shortcomings, and whose application reduces many complex processes to a single figure. Like so many other indexes and global measures, it does tell us something important, in this case about the human claim on the biosphere's primary resource, but the published HANPP values are too uncertain, too dependent on the concept's definition, and, perhaps most important, entirely devoid of any qualitative connotation to offer a special insight or guidance.

Rather than using the latest data to produce a new version according to one of the published HANPP procedures or creating yet another definition of appropriation and offering a new range of global values, I will devote the following two sections to fairly straightforward quantifications of phytomass harvest and will do it in two ways, by presenting rather comprehensive accounts of phytomass harvests at the beginning of the twenty-first century and then by making more approximate historical estimates in order to uncover some long-term trends, stressing the data limitations as I proceed. For practical reasons I take up this exercise in two topical sections, first presenting a detailed account of the phytomass needed to produce food for seven billion people and then aggregating the best available information on the harvests of woody phytomass.

Evolution of Terrestrial Food Harvests

Rough estimates of crop phytomass harvested on national and global levels are not difficult to make and, given a number of unavoidable approximations that must be made (even in the case of countries with the best available statistics), they may suffice to make an effective assessment of major phytomass flows and their long-term trends. The FAO's annual global collation of production statistics provides a readily accessible basis for global calculations whose overall error for major crop categories is unlikely to be greater than 5%–10%. In the year 2000, the global harvest of cereal food and feed grains (dominated by wheat, rice, corn, sorghum, and barley) was nearly 2.1 Gt. Sugar crops (beet roots and sugarcane stems) came second, with 1.5 Gt; the vegetable harvest (about 750 Mt) was more than 50% higher than fruit production (about 470 Mt), while tubers (mainly white and sweet potatoes and cassava) added about 700 Mt, oil crops (mostly soybeans, peanuts, sunflower, rape and sesame, olives, and oil palm fruits) about 100 Mt, and leguminous grains (led by beans, lentils, and peas) more than 50 Mt (FAO 2011d).

FAO-listed food and feed crops harvested from fields as annual crops or gathered from permanent plantations thus added up to nearly 5.7 Gt of fresh phytomass. Fiber crops (cotton, flax, hemp, jute, ramie, and sisal) are a minor addition: their fresh mass (about 25 Mt in the year 2000) is much lower than a reporting error for tubers or cereals. Agricultural land is also used to grow forages, including silage (mainly corn and sorghum) and leguminous cover crops (mainly alfalfa and clover in North America and Europe and vetches in Asia), often in rotation with other feed or food crops.

The FAO does not publish any regular statistics regarding these harvests, and annual outputs have to be pieced from data published by major agricultural producers and then extrapolated in order to account for production in countries that publish no relevant information: consequently, all of these annual rates have large (more than 25%) margins of uncertainty. American corn silage production has been on the order of 100 Mt/year, France (Europe's greatest producer) every year grows about 20 Mt of corn and grass silage, and the worldwide output has been close to 250 Mt/year. National totals for hay produced on arable land in the year 2000 were about 140 Mt of fresh weight in the United States and close to 20 Mt in France, and the global output has been on the order of 1 Gt/year.

And then there is a specific accounting problem posed by green manures. These leguminous cover crops (alfalfa, clover, vetch, beans) are used in rotations (often also as forages) and then worked into soil. This recycling takes place usually 60–120 days after their planting, and in temperate latitudes it can add enough nitrogen to produce a good summer cereal crop. This practice was common in all traditional intensive agricultures: in China it peaked as recently as 1975 with plantings of nearly 10 Mha of green manures before the availability of affordable synthetic nitrogenous fertilizers brought its rapid decline (Smil 2004). How should the phytomass production of green manures be counted? Obviously, their phytomass is not directly appropriated, and the nitrogen they fix is available not only to a subsequent nonleguminous crop but also to soil bacteria and soil invertebrates.

The only adjustment that has to be done for all production data to make them comparable across many species is to convert them from fresh-weight basis to absolutely dry weight. Arithmetically this is just a matter of simple multiplication—but also another occasion for introducing inevitable errors. But these errors are bound to be considerably smaller than those arising from the often questionable reporting of many national crop harvests or their estimates made in the FAO's Rome headquarters, or from the necessity to make nothing better than rough estimates of animal fodder produced from agricultural land. For food and feed crops the conver-

Box 11.1
Moisture Content of Harvested Crops

There are obviously large differences between the water content of oilseeds (often less than 5%), staple grains (in excess of 10%), tubers (more than 70%), green forages (around 85%), and vegetables, with many cultivars having more than 90% water (Watt and Merrill 1963 Misra and Hesse 1983; Barreveld 1989; NRC 1996; Bath et al. 1997). To prevent losses during transportation and storage, cereals should not be marketed with a water content exceeding 14.5%, but highly variable weather conditions mean that many cereal crops are harvested when their moisture content is closer to 20% than to 15%, while others enter commercial exchange with a water content just above 10%. All FAO production statistics refer to fresh (field) weight, and hence there can be no universally valid specific conversion factors. A good approximation would be to use an average moisture content of 12% for grains and oilseeds; 75% for tubers, roots, sugar beets, and sugarcane; 85% for fruits and green fodder crops (alfalfa, clover); and 90% for vegetables.

When using these shares the most likely error would be to overestimate the actual total of dry phytomass by at least 5%, as the typical moisture content for marketed cereals may be 13% rather than 12% and white potatoes, the most important tuber crop, contain 80% water (sweet potatoes and cassava have about 70%). When these approximate conversion ratios are applied to the harvest in the year 2000, cereals dominate the dry phytomass, at about 1.8 Gt, and they are also the leading harvest in all countries except some tropical nations. Among the major food producers they account for between 70% of the total in China and 55% in the United States. Sugar crops come second in the global ranking (nearly 400 Mt), followed by tubers and roots (175 Mt), oil crops (about 100 Mt), and fruits and vegetables (each around 70 Mt).

sions are fairly accurate: their total global mass in the year 2000 was about 2.7 Gt, a bit less than half the fresh weight (box 11.1). For comparison, Wirsenius (2000) ended up with a very similar total of 2.54 Gt of dry phytomass as an average for 1994–1996.

My approximate reconstructions of past agricultural harvests (all in dry weight) show them rising from about 400 Mt in 1900 to 600 Mt in 1925, 800 Mt in 1950, 1.7 Gt in 1975, and 2.7 Gt in 2000, or a nearly sevenfold increase in 100 years, most of it (as already noted) due to increased crop yields. Between 1900 and 1950, wheat yields were up only by about 20% in the United States and 38% in France. Larger U.S. harvests (up 47% by 1950 compared to 1900) came largely because of the steadily expanding area planted to wheat (Dalrymple 1988). Significant yield gains began only during the late 1950s, when the U.S. area planted to short and semidwarf cultivars surpassed 40% of all wheat fields and when sharply rising fertilizers inputs made it possible to realize more of their inherent yield potential.

Analogical trends were seen in all other wheat-growing countries, the only difference being that many of them, including France and the UK, started their post-1950s yield rise from higher average levels. Average U.S. wheat yields in the year 2000 were 3.1 times the 1900 mean, and the analogical multiples are 5.8 for France and 3.8 for China. The average yield of U.S. corn, the country's most important field crop, rose more than fivefold during the twentieth century, from 1.6 t/ha to 8.5 t/ha, and by 2009 even the nationwide mean (and not just the harvests in key Corn Belt states) had surpassed 10 t/ha. In contrast, China's average corn yield had only doubled, to 2 t/ha. Japan's average rice yield, already relatively high in 1900 (2.2 t/ha), increased nearly three times before reaching a plateau between 5.8 and 6.5 t/ha during the 1980s. Nationwide and regional yield maxima for rice are at 6–8 t/ha in East Asia and in California. Chinese and Japanese soybean yields grew about 1.7 times during the twentieth century, while the U.S. yield tripled between 1925 and 2000 (no soybeans were cultivated in North America in 1900).

I have also reconstructed specific production shares for major crops grown by the largest agricultures. The basic pattern of U.S. cropping remained relatively constant during the twentieth century. Corn retained its primacy and output share: in 1900 its total phytomass accounted for a bit over 40% of the entire U.S. crop production, and essentially the same share prevailed 100 years later, when the crop's total output had increased just over threefold compared to 1900. Wheat retained roughly one-seventh of all output after a 2.7-fold increase in production and remained the second most important crop until the last decade of the century, when its total phytomass was equaled and perhaps even slightly surpassed by soybeans.

The rise in soybeans and larger hay harvests (roughly tripling in a century) were the two primary reasons for the declining share of cereal phytomass in the total output, from nearly 75% in 1900 to less than two-thirds in 2000. No other crop has seen such a spectacularly rapid diffusion as soybeans: their American plantings amounted to a few thousand hectares in the early 1930s, but since the early 1970s they have been sown annually on more than 20 Mha, producing more than 50 Mt of seeds a year. The increase in Brazilian soybean production has been even faster, from a negligible harvest in the early 1960s to more than 20 Mt in the early 1990s. The global share of sugar crops doubled, to 2% of total phytomass, while the share of oil crops was nearly halved, to less than 3%.

To find the total phytomass harvests, the crop-yield figures must be adjusted for the production of crop residues. While the typical moistures of harvested grains have not changed, specific harvest indices (HI) were (as already explained) transformed during the twentieth century. In comparison, changes in the production of other aboveground residues (whether legume and oilseed crops or the vines and stems of tuber and root crops) have been generally much smaller. The HI values of traditional cultivars were most often below 0.3. Traditional wheats had an HI as low as 0.25–0.30, producing up to three times as much residual phytomass as grain (Singh and Stoskopf 1971). In contrast, Mexican semidwarf cultivars introduced during the 1960s measured no more than 75 cm, and their HI was around 0.35; by the late 1970s many short-stalked wheat cultivars had an HI close to 0.5, producing as much grain as straw (Smil 1987). Typical HI averages are now between 0.40 and 0.47 for semidwarf wheats.

Similarly, traditional barley cultivars had HI values below 0.35, while the improved varieties grow to just 70–80 cm and their HI range is between 0.45 and 0.50 (Riggs et al. 1981). Traditional low-yielding, long-stalked rices had HI values between 0.12 and 0.20, medium-yielding cultivars could reach 0.3–0.4, and only the best cultivars grown under optimal conditions had an HI in excess of 0.4. The first short-stalked *Oryza japonica* varieties were bred in Japan early in the twentieth century; a similar *indica* cultivar (TN1) was released in Taiwan in 1956. A major breakthrough came only in 1966, when the International Rice Research Institute introduced IR8, the first in a sequence of new cultivars that transformed Asian rice farming (Donald and Hamblin 1984; Smil 1987). Under optimal conditions these plants have an HI between 0.5 and 0.6, and their typical HI is between 0.45 and 0.50. Soybeans have an HI around 0.5, up from 0.4 or less for traditional cultivars (Weilenmann de Tau and Luquez 2000). Sweet potato is the only major crop whose HI is now often at or even slightly above 0.60. The HI values of vegetables range from less than 0.3 for a poor crop of tomatoes to more than 0.7 for red beets.

The overall effects of a higher HI can be shown by using a national example. The average yield of French winter wheat rose from 1.3 t/ha in 1900 to 7.5 t/ha in the year 2000, and during the same time the total aboveground phytomass increased from 4.3 t/ha to 15 t/ha. With an unchanged HI of 0.3, the grain yield would have reached 4.5 t/ha in the year 2000, a gain of 3.2 t/ha compared to the 1900 harvest, while with the HI rising to 0.5, this increased phytomass productivity brought a gain of 6.2 t/ha. The difference between the two gains (6.2 − 3.2) is attributable to the HI gain, and in this case it accounted for nearly half (48%) of the higher grain yield. Many HI values have been published for major field crops, but there are no comprehensive national data sets of typical rates, and estimating past global averages is even more uncertain. As a result, the actual production of crop residues could have departed by as much as ±20% from the rough pre-1950 estimates, and the

Box 11.2
Crop Residues

A rough global approximation for the recent production of crop residues can be done by assuming an HI of 0.4 for cereals, pulses, and oilseeds, 0.5 for tubers, 0.6 for sugar crops, and 0.1 for fruits and vegetables. Errors can be minimized by using different specific HI averages for major producing regions, or at least for the continents: the differences are largest for food and feed grain crops, with low-yielding African wheat, corn, sorghum, and barley having HI values of just 0.25–0.3, compared to HI values of 0.45–0.50 for comparable crops grown in Europe and the United States. The variability of environmental and agronomic factors precludes any accurate calculation of annual crop residue production, and well-documented fluctuations in cereal straw output make a particularly large difference. For example, using an average nationwide HI of 0.42 instead of 0.45 in calculating the total harvest of the U.S. cereal phytomass results in about 50 Mt of additional straw, a difference larger than all the residues produced annually by all of the country's tuber, oil, and sugar crops!

My historical reconstructions show that in 1900, crop residues accounted for nearly 75% of the total harvest of 1.5 Gt, and that share was still about 70% 2.7 Gt in 1950, but it fell to 65% by 1975 (about 3.2 Gt out of the total harvest of 4.9 Gt) and to 58% by the year 2000. Two accounts for the mid-1990s used disaggregated HI data for major crops or crop categories and ended up with very similar results. I concluded that crop residues incorporated 3.5–4 (most likely about 3.75) Gt of dry phytomass, or 55%–60% of the total agricultural harvest; cereals (stems, leaves, and sheaths) accounted for two-thirds of all residual biomass, and sugarcane tops and leaves were the second largest contributor (Smil 1999b). Wirsenius (2000) ended up with 3.46 Gt, including 2.7 Gt in cereal straws and corn stover, 400 Mt in oil crop residues, and about 200 Mt in sugarcane and beet residues.

---

most likely error is at least 10% even for the latest estimates. Only approximations can be offered as to the postharvest fate of crop residues (box 11.2).

Given the enormous uncertainties in estimating the annual production of forage crops grown on agricultural land, I will not offer detailed outlines of alternative extrapolations and will simply assume an annual output on the order of 250 Mt of fresh silage and just over 1 Gt of hay. These rates would translate (with typical moisture contents of respectively 75% and 10%) into roughly 1.2 Gt of dry phytomass in the year 2000, making the total harvest of about 7.6 Gt (2.7 Gt in field crops, 3.75 Gt in their residues, 1.2 Gt in forages) harvested from fields and permanent plantations. This figure represents an annual aggregate; the peak global preharvest phytomass is considerably lower, owing to differences in harvest times. In northern temperate latitudes, harvests extend over a period of about six months, starting in mid-May for winter wheat, peaking in August and September for major spring-planted grains, and ending in October and November for many root and vegetable crops.

A second reason is that the global multicropping ratio is now close to 1.5, ranging from single crops harvested annually in high temperate latitudes (such as Canadian or Russian spring wheat) to triple-cropping (and for vegetables even harvesting as many as five crops) in some of the world's most intensively cultivated regions in China, Indonesia, or California. As a result, the maximum global preharvest phytomass (during the late summer in the Northern Hemisphere) is at least a third smaller, no more than about 5 Gt (2.5 Gt C), a small fraction of the terrestrial phytomass. Phytomass that is directly consumed by humans was about 1.2 Gt, while the phytomass consumed by animals—about 1.2 Gt of feed crops and crop-processing residues, 1.2 t of forages, and 1.6 Gt of crop residues—added up to 4 Gt of dry matter. Fiber crops and their residues incorporated only about 50 Mt of phytomass.

For comparison, Haberl et al. (2006) estimated the global crop harvest for the years 1992–1994 at 6.28 Gt of dry matter, composed of 2.53 Gt of crops intended for human consumption, 1.04 Gt of animal forage crops, and 2.71 Gt of crop by-products, and Wirsenius (2000) presented the following global flows (all totals also in dry weight) for the years 1992–1994: food crops, 2.54 Gt (dominated by 1.65 Gt of cereals); their by-products, 3.46 Gt; animal forages, 1.15 Gt; and 5.82 Gt of feed from pastures. Imhoff and Bounoua (2006) calculated that in 1995, 4.09 Gt C (median estimate, with the low-high range of 2.83–5.85 Gt C) of phytomass were required to produce the global food supply: 1.73 Gt C were in crops destined for direct food consumption and 2.36 Gt C in phytomass for the production of animal foods, with meat accounting for just over 80% of the phytomass used for feeding.

All of these totals (with the obvious exception of root crops) refer to the aboveground dry matter of harvested crops and their residues. To express them as total crop phytomass, they should be enlarged by 20% for annual cultivars, and approximately doubled for all perennial forages. This aggregate would indicate the total standing phytomass at (or shortly before) the time of harvest, but it should not be used in any accounts of human appropriation, as the root biomass is usually completely recycled during regular plowing of annually planted fields or is left untouched for years to decades in no-till fields and in multiyear or perennial cover crops.

My reconstructions show that the total phytomass produced by American cropping tripled during the twentieth century, with more than 75% of the gain taking place after 1950 and more than 40% of it realized after 1975. The country's phytomass output per hectare of cultivated land increased about 2.4 times, from

3.2 t/ha in 1900 to almost 7.8 t/ha in the year 2000. The productivity of cereal phytomass more than tripled, from about 3.4 t/ha to 10.6 t/ha, and hay productivity doubled, from 2.7 to 5.5 t/ha. Lobell et al. (2002) used a biogeochemical model driven by satellite observations to estimate the productivity of U.S. agriculture between 1982 and 1998, and their results (expressed in carbon units) are in close agreement with my calculations. The total NPP for 48 states was 620 Mt C/year, or about 20% of total NPP, and there was an average annual increase of nearly 4 Mt C during the studied period. With annually harvested area averaging 126 Mha, this prorates to nearly 5 t C/ha, with rates ranging from 2.5 t C/ha for Kansas wheat to 6.5 t C/ha for Iowa corn.

Similarly, China's total crop phytomass expanded 3.2-fold during the twentieth century. Cereal phytomass accounted for nearly 85% of the total in 1900 and, after a 2.7-fold increase in the production, its share was down to just over 70% in 2000. The share of oil crop phytomass doubled and that of sugar phytomass rose more than sixfold. Vegetables and fruits recorded remarkable gains after 1980 under privatized farming, while the cultivation of green manures declined sharply as synthetic fertilizers replaced the traditional practice of rotating leguminous cover crops with food or industrial crops. The total phytomass productivity was almost 11 t/ha in 2000, when China's multicropping index stood at 1.2. By the year 2000, French agriculture produced 2.5 times as much cereal phytomass as it had in 1900, and the total phytomass production had expanded 2.2-fold. Notable trends included halving of the tuber phytomass, a large expansion of the legume (nearly 10-fold) and oil crop (about 50-fold) phytomass output, and a declining hay harvest. The aggregate French crop phytomass productivity rose from about 3.8 t/ha of cultivated land in 1900 to 10 t/ha in 2000, a 2.6-fold increase.

But it would be wrong to put all crop phytomass in the category of human appropriation because not all crop residues are removed from agroecosystems: a considerable share remains available in fields for microbes and other heterotrophs. Quantifying the postharvest flows of the residual phytomass is very difficult, but rough approximations indicate the extent to which the harvested phytomass is not removed from (or, to a lesser extent, is returned to) fields. The estimate will obviously be most influenced by apportioning the fate of cereal crop residues, whose eventual fate ranges from direct recycling and in situ burning to removal for feed and bedding, household fuel, and construction material.

The Intergovernmental Panel on Climate Change estimated that in low-income countries, about 25% of all residues were burned, and that the corresponding share in affluent nations was just 10% (IPCC 1995). Actual rates are higher, especially when one includes the use of residues for fuel. Even in affluent nations, data on average burn fractions indicate much higher regional rates both for field and for orchard crops (Jenkins, Turn, and Williams. 1992). Yevich and Logan (2003) estimated that in the mid-1980s, 400 Mt of crop residues were burned in fields, ranging from just 1% in China to 70% in Indonesia. In deforested regions most of the residues ended up as household fuel: as recently as the late 1980s roughly 75% of China's crop residues, including more than two-thirds of all cereal straws, were burned in cooking stoves (Smil 1993). The modernization of fuel supply has reduced that demand throughout Asia, and the use of wheat straw as fuel is minimal in Europe and nonexistent in North America. The demand for bedding straw has also decreased with modern milk and beef production arrangements, but postharvest straw burning in fields remains common around the world.

I put the overall share of recycling during the late 1990s at no less than 60% of the total residual phytomass in high-income countries and at 40% of the total in low-income countries, resulting in the most likely global mean of just below 50% (Smil 1999b). In addition, a significant share of residues used in ruminant feeding and bedding is returned to fields as recycled manure. This means that close to 2 Gt of dry-matter phytomass in recycled residues and in undigested feed and used bedding belongs to a special category. This phytomass either is not removed from fields or is rather promptly returned to them, and hence it should not be added to the human appropriation total—or it should be counted only with qualifications: once it becomes available for other heterotrophs, its recycling serves to replenish organic matter in agricultural soils and hence to enhance the production of croplands, not of natural, unmanaged ecosystems. There is no obvious unequivocal solution to this accounting problem.

Even greater challenges arise with the inclusion of phytomass consumed by domesticated animals either grazing on pastures created by the conversion of natural ecosystems or as they share natural grasslands with wild herbivores, a situation common in Africa and parts of Asia and South America. A theoretical maximum can be found easily by assuming that all annual production of all permanent meadows and pastures (3.35 Gha, according to the FAO) is consumed by domesticated herbivores to be converted to meat and milk: that was, indeed, the method used by Vitousek et al. (1986) and by Rojstaczer, Sterling, and Moore (2001). Even with a conservative productivity average of 5 t/ha, that would amount to nearly 17 Gt/year.

This is an obviously unrealistic assumption: even in those pastures that are heavily overgrazed some phytomass remains, and in well-managed, fully stocked

European and North American pastures no more than 25%–50% of their annual primary productivity is eaten by grazing livestock (Detling 1988; Hodgson and Illius 1996; Lemaire et al. 2000). Assuming that between 30% and 40% of grassland NPP (1.5–2 t/ha) is consumed by domesticated grazers would yield a total phytomass removal of at least 5 and as much as nearly 7 Gt. More elaborate alternatives to estimate the grazing claim are possible, but all of them depend on concatenated assumptions and may not yield any superior results.

Detailed reconstructions start with average body masses of grazers (they vary substantially even for the same species, depending on breeds and typical feeding), their typical metabolic rates, take-off rates (animal shares killed annually), average ages at slaughter, and typical meat yields per carcass (or typical lactation rates). Wirsenius (2000) acknowledged these complications but undertook such a detailed examination on the global and regional levels: he concluded that during the mid-1990s, domesticated grazers consumed about 5.8 Gt of dry phytomass (23.3 Gt of fresh weight). Comparisons with other estimates of phytomass "appropriated" by grazing show the extent of unavoidable uncertainties: the totals were 10.6 Gt in Vitousek et al. (1986) and 8.89 Gt (assuming twice the total of 4.45 Gt C) in Haberl et al. (2007), compared to the 5–7 Gt I estimated by doing just two multiplications.

Aquatic Species

Because marine species now provide about 15% of the world's dietary protein and because much of the world's ocean either has been exploited to capacity or is now overexploited, it is important to estimate the evolution of global fish and invertebrate harvests. In contrast to the relatively abundant research on terrestrial harvests, little has been published on marine catch and the ocean's primary production. Most notably, all recent studies of the HANPP have chosen to ignore the marine share of zoomass harvests. But Vitousek et al. (1986) estimated the human claim on marine phytomass by converting 75 Mt of fresh-weight landings of fish and aquatic invertebrates to 20 Mt of dry weight and then assuming that the average catch takes place at the second trophic level and that the transfer efficiencies are 10% in each case. This would, obviously, result in about 2 Gt/year of oceanic phytomass consumed by the aquatic zoomass harvested by humans, and Vitousek et al. (1986) put that rate at 2.2% of the total oceanic NPP.

The assumption of an average energy transfer efficiency of 10% has been commonly made when calculating interlevel transfers ever since Lindeman's (1942) pioneering research on energy transfers in Wisconsin's Lake Mendota. Moreover,

Pauly and Christensen (2002) used the 10% figure not because they assumed it but because they found that it comes very close to the mean of more than 100 estimates they examined for their study. Given the enormous variety of aquatic species and habitats, it is highly unlikely that we could ever find the precise transfer mean, but we know that actual transfer efficiencies depart significantly from the 10% rate (Smil 2008). In upwelling areas, which occupy only about 0.2% of the world's oceans but produce between 15% and 20% of all harvests (Pauly and Christensen 1995), the average efficiencies may be close, or even above, 15%: Cushing (1969) found rates between 13.9 and 18.5% in some Pacific upwelling areas, and a recent study of marine food webs by Tsagarakis et al. (2010) used mean values of 10% for the Adriatic but 12.6% for the southern Catalan Sea and 17.4% for the northern Aegean Sea.

The average trophic level of marine harvests are much easier to address, as several studies have used fractional data for a large number of species groups in order to calculate the historical changes in global fishing effort. As a result, we know that Vitousek et al. (1986) made a serious underestimate when they chose a mean trophic level of two: since the early 1980s the actual trophic levels of exploited fishes have been between 3.2 and 3.3 (Zeller et al. 2009). But to which total should be these multipliers applied? While crop production statistics are among the most accurate figures used to assess the human claim on phytomass, and while for many countries they are available for the entire twentieth century (and even further back), there are significant problems even with the latest fishery statistics, and only a few reliable national series and a few global approximations exist for the pre–World War II period. Moiseev (1969) made educated guesses for the global catch in 1800, 1850, and 1900, with the totals remaining very low, at, respectively, 1 Mt, 1.5 Mt, and 2 Mt.

The FAO began its post–World War II data series in 1947, when it also estimated the global catch for the last prewar year, putting the 1938 landings at 22 Mt. But Sarhage and Lundbeck (1992) put the worldwide fish landings at just 10 Mt during the late 1930s. If the former total is correct, then the record Japanese prewar marine harvest would have accounted for only about 20% of the worldwide total; if the latter is closer to the real value, then Japan's share would have been more than 40% of global seafood landings. The FAO's data series (in yearbooks since 1947, online since 1961) shows the global marine catch rising from about 18 Mt in 1950 to just over 80 Mt in 1988 and 1989, then falling back a bit before getting to about 85 Mt in 1996 and 1997, then slightly declining again, a record that amounts to two decades of stagnation (FAO 2011a).

Global landings remained around 90 Mt during the second half of the first decade of the twenty-first century, falling slightly from 94 Mt in 2005 to 92 Mt in 2006 and then stabilizing at just below 90 Mt (FAO 2011a). Freshwater landings added up to about 10% of this total (9.4 Mt in 2005, 10.2 Mt in 2008), with fish species accounting for nearly 90% of harvests in inland waters. Marine landings have not departed far from 80 Mt/year, declining from 86.8 Mt in 2000 to 81.5 Mt in 2003 and to 79.5 Mt in 2008. About 85% of those totals were marine fishes, with mollusks contributing about 8% and crustaceans about 6%. Worldwide fish catches have been dominated by the anchoveta (Peruvian anchovy), with landings just above 10 Mt/year until 2005 and above 7 Mt/year afterward. Alaskan pollock comes next (between 2.5 and almost 3 Mt/year), followed by roughly equal masses (each about 2.5 Mt/year) of Atlantic herring (*Clupea harengus*) and skipjack tuna (*Katsuwonus pelamis*) and chub mackerel (*Scombrus japonicus*), with nearly 2 Mt/year.

These top five species now make up about 25% of all marine fish landings. In national terms China has been the largest producer, with total landings leveling off at between 14 and 15 Mt/year. An anchoveta bounty puts Peru in second place, with landings fluctuating between 6 and more than 9 Mt/year. Indonesia, the United States, and Japan (each with more than 4 Mt/year) complete the top five. In regional terms, the Northwest Pacific has recently yielded about 20 Mt/year, and the western Central Pacific and Southeast Pacific are nearly equal second and third most bountiful fishing areas, each with about 11 Mt/year. The Northwest Atlantic still ranks fourth, ahead of the eastern Indian Ocean.

Accounting for global aquatic harvests gets even more complicated because aquacultural production is now of the same order of magnitude as are marine catches. Whereas in 1950 freshwater and marine aquaculture (including seaweeds) contributed less than 15% of all landings, by the year 2000 its global output of 35 Mt had reached about 27% of the total, and in 2008 aquaculture added more than 52 Mt, to nearly 90 Mt of captured species, or 37% of all fishes, invertebrates, and aquatic plants used for food, feed, and fertilizer (FAO 2011a). Unlike the capture of wild species, aquacultural production comes largely from inland waters and is dominated by three carp species—silver carp (*Hypophthalmichthys molitrix*), grass carp (*Ctenopharyngodon idella*), and common carp (*Cyprinus caprio*)—with salmonids in a distant second place, and its growth has been disproportionately concentrated in China (now accounting for about 70% of the global production).

While the aquacultural statistics may be fairly accurate, there are major concerns about the quality of the FAO's data on global and national landings of fish and marine invertebrates. Catch reconstructions for various maritime nations showed that the official statistics underestimate their likely true catch—that is, reported landings plus illegal, unreported, and unregulated (IUU) catches—by as much as a factor of two or more (Zeller and Pauly 2007). And these discrepancies were found not only for low-income countries but even for Greece, a member of the EU: Tsikliras, Moutopoulos, and Stergiou (2007) found that between 1964 and 2003, the FAO data underestimated actual landings by an average of 35%, with an error range of 10%–65%.

The single most important reason to question recent global FAO landings data during the last decades of the twentieth century was the massive overreporting of Chinese fish and invertebrate landings. Watson, Pang, and Pauly (2001) showed that while the official reports of Chinese marine catches kept on rising between the mid-1980s and 1998, other evidence pointed to declines averaging 360,000 t/year; as a result, the total 1998 catch in China's exclusive economic zone and in distant waters was not 15 Mt but perhaps as little as 5 Mt. The FAO had also acknowledged these Chinese overestimates. At the same time, there has also been a great deal of illegal fishing, and the FAO does not make any attempt to estimate it.

The first worldwide analysis of IUU catches concluded that their recent rate has been between 11 and 26 Mt/year (Agnew et al. 2009). On the regional level, West African waters (the eastern Central Atlantic) have had the highest level of illegal fishing, amounting to nearly 40% of the reported catch, and areas with illegal fishing rates above 30% of the acknowledged landings are the western Central Pacific (34%), the Northwest Pacific (33%), the Southwest Atlantic, and the eastern Indian Ocean (both 32%). In specific term, the highest illegal catches are for demersal fishes, salmons, trouts, and smelts (mostly over 40%), the lowest (less than 10%) for flounders, halibuts, and soles.

Officially tabulated landings also exclude discarded by-catch, the zoomass of all unwanted species caught alongside the targeted catch, including fish and invertebrates, marine mammals (most often dolphins), reptiles (most often turtles enmeshed in fishing nets), and even seabirds, including gannets, fulmars, puffins, albatrosses, and petrels killed by tuna drift nets and longline fisheries (Gales, Brothers, and Reid 1998; Rogan and Mackey 2007). But it also includes desirable commercial species that are smaller than the allowable size or are over the permissible catch quota, and females that are protected for breeding. By-catch is simply discarded overboard, and both its share in specific fisheries and the survival rates of discarded species vary widely. Discards can amount to a small fraction of actual landings or (as in tropical shrimp trawling) can be an order of magnitude greater than the catch itself.

The most comprehensive review of by-catch, based on some 800 studies, showed its worldwide mean to be about 35%, with specific rates ranging from less than 10% for cephalopods, more than 60% for redfish and basses, 75% for flounders and soles, more than 80% for eels, and nearly 250% for crabs to more than 500% for shrimp (Alverson et al. 1994). The great variability of these estimates precludes coming up with a satisfactory mean discard rate. Discard mortality is also highly variable, with lows of just a few percent and highs of more than 80% or even 100% for such species as halibut, king crab, and salmon (Alverson et al. 1994).

Published totals of global discard were estimated to be as high as 27 Mt/year for the late 1980s (range of about 17–40 Mt) by Alverson et al. (1994) and about 15 (10–20) Mt/year (Alverson 2005) and as little as 7.3 (6.8–8.0) Mt/year (Kelleher 2005) for the beginning of the twenty-first century. Zeller and Pauly (2005) published a reasoned reconstruction of the past discards, and their best estimates indicate rising totals until the late 1980s (from 4 Mt during the late 1970s to nearly 30 Mt by 1990), followed by declines (around 15 Mt by 1995, and down to 8 Mt by the year 2000). While this is a welcome news, they also point out that it also indicates even faster declines in actual landings than was believed previously (by at least 1.5 Mt and perhaps up to 2.3 Mt/year, compared to the previously estimated 0.39 Mt/year (Watson and Pauly 2001).

While illegal and unaccounted-for catches should obviously be added to reported landings to get the total of annually harvested marine zoomass, the treatment of discards is problematic. Those fish and invertebrates that are dumped back into the ocean and survive the experience obviously should not be counted as a part of harvest, and the discarded by-catch that dies serves to support wild marine heterotrophs, and it should not be counted as human-appropriated either. At the same time, there can be no doubt that annual removal followed by the prompt dumping of millions of tons of live and dead marine organisms amounts to a serious disturbance and degradation of aquatic food webs, particularly in the case of fisheries with very high by-catch rates.

With all of these complications and caveats in mind, I will offer the best estimates of phytomass needed to support recent levels of fish catches and aquacultural production. Pauly and Christensen (1995) added 27 Mt of discarded by-catch to the mean 1988–1991 catch (94.3 Mt), assigned fractional trophic levels to 39 species groups, and used an average interlevel energy transfer efficiency of 10% to estimate that about 8% of the world's primary aquatic production was needed to sustain commercial fishing in the early 1990s, with the shares ranging from about 2% for the open ocean to 24% for tropical and 35% for nontropical shelf ecosystems. I use the total catch of 93 Mt (average of 1998–2002 rates) augmented by about 17 Mt (18% of the reported total) representing illegal landings, and 8 Mt of discarded by-catch. When conservatively converted (12% C in fresh weight), 118 Mt of aquatic zoomass are equal to about 14 Mt C, and with the average trophic index at 3.3 and the mean transfer efficiency at 10%, the production of 118 Mt of aquatic zoomass would have needed at least 2.8 Gt C of phytoplankton and aquatic plants. With a global aquatic NPP of about 50 Gt C, this would have equaled nearly 6% of the primary marine productivity.

Calculating the global claim made by the rising aquacultural production would be fairly simple if the output were limited to the traditional pond cultures of Asian and European cyprinids feeding on natural phytoplankton and detritus or the maricultures of mollusks capturing their feed from coastal waters by filtration and hence feeding near the bottom of their respective trophic webs. The dominant species of cultured carps (including grass and silver carp) are strict herbivores (trophic level 2), while the common carp feeds at levels ranging from 2.1 to 2.8 and the most popular species of mollusks (mussels, oysters, clams, scallops) are herbivores and detritivores (level 2). But modern aquaculture uses mass-produced commercial plant-based feeds to increase the efficiency of herbivorous fish production and has been producing increasing amounts of carnivorous species (above all salmon and shrimp, in the Mediterranean mainly sea breams and sea basses) whose feeding requires protein derived from captured wild fish and marine invertebrates processed into fish meals and oils.

As a result, the mean trophic level of the Mediterranean marine farmed species rose from 2 during the 1970s and 1980s to about 3 by 2005 (Stergiou, Tsikliras, and Pauly 2008). Moreover, some producers used to increase the weight of herbivorous and omnivorous species by supplementary (up to 15% of all energy) rations of fish meals and oils (Naylor, Williams, and Strong 2000). But both of these feed streams have already been accounted for: plant-based aqua-feeds claim just a small share of cereal, leguminous, and oil crop harvests, while fish-based feeds claim a significant share of fish landings (particularly such small pelagic species as anchoveta and sardines) that are not eaten by humans.

According to the FAO (FAO 2011a), 2006 was the first year when worldwide aquaculture produced more than 50 Mt of fish and marine invertebrates (it was about 35 Mt in 2000), and nearly 75% of that output was as species that consume feeds. As a result, the production of aqua-feeds increased from just 4 Mt in 1994 to about 20 Mt by 2006, and Tacon (2008) estimated that in that year, nearly 70% of all fish meal and almost 90% of fish oil were consumed by farmed species (about

20% of fish meal went to pigs, the rest to poultry). Although aquatic ectotherms convert this feed with higher efficiencies than land animals—typical feed-to-product ratios are 1.9–2.4 for salmon and 1.2–1.5 for shrimp (Jackson 2009)—aquafarming of carnivorous species fed fish-derived protein and oils will always result in a net loss of digestible protein, arguably justified by the fact that people prefer to consume salmon and shrimp rather than anchoveta.

The only unaccounted-for claim of aquaculture on aquatic primary production is phytoplankton, zooplankton, and detritus captured by farmed herbivorous fish and invertebrates. The upper limit of this requirement can be estimated by assuming (contrary to actual practices) that no cyprinids (trophic level 2) receive any supplementary feeding: producing some 15 Mt (1.8 Mt C) of these fishes in 2008 would have required about 18 Mt C in aquatic phytomass, an insignificant total compared to inherent inaccuracies in calculating the feeding requirements of wild fish. The previously calculated share of about 6% of the worldwide aquatic NPP can thus represent the total claim by all harvested marine and freshwater species, caught or cultured.

This seems to be a comfortably low rate, but it is yet another perfect illustration of the limits of using the human-appropriated NPP shares as supposedly revealing measures of human impact. This particular rate ignores the changing primary productivity base, and it entirely misses three critical factors affecting modern marine catches that cannot be captured by a simple tracing of total landings and their conversion to primary production requirements. The first one is a long-term decline in phytoplankton production, the second one is the extent of overfishing, and the third one (less assuredly) is the shift in average trophic level of harvested species.

Satellite-derived observations of phytoplankton concentrations have been available only since 1979, but numerous standardized ocean transparency measurements (made using a Secchi disk) have been available since 1899, and they can be used to calculate surface chlorophyll concentration and augmented by direct in situ optical measurements. Boyce, Lewis, and Worm (2010) used this information to estimate a century of phytoplankton changes on regional and global scales and found significant interannual and decadal phytoplankton fluctuations superimposed on long-term declines in eight out of ten ocean regions, a worrisome trend whose continuation would have significant effects on future catches.

The second concern, the increasing extent of overfishing, has been masked for a long time by higher landings. Worm et al. (2006) concluded that the rate of fisheries collapses (when the catches drop below 10% of the recorded maxima) has been accelerating, and that by 2003, the stocks of 29% of all currently fished species had to be considered collapsed. Moreover, that study also found that despite substantial increases in fishing effort, cumulative yields across all species and large marine ecosystems had declined by about 13% after reaching a maximum in 1994. Not surprisingly, these declines have affected most of all the largest predators, tuna, billfish, and swordfish.

Using the logbook data from Japan's worldwide fishing activities, Worm et al. (2005) demonstrated that between 1950 and 2000, the predator species density showed gradual declines of about 50% in both the Atlantic and Indian Oceans and about 25% in the Pacific. And Pauly (2009) concluded that commercial fishing has reduced the zoomass of traditionally harvested large demersal and pelagic species (cod, tuna) by at least an order of magnitude. Despite our surprisingly poor understanding of actual fish stocks and their dynamics, there is no doubt that most of the traditionally targeted species and most of the major fishing areas now belong to three unwelcome categories: those whose the stocks have already collapsed or are close to doing so, those that continue to be overfished, and those that are fished to their full capacity.

The declines have affected many species that have been traditional food choices for millennia and whose catches switched from abundance to scarcity or even to total collapse within often very short periods of time. The already noted collapse of the Canadian cod fishery and the sharp declines in bluefin tuna have attracted a great deal of attention: Mediterranean and Atlantic populations are particularly endangered, and even the larger Pacific tuna population may crash at any time. But there are many less reported declines, including the low counts of one of Western Europe's well-liked species: the numbers of European eels (*Anguilla anguilla*) have declined sharply since the 1970s, perhaps by as much as 99%, while the effectiveness of restocking efforts remains uncertain (Vogel 2010).

The third notable change that is not captured by the seafood-NPP appropriation ratio has been a gradual decline in landings of larger, slow-growing, and more valuable carnivorous top predator species (tuna, cod, halibut) and the rising harvests of larger amounts of smaller, faster-growing, less valuable species that either are herbivorous (primary consumers) or feed largely on zooplankton (Pacific anchoveta and herring). Pauly et al. (1998) traced these specific changes during the latter half of the twentieth century and discovered that the mean trophic level of reported landings declined from slightly more than 3.3 during the early 1950s to less than 3.1 in 1994, with the index as low as 2.9 in the northwestern and west-central Atlantic.

Although the methodology and findings of these studies were questioned (Caddy et al. 1998), slowly declining trends in mean trophic levels (the phenomenon known as fishing down marine food webs) have been subsequently confirmed by many studies using detailed local data, including fishing around Iceland, in the Gulf of Thailand, in both eastern and western Canadian waters, in the Chesapeake Bay, and in the Celtic Sea (Zeller et al. 2009). Even if the continuing declines were to amount to just 0.1 of trophic level per decade, such an apparently small decrease would lead inevitably to more fishery collapses, and within a century, there would be no carnivorous fishes available for large-scale commercial capture.

But such conclusions were questioned once again by Branch et al. (2010) whose comparison of catch data, trawl surveys, and stock assessments showed no discernible fishing-down the food web on the global scale and concluded that marine organisms at all trophic levels (from mollusks to tunas) are being harvested in ever-higher quantities. But this is hardly good news: it means that intensifying fisheries can collapse even as the mean trophic levels are stable or increasing. In turn, these conclusions were questioned by the authors of original fishing-down reports, offering yet another illustration of our uncertain understanding of human impacts on the biosphere. Moreover, Smith at al. (2011) argue that even the fishing of low-trophic-level species at conventional maximum sustainable yield levels can have large impacts on other parts of marine ecosystems and that reducing such rates would be desirable.

Harvesting Woody Phytomass

Reconstructing past wood harvests (at any scale) is to engage in guesses and crude approximations. As already shown (in chapter 2), historical data on wood consumption as household and industrial fuel and as a raw material in construction are scarce, highly discontinuous, and hard to interpret. Modern statistics, some going back to the late nineteenth century, offer fairly good accounts of commercial wood uses for lumber and pulp, but as far as the use of wood as household fuel in low-income countries is concerned, they too are largely estimates. In this section I first present the global estimates of fuelwood combustion and then review the best available data on the use of wood as a raw material.

Estimates of aggregate fuelwood consumption during the medieval period or in antiquity would be nothing but guesses, not only because of our poor knowledge of what constituted typical usage rates but also because a complete lack of information about the actual extent of seasonal heating and the frequency and duration of everyday cooking, as well as because of uncertain population totals. Estimates become more defensible for the modern era, and my choice of a plausible global approximation is an average annual fuel wood consumption of about 20 GJ/capita in 1800 (Smil 2010b). With a global population of one billion, that would yield a worldwide woody phytomass combustion of about 20 EJ (about 1.3 Gt of air-dried wood). Crop residues used as fuel added most likely no more than another 2 EJ, within the conservative error range of ±15%. By the mid-nineteenth century the total mass of wood and crop residues used for fuel surpassed 25 EJ, but then the rapid shift to coal lowered the phytomass contribution to a level only slightly higher than in 1800.

As a result, the nineteenth century's cumulative combustion of wood-dominated solid biofuels added at least 2.4 YJ, while all fossils fuels (until the early 1860s only coal and peat, afterward also small volumes of crude, and just before the century's end also small volumes of natural gas) consumed between 1800 and 1900 contained only about 0.5 YJ (Smil 2010b). On the global scale, wood-dominated phytomass combustion thus provided no less than 85% of all fuel burned during the nineteenth century, making it—contrary to a common impression that it was the first century of fossil fuel (more accurately, coal) era—still very much a part of the long wooden (or, more accurately, wood-and-straw) age.

Fernandes et al. (2007) offer a slightly different account: they put the global biofuel use at 1 Gt in 1850 and about 1.2 Gt in 1900; these totals would translate to about 18 (17–20) EJ/year, somewhat lower than my estimates. In most national cases accuracy of fuelwood consumption accounts does not improve significantly during the twentieth century as approximations are unavoidable even when quantifying a relatively high dependence on phytomass fuels that lasted in the world's most populous Asian nations (China, India, Indonesia) well into the second half of the twentieth century and that still dominates rural energy use in Brazil, the most populous country in Latin America and the world's eighth-largest economy. Perhaps the best illustration of the data problem is the fact that Fernandes et al. (2007) put the uncertainty range of their global biofuel estimate in the year 2000 (2.457 Gt) at ±55%, no smaller error than for 1950 (their uncertainty range for 1850 was ±85%).

My reconstruction puts the global combustion of fuelwood and crop residues at 22 EJ (1.45 Gt) in 1900, 27 EJ (1.8 Gt) in 1950, and 35 EJ (2.3 Gt) in 1975. The lowest total of global fuelwood combustion at the end of the twentieth century is the FAO's 1.825 Gm^3 of coniferous and nonconiferous "wood fuel" (the latter category accounting for nearly 90% of the total), as well as 75 Mm^3 of "wood residues" and 49.2 Mt of "wood charcoal" (FAO 2011e). Charcoal production (using the FAO's recommended conversion factor of 6.0) required 295 Mm³ wood, resulting in a grand total of 2.195 Gm³ of wood. Assuming averages of about 0.65 t/m³ and 15 GJ/t of air-dried wood, this converts to approximately 21.5 EJ or 1.43 Gt. As already noted (in chapter 2), in many countries most woody phytomass does not come from forests, and that is why the FAO's total greatly underestimates the actual use.

My estimate for the year 2000 is about 2.5 Gt of air-dried wood (about 2 Gt of absolutely dry matter containing about 35 EJ, including wood needed to make charcoal), and the burning of crop residues in fields and their recycling and feeding to animals left about 20% of the entire residual phytomass (about 10 EJ in the year 2000) to be used as household fuel. My most likely estimate of phytomass combustion is thus 45 EJ in the year 2000, an equivalent of 3 Gt of air-dried wood (Smil 2010b). This compares to 2.06 Gt (nearly 31 EJ) of all fuel phytomass (of which 1.32 Gt was wood) estimated only for the low-income countries in the year 1985 by Yevich and Logan (2003) and to a range of 45 ±10 EJ estimated by Turkenburg et al. (2000). The most recent (and the most detailed) estimate, based on the compilation of many national totals, ended up with 2.457 Gt of solid phytomass fuels in the year 2000, with about 75% as wood (1.85 Gt), 20% (0.5 Gt) as crop residues and the rest as dung and charcoal, and with 80% of the total consumed by households and 20% by workshops and industries (Fernandes et al. 2007).

Their aggregate solid biofuel consumption translates to at least 37 EJ. Published estimates of solid phytomass combustion thus converge to around 40 EJ, or around 2.5–2.7 Gt of all solid phytomass fuels in the year 2000. Given the many inherent uncertainties in estimating this supply, plausible totals for the year 2000 range between 35 and 45 EJ, or between 2.3 and 3.0 Gt of wood equivalent, of which wood accounts for 30–35 EJ, or between 2 and 2.35 Gt. This range implies an increase of nearly 70% between 1950 and 2000 and a doubling of solid biofuel harvests during the twentieth century. But this absolute growth has been accompanied by substantial declines in per capita uses in all high-income countries, with a rising or stable per capita supply found only in some countries of sub-Saharan Africa and in rural areas of low-income economies in Asia.

The global harvest of 40 EJ of solid biofuels in the year 2000 was just over 10% of all primary energy supply when we compare the initial energy contents of fuels and primary electricity: that year's total was about 382 EJ, with fossil fuels accounting for about 305 EJ and hydroelectricity and nuclear electricity generation contributing respectively about 10 EJ and 25 EJ. But because of much higher conversion efficiencies of fossil fuel combustion—the differences between simple wood stoves and large coal or hydrocarbon-fired boilers, household gas furnaces, and various internal combustion engines burning liquid fuels or natural gas are commonly two- to threefold—biofuels contributed less than 5% of the world's final (useful) energy supply.

Information about wood harvests for lumber and pulp has much smaller margins of uncertainty. The FAO database lists roundwood consumption (a category that includes all saw logs, veneer logs, and pulpwood) of about 1.7 Gm^3 in 1950, 2.88 Gm^3 in 1975, and 3.43 Gm^3 in 2000 (FAO 2011e). These volumes are for natural (as felled) wood, and the conversion to dry matter would yield—assuming, as previously noted (Glass and Zelinka 2010), an average specific gravity of 0.5 and an average moisture content of 50%, that is, a weight of 750 kg of oven-dry weight per cubic meter—nearly 2.6 Gt. Once again, these totals underestimate the actual impact of harvesting because they do not account for woody phytomass that is not included in the reported volumes of commercial harvests.

The first omission, and the one that is most difficult to quantify, is wood obtained by common illegal cutting. In 1999 it was estimated that as much as 27% of German and 62% of British tropical timber imports could have been from illegal sources (Forest Watch 2007). A very conservative assumption based on data, reports, and estimates by Global Timber (2011) would be to increase the global total by 15% to about 3 Gt. An omission that is easier to quantify is that of the exclusion from harvest data of all small trees that are destroyed during forest cutting: generally only live sound trees of good form with a diameter at breast height of more than 12 cm and a 10 cm diameter at the top (measured outside the tree bark) count.

U.S. statistics also exclude live sound trees of poor form: such cull trees usually account for about 6% of the total forest phytomass (USDA 2001). More important is the omission of all phytomass that does not become part of the merchantable bole (stem, trunk): stumps and roots, branches, and treetops. Various formulas convert merchantable stemwood volumes to aboveground phytomass (Birdsey 1996; Penner et al. 1997). Species-specific U.S. multipliers from timber to total phytomass show a range from about 1.7 (southeastern loblolly pine) to just over 2.5 (spruce and fir), with a median at about 2.1. An inventory of Canadian forests showed multipliers of 2.17 in Alberta and 2.43 in British Columbia, and a nationwide mean of 2.56 (Wood and Layzell 2003). Actual shares in large-scale harvesting will depend not only on specific tree composition but also on harvesting methods. A conservative conversion would be to double the total harvested mass of 3 Gt to yield the annual roundwood harvest and removal of at least 6 Gt, and an upper bound of adjustment would be to multiply by 2.5 and raise the total to 7.5 Gt of dry weight.

This correction should also apply to at least a quarter of all fuelwood harvests that are done in commercial fashion rather than by children and women collecting woody litter; this would add 1–1.3 Gt of dry wood (2.0/4 = 0.5 × 2 = 1 Gt; 2/4 = 0.5 × 2.5 = 1.25 Gt). These adjustments would mean that overall harvest of woody phytomass in the year 2000 would have been 2.6–3.3 Gt of fuelwood, charcoal, and associated losses, and between 6 and 7.5 Gt of roundwood removed from forests, destroyed during harvesting, or abandoned after harvests, for a grand total of 8.6–10.8 Gt of absolutely dry weight. These totals should be enlarged by an average of 25% to account for root biomass (Cairns et al. 1997), raising the global woody phytomass appropriation to 10.8–13.5 Gt.

Further additions could be made by considering a variety of other human impacts. Forest disturbances resulting from construction (road and dam building), settlements, and logging will lead to reduced productivity that would have to be quantified as a difference between potential and actual photosynthesis. In some areas this impact may be of the same order of magnitude as tree clearing: an analysis of satellite imagery found that between 1972 and 2002, 15% of Papua New Guinea's tropical forests had been cleared and another 8.8% degraded through logging (Shearman et al. 2009). A very conservative global estimate of such degradations is at least 5 Mkm², and the real extent may be twice as large. Even if the degradation were to reduce average NPP by no more than 10%–15%, the annual productivity loss would be on the order of 1–1.5 Gt of dry phytomass, pushing the annual grand total to 11.8–15 Gt of dry mass in the year 2000. Consequently, a broad definition of human claims on woody phytomass in the year 2000 would bring the most likely total to about 13.5 (12–15) Gt of dry matter. Table 11.2 summarizes the derivation of that annual rate.

More phytomass will be lost to anthropogenic fires as commercial logging opens up previously inaccessible areas to settlement. And tree growth will change (the shift can also be positive, at least for a period of time) as a result of the long-term effects of atmospheric deposition, above all from soil acidification in some regions and nitrogen enrichment (from the deposition of airborne ammonia and nitrates) in others. The latter process, well demonstrated in temperate regions, is now affecting even some tropical forests (Hietz et al. 2011). None of these effects can be quantified without resorting to concatenated and uncertain assumptions, but the inclusion of all of these factors could raise the aggregate claims by at least 10%.

Table 11.2
Derivation of the Global Harvest of Woody Biomass in 2000 (in Gt of dry weight)

Category	Range of Estimates	Best Estimates
Fuelwood removal	1.6–2.0	2.0
With phytomass losses	2.6–3.3	3.0
Roundwood removal	(3.43 Gm³)	2.6
With illegal cutting		3.0
With phytomass losses	6.0–7.5	6.5
Fuelwood and roundwood	8.6–10.8	9.7
Including roots (+25%)	10.8–13.5	12.0
Including disturbances	11.8–15.0	13.4

In light of these multiple assumptions and corrections, it is not surprising that the appropriation studies did not handle this in any uniform manner. A narrowly defined removal account by Vitousek et al. (1986) ended up with 2.2 Gt/year during the late 1970s. Rojstaczer, Sterling, and Moore (2001) added 0.9 Gt of fuelwood and 1.5 Gm³ of timber (the equivalent of 0.83 Gt, using their conversion mean of 560 kg/m³) to get about 1.75 Gt of removed phytomass during the late 1990s, and multiplied the total 2.7 times to convert merchantable wood to total forest biomass: that adjustment, if applied to all wood removals, would yield an annual appropriation of about 4.75 Gt. In contrast, Imhoff and Bounoua (2006) put wood product harvests (fuel, construction paper) at about 7 (4.84–8.53) Gt C in 1995, with fuelwood accounting for about 60% of that total. Assuming a 50% carbon content, their aggregate appropriation comes to about 14 Gt of dry matter, a total that is in an excellent agreement with my liberal estimate that includes all below- and aboveground woody phytomass removed, lost, and degraded in the course of wood harvests (see table 11.2).

Expressing wood harvests as a share of standing forest phytomass or as a fraction of global forest NPP is subject to substantial errors. The forest inventories of rich nations are fairly reliable, and hence the calculations of harvest shares will be close to actual values. For example, the end-of-twentieth-century U.S. forest inventory listed 23.65 Gm³ of growing stock (about 14 Gt) in the country's timberlands and an annual growth rate of 670 Mm³ (USDA 2001), compared to nearly 490 Mm³ (that is, 250–290 Mt, including all timber, pulp, and fuelwood) cut in the year 2000 (USDA 2003). U.S. wood removals were thus equal to 2% of the standing stock in all timberlands and 73% of the annual growth of all merchantable timber. But during the mid-1990s the total standing phytomass of all U.S. forests (timberlands, reserved forests, and other wooded areas) was at least 25 Gt, and their annual NPP was about 4.5 Gt (Joyce et al. 1995). This means that the removal of all woody phytomass was equal to only about 1% of the existing woody phytomass and on the order of 6% of the annual forest NPP.

Analogical calculations have been relatively easy to make for Canada, Germany, France, or Japan, but only the latest monitoring advances have resulted in fairly reliable stocktaking for the tropical countries. When Gibbs et al. (2007) surveyed the totals of national carbon stock estimates in tropical forests based on compilations of harvest data as well as on forest inventories published between 2002 and 2007, they found that the differences between the lowest and highest estimates were commonly two- or even threefold. These estimates included 1.9–5.1 Gt C for Mozambique, 3.6–11.8 Gt C for Angola, 2.5–9.2 Gt C for Bolivia, and 2.5–11.5 Gt C for Colombia. For the three countries with the largest areas of tropical forests, Brazil, Congo, and Indonesia, the differences were respectively 1.5-, 1.8-, and 2.5-fold, and in absolute terms the aggregate difference amounted to more than 60 Gt C.

These uncertainties have been greatly reduced by a new mapping of phytomass stored in 2.5 Gha of tropical forests (Saatchi et al. 2011). The study combined data from nearly 4,100 inventory plots with satellite Lidar (light detection and ranging) samples of forest structure and with high-resolution (1 km) optical and microwave imagery to estimate the global tropical forest carbon stock of 247 Gt C, with nearly 80% (193 Gt C) aboveground and the rest in roots. Latin American forests store 49% of the total, sub-Saharan Africa accounts for 25%, and Southeast Asia accounts for 26%. Phytomass density for growth with 30% canopy cover (with the global area of 1.67 Gha) averaged 124 t C/ha, ranging from just 80 t C/ha in Angola to 180 t C/ha in Malaysia.

12

Long-Term Trends and Possible Worlds

Our species has evolved to become the planet's dominant heterotroph in what has been (when measured on the biospheric time scale of more than three billion years) a very brief period of time. Less than 2.5 million years have elapsed since the emergence of our genus (with *Homo habilis*), our species became identifiable about 200,000 years ago, and shortly after the end of the last glaciation (less than 10,000 years ago) various societies began to move from subsistence foraging to settled existence energized by harvesting cultivated plants and domesticating animals. Afterward our capacities for expansion, extraction, production, and destruction began to grow rapidly with the emergence of the first complex civilizations.

After millennia of slow gains during the Pleistocene and the early part of the Holocene, global population began to multiply as humans commanded increasing flows of energy and potentiated their impacts, thanks to many technical and social innovations that improved the quality of life, whether it is measured in physical terms (a much greater average life expectancy) or in economic accomplishments (per capita economic product or average disposable income). Reconstructions of all of these fundamental long-term trends are uncertain but are good enough to capture the magnitude of specific advances and their relentless growth (table 12.1).

Five thousand years ago, when the first complex civilizations began to leave written records, there were, most likely, fewer than 20 million people; at the beginning of the Common Era the total was about 200 million; a millennium later it rose to about 300 million; in 1500, at the onset of the early modern era, it was still less than 500 million. The billion mark was passed shortly after 1800. In 1900 the total was about 1.6 billion, in 1950 2.5 billion, in 2000 6.1 billion, and in 2012 it surpassed 7 billion. Consequently, there has been a 350-fold increase in 5,000 years, more than a 20-fold gain during the last millennium, and roughly a quadrupling between 1900 and 2010.

Perhaps nothing illustrates the extent of the anthropogenic Earth better than the composite of nighttime satellite images by NASA. The file can be downloaded at http://eoimages.gsfc .nasa.gov/images/imagerecords/0/896/earth_lights_lrg.jpg.

Table 12.1
Some Important Global Trends

Year	Population (million)	Energy Use (GJ/capita)	Economic Product (2000, $/capita)	Life Expectancy (years)	Global Phytomass Stock (Gt C)
5,000 BP	20	<3	<100	20	>1,000
0	200	<5	500	<25	1,000
1000	300	<10	500	<30	900
1800	900	23	600	35	750
1900	1,600	27	1,200	40	660
2000	6,100	75	6,500	67	600
2010	6,900	75	7,500	69	<600

Note: All of these values (with the exception of the post-1900 population, energy, and life expectancy) are just approximations. Economic product and phytomass stock series have the highest (at least ±15%) uncertainty. Population series are available in McEvedy and Jones (1978) and HYDE (2011). Average per capita energy use is based on Smil (2008, 2010b). Economic product estimates are based on Maddison (2007). Global phytomass stocks are based on data in table 2.4.

Fuel use in the earliest complex civilizations was limited to burning wood and crop residues, and even during the first centuries of the Common Era the average per capita energy consumption in the Roman Empire, whose power was unmatched in the West and whose territory extended to three continents, was no higher than 10 GJ per capita (Smil 2010c). By 1800 the British mean, the world's highest, had reached about 50 GJ per capita (Warde 2007), and in 1900 the average U.S. per capita energy (fossil fuels and wood) supply had surpassed 130 GJ (Schurr and Netschert 1960). A century later the largest EU countries were, much like Japan, at about 170 GJ, while the U.S. and Canadian per capita supply of primary energy was running at twice that rate (BP 2011). And these are comparisons of gross inputs: because of vastly improved energy conversion efficiencies, the gains in terms of actually available useful energy were at least three times higher.

Life expectancy at birth among the citizens of the Roman Empire was less than 25 years, and it was not until 1900 that the average for both sexes surpassed 50 years in the United States and various European countries; by 2010 it stood at around 80 years in the world's most affluent nations—and more than 70 years even in China (UN 2011). And while per capita GDP is an imperfect measure of economic well-being, its reconstructions for the Roman Empire (Maddison 2007; Scheidel and Friesen 2009) yield only $500–$1,000 in today's monies, similar to the rates now prevailing in the poorest countries of sub-Saharan Africa, while the 2010 averages of large economies ranged from more than $40,000 for the United States, Japan, and the EU's richest countries to about $4,000 for China (IMF 2010).

These indicators make it clear than when judged by its technical advances and by its success in extending a comfortable standard of living to an increasing part of global population, ours is an admirably accomplished civilization. In its quotidian mental detachment from nature it sees its fate, not incorrectly, as highly dependent on incessant and affordable supplies of modern energies in general and fossil fuels in particular (hence the constant worries about "running out" or "peak oil") and on the availability of a wide range of nonenergy minerals. But first things first: photosynthesis remains the most important energy conversion in the biosphere, and without its products no heterotrophic life—including human civilization—would be possible. Harvesting the biosphere is as quintessential for us as it was for our primate ancestors, for the Paleolithic hominins, for the first sedentary societies engaged in perfecting annual cropping eight millennia ago, or for the first great empires of antiquity.

All of the intervening technical and scientific innovations and the higher use of energy would have been impossible without greatly expanded claims on the biosphere's photosynthetic productivity, without rising harvests of cultivated and wild phytomass, and without increasing contributions of animal foods produced by domesticated species or by hunting wild mammals, birds, and fishes. The biosphere has paid a considerable price for these great human gains as both its total stock of standing phytomass and its overall primary productivity have declined by significant amounts—as have the stocks of wild terrestrial and marine zoomass. In closing this book, I will review these long-term trends before making some observations about the anthropogenic Earth and raising some possibilities, and many more uncertainties, regarding the future of life on the third planet from the Sun.

Biomass Changes

As I have noted repeatedly, all but a few quantifications of global stores of biomass rely on multiple assumptions, and such exercises can produce only more or less acceptable approximations. Advances in remote sensing have greatly improved our ability to monitor land-use changes, to map specific land-cover categories, and hence, in conjunction with field studies, to quantify the phytomass stocks of biomes and ecosystems. Even so, estimates of total terrestrial phytomass at the end of the twentieth century span an unhelpfully wide range between about 470 and 780 Gt C, and we cannot be sure if the most likely total is 550, 600, or 650 Gt C. Not surprisingly, uncertainties regarding the evolution of phytomass stocks are even greater.

Only a qualitative narrative is fairly clear: the most recent ice age reduced the Earth's plant cover and hence the global phytomass stocks, which rebounded with deglaciation; global storage peaked sometime during the mid-Holocene before the more extensive human interferences (in the form of shifting and permanent cultivation, the grazing of domestic animals, a higher incidence of fires, and the extension of settlements) began to change the natural land cover and reduce the phytomass stores. These processes accelerated in the past two centuries, and even the substantial post-1950 return of temperate forests has not been able to eliminate the overall net loss of woody phytomass.

Quantifying all of this is another matter. The best conclusion is that during the last glacial maximum, the land plants stored up to 200 Gt less carbon than they did before the glaciation. A substantial gain during the Holocene—doubling does not seem excessive, as the total area of tropical rain forest had roughly tripled between 18,000 and 5,000 years before the present and that of cool temperature forests had expanded more than 30-fold (Adams and Faure 1998)—could have raised the stocks to more than 1,000 Gt C, with subsequent land-use changes reducing it to no more than 900 Gt C by the eighteenth century, with a total of 750–800 Gt C the most likely value. Plant carbon losses during the last two centuries amounted most likely to 150–200 Gt C, lowering the late twentieth-century terrestrial stocks to no more than 650 Gt C, and perhaps even below 600 Gt C.

Human actions have thus reduced the biosphere's phytomass by as much as 35%–40% from its preagricultural level. During the twentieth century the net reduction in global phytomass was about 110 Gt C, or about 15% of the 1900 total—but, concurrently, the phytomass of field crops increased fivefold.

Accurate knowledge of long-term changes of the terrestrial phytomass would make it possible to quantify with greater confidence the partitioning of carbon emitted from fossil fuel combustion, as well as the magnitude of potential terrestrial carbon sinks and the interactions of carbon and nitrogen cycle. And, obviously, our assessment of cumulative anthropogenic losses of biodiversity would be very different if we could assert that we have lost only 25% rather than 40% of the peak postglacial phytomass.

We are on firmer ground when appraising the conversion of natural ecosystems to fields and the global expansion of cropping driven by growing populations and by the universal dietary transition from an overwhelmingly vegetarian diet to one containing a higher share of animal protein. By the middle of the eighteenth century the global area of farmland was still only about 350 Mha; by 1900 it was 850 Mha, and by 2010 the land used for annual and permanent crops had surpassed 1.5 Gha, claiming about 12% of all ice-free land, but the peak seasonal preharvest phytomass of global cropland is less than 0.5% of all plant mass. The obverse of these gains was a major loss of temperate grasslands and tropical forests. After 1850, most of North America's and Russia's new cropland came from plowing up grasslands; in the tropics, most new fields came from deforestation. In total, ecosystem conversions led to the loss of at least 150 Gt of plant carbon between 1850 and 2000 (Houghton 2003).

But perhaps the most impressive way to illustrate the extent of human impacts on the stocks of global organic matter is to trace the gains and losses of mammalian biomass, that is, the increasing mass of humanity (anthropomass) and its domesticated animals and the declining zoomass of wild terrestrial animals, particularly of the largest herbivores and anthropoid primates. Again, the need to resort to chained assumptions precludes a high degree of accuracy, but conservatively made calculations not only reveal the magnitude of indisputable secular trends but can

Box 12.1
Calculating Global Anthropomass in the Year 2000

When calculating the total mass of humanity, it is imperative to take into account substantial differences in age compositions of the constituent populations and the disparities in their average body weights. Low-income countries have much higher shares of children and young teenagers (whose average body weights are fractions of the typical adult mass) than do affluent countries. For example, in 2010, 40% of Africa's population was younger than 15 years, and the continent's median age was 19.7 years, while the corresponding numbers for Europe were 15% and 40.2 years (UN 2011). The great economic divide is also reflected in average body weights: five-year-old children in the United States are 3–4 kg heavier than in India, and by the age of 15 the difference is twice as high (Ogden et al. 2004; Sachdev et al. 2005). But national means differ even within the same economic group. Most notably, in 2005 the average obesity rates (defined as a body mass index higher than 30) were as low as 3.9% in Japan and as high as 33% in the United States, with the European shares ranging from 10% in Italy to about 23% in England (NOO 2009).

That is why, in calculating the global anthropomass in the year 2000, I have used four different weighted means of body averages: for North America with its overweight population of more than 300 million people; for all the other high-income countries (about 800 million, dominated by Europe); for modernizing nations (4.2 billion, dominated by China and India); and for the world's poorest economies (about 700 million, mostly in Africa). Age and sex structures are available for these four population categories (UN 2011), and I used average body masses derived from anthropometric studies and growth curves for populations of four representative countries, the United States, Germany, China, and India (Schwidetzky, Chiarelli, and Necrasov 1980; Sachdev et al. 2005; Zhang and Wang 2010). This procedure resulted in a weighted mean of about 50 kg, which means that the total live weight of the global anthropomass of 6.1 billion people was about 300 Mt in the year 2000. With human body water content averaging 60% (Ellis 2000) and with 45% of carbon in the dry mass, that total yields about 55 Mt C. In contrast, Barnosky (2008) used an average live weight of 50 kg up to about 400 years ago, and 67 kg afterwards, both being clear exaggerations.

also produce specific rates for some surprising comparisons of biomass densities, which must start with calculating the global anthropomass.

Because better diets (brought about by the dietary transitions of the late nineteenth and twentieth century) have resulted in higher average body weights—for example, Japanese records show that average male weight at age 20 rose from 53 kg in 1900 to 65.4 kg in 2000 (SB 2010)—the total biomass of our species has risen at a faster rate than did the overall population, which was about 3.7 times greater in 2000 than it was in 1900. Using the weighted global body mass mean of 45 kg and an approximate population total of 1.65 billion results in 13 Mt C in human biomass in 1900: the global anthropomass thus more than quadrupled during the course of the twentieth century. What is much more surprising is the extent to which the anthropomass and, even more so, the domesticated zoomass increased when compared not only to the zoomass of wild vertebrates but even to the much more abundant zoomass of soil invertebrates, whose activities maintain productive agricultural soils.

Even the largest wild terrestrial vertebrates now have an aggregate zoomass that is only a small fraction of the global anthropomass. The pitiful remnants of once enormous herds of bison, America's largest surviving megaherbivore, now add up only to about 40,000 t C (fewer than 400,000 animals, with an average body mass of 500 kg and a water content of 55%), an equivalent of the anthropomass of a city of four million people. The latest count of African elephants estimated 470,000 individuals in 2006 (see box 7.3): with an average body mass of 2.6 t per individual, this equals only about 1.2 Mt of live weight and (at 55% water and 45% C in dry matter) only about 250,000 t C, equivalent to about 0.5% of the global anthropomass. Given this huge disparity, it also seems very likely that today's anthropomass exceeds the global biomass of any other extinct large vertebrate with which our species shared the biosphere (it is unlikely that the peak mammoth zoomass would have been more than 200 times larger than today's global elephant biomass).

And even a liberal estimate of the total zoomass of wild terrestrial mammals at the beginning and end of the twentieth century—assuming averages of 1 kg/ha in croplands, 2 kg/ha in low-productivity ecosystems (in both cases dominated by rodents), and 5 kg/ha (dominated by large herbivores) in the richest grasslands and forests, and using the relevant historical land cover data (HYDE 2011)—yields no more than about 40 Mt of live weight (about 8 Mt C) in 1900 and 25 Mt of live weight (about 5 Mt C) in the year 2000, a decline of 35%–40%. In contrast, during the same time period the global anthropomass rose from roughly 13 Mt C to 55 Mt C. This means that the global anthropomass surpassed the mammalian terrestrial zoomass sometime during the first half of the nineteenth century, by 1900 it was at least 50% higher—and by the year 2000 the zoomass of all land mammals was only about a tenth of the global anthropomass (see table 12.2)!

In contrast, Barnosky (2008) concluded that the human biomass had already surpassed the global megafaunal zoomass about 3,000 year ago, and that in recent years anthropomass was about 80 times greater than the mass of nonhuman megafauna (all of his comparisons were in terms of live weight). His conclusions are based on two indefensible assumptions. First, Barnosky assumes that the megafaunal

Table 12.2
Anthropomass and Zoomass of Wild and Domesticated Animals, 1900–2000 (in Mt C)

Year	Anthropomass	Wild Mammalian Zoomass	Elephants	Domesticated Animals	Cattle
1900	13	10	3	35	23
2000	55	5	0.3	120	80

Note: These values are only approximations of global totals; those for the anthropomass and the zoomass of domesticated animals and cattle in the year 2000 are relatively the most accurate.

biomass has remained constant for the past 10,000 years—an obvious error given the intervening slaughter of elephants, bison, and other large herbivores. And his estimate of megafauna biomass—just 4 Mt live weight for the past 10,000 years—is much too low: for example, in 1800 the live mass of 50 million North American bison was alone (even when assuming a conservative herd average of 400 kg/animal) about 20 Mt, and the mass of African elephants would have at least doubled that total.

My calculations show that the differences between average densities of humans and wild animals are similarly large, as anthropomass densities, supported by modern intensive farming, have far surpassed the highest possible densities of wild mammals and have risen orders of magnitude above those of anthropoid primates. The chimpanzee zoomass (live weight) of some communities surpasses 1 kg/ha but is typically less than half that rate. The densities of many early foraging societies were similar (at less than 0.5 kg/ha), but the most productive traditional agricultures could eventually support more than five people, or more than 200 kg/ha of arable land. By the year 2000 the most intensively farmed temperate climate regions could feed more than 15 people/ha, or in excess of 250 kg of dry-weight anthropomass per hectare. In contrast, the total dry-matter zoomass of soil fauna in temperate fields is usually less than 100 kg/ha (Coleman and Crossley 1996). Remarkably, the normal trophic order that characterized agroecosystems for millennia has been transformed, as the primary production (crops) in many agricultural regions now supports a mass of people larger than the mass of all soil invertebrates.

And the zoomass of wild vertebrates is now vanishingly small compared to the biomass of domestic animals. In 1900 there were some 1.6 billion large animals, including about 450 million head of cattle; a century later the count of large domestic animals had surpassed 4.3 billion, including 1.65 billion head of cattle and water buffaloes and 900 million pigs. Calculations using these best available head counts and average body weights (they have increased everywhere during the twentieth century, but the difference between typical body masses in North America and Europe, on the one hand, and in Asia, Africa, and Latin America on the other persists) results in at least 170 Mt (35 Mt C) of domesticated zoomass in 1900 (more than four times greater than the total zoomass of all wild land mammals) and at least 600 Mt of live weight (120 Mt C) in the year 2000, a 3.5-fold increase in 100 years and roughly 25 times the values for the total wild mammalian zoomass. The cattle zoomass alone (about 80 Mt C) is now at least 300 times greater than the zoomass of all surviving African elephants, whose biomass is now less than 2% of the zoomass of Africa's nearly 300 million bovines.

In some countries the presence of domestic animals has reached unprecedented densities. In 2009 the Netherlands had nearly 4 million heads of cattle, more than 12 million pigs, and 1.1 million sheep and goats (PVE 2010), whose live-weight zoomass added up to about 1.3 t/ha of crop and grazing land, three times the average anthropomass per hectare—and in some parts of the country the difference was twice as high. Even more remarkably, this high density of domesticated zoomass was an order of magnitude greater than the combined biomass of all soil invertebrates and was surpassed only by the mass of soil bacteria. Even the extraordinarily high Dutch crop yields could not support such densities, and the country is a major importer of animal feed (Galloway et al. 2007).

With 500–800 Gt C, the biosphere's phytomass binds an equivalent of between 65% and 102% of the element's current atmospheric content (831 Gt C, corresponding to 390 ppm CO_2). In terms of penetration of the Earth's physical spheres, the eukaryotic biomass constitutes a mere 7×10^{-9} of the ocean's volume and less than 0.01% of all the carbon in the ocean. Subterranean and subsea prokaryotes, even when credited with very high cell totals, amount to just 6×10^{-7} of the top 4 km of the crust, the generally accepted depth limit for the survival of extremophiles. The uniform distribution of dry terrestrial phytomass over ice-free land would produce a layer about 1 cm thick; the same process in the ocean would add a mere 0.03 mm of phytoplankton (in both cases I am assuming the average biomass density to be equal to 1 g/cm^3). I know of no better examples to illustrate the evanescent quality of life.

Productivities and Harvests

During the millennia of prehistoric evolution the human impact on the biosphere, however profound locally or regionally, was limited from spreading to a global scale by low population numbers and low technical capacities. Foragers had a clearly discernible but rarely a highly destructive influence. Harvesting to complete extermination might seem to be an exceedingly difficult if not impossible task to achieve with organisms whose natural populations numbered in millions of individuals. As I have tried to demonstrate, mammoth extinctions do not belong to that category: hunting was clearly a factor, but hunting alone did not eliminate the species.

But that incredible feat of complete extinction was accomplished in the case of an animal whose numbers were in the billions. Audubon, after years of witnessing the extensive slaughter of passenger pigeons, thought them to be safe because they commonly quadrupled their numbers every year, and always at least doubled it— but they did not make into the twentieth century (Jackson and Jackson 2007). Many less numerous species were eliminated by hunting since the beginning of the early modern era. Near extinctions, in which the original overwhelming richness has been reduced to tiny, isolated remnants, have been numerous, affecting species ranging from the American bison (essentially gone by the 1890s) to the Newfoundland cod (impossible to catch in large quantities by the 1980s).

And these near extinctions continue to unfold, with the gravest threats (despite the continued tropical losses) not on land but in the ocean because of the combination of persistent overexploitation of all commercially valuable species and frenzied competition to catch the most valuable specimens. Only the great whales have become exempt from this global hunt, and Schneider and Pearce (2004) have shown that the dismal economics of whale hunting, rather than the public opinion in most Western nations, was the main reason for the 1986 moratorium: declining stock and the resulting high unit catch costs, and higher incomes that lowered rather than increased demand for whale products, were the decisive factors.

There is no shortage of analyses and reviews explaining the follies of modern fisheries (none as irrational as the fact that the continuing overfishing has been heavily subsidized by virtually all governments of maritime nations) and suggesting steps toward ending the overexploitation, rebuilding endangered stocks, and making a transition to truly sustainable catches (Hilborn et al. 2003; Worm et al. 2009; M. D. Smith et al. 2010). But political pressures and narrow national interests have made it very difficult to translate these rational precepts into everyday actions. Similarly, it would be naïve to expect that the expanding aquaculture will suffice to prevent further deterioration of wild stocks; indeed, the culturing of carnivorous species may only accelerate the overall decline.

The most worrisome hunting on land affects the two remarkable mammalian families, anthropoid primates, which are killed for their meat, and elephants, which poachers kill for their tusks. No effective solutions are at hand. The largest ivory seizure since the 1989 trade ban, made in Singapore in 2002, contained 532 tusks with an average weight of more than 11 kg, substantially larger than the usually traded tusks (Wasser et al. 2008). The scale of the ongoing illegal ivory trade makes it clear that most African countries still cannot protect their elephants, and periodic sales, often touted as "one-off" affairs, do not help improve that dismal situation (Wasser et al. 2010). Loss of forest habitat and the killing of anthropoid primates have reduced the distribution of gorillas and chimpanzees to ever-smaller disconnected patches and brought their numbers to levels that justify their inclusion in the critically endangered category for the mountain gorilla (*Gorilla gorilla* and *G. beringei*) and to a somewhat less dire extent the Western as well as endangered listing for both chimpanzee species, *Pan troglodytes* and *P. paniscus* (IUCN 2011).

For millennia, the greatest human impact on the biosphere was not caused by actual harvests of food, feed, and wood but by the alteration and destruction of ecosystems to create more open treed landscapes, expand agriculture and grazing, and build terraces, settlements, and roads. We now know that such changes affected even places that were seen until very recently as paragons of intact purity. The pre-1492 Americas were not pristine wilderness but contained cultural landscapes, including the fields in the Great Lakes region, terracing and irrigation in Mesoamerica, and settlements in the forests of Amazonia (Denevan 1992). Similarly, in Central Africa van Gemerden et al. (2003) found that today's tree composition of a species-rich rain forest still shows signs of historical disturbances most likely caused by human use three to four centuries ago, a reality that makes the commonly used categories of "old growth" and "secondary forest" less clear than is usually assumed. This has an important implication for the identification of preindustrial or preagricultural natural (intact, pristine) vegetation baselines.

Advancing deforestation and the conversion of grasslands and wetlands to agricultural land or settlements led to declines in the biosphere's total net primary production (NPP), but this loss was relatively much lower than the accompanying loss of standing phytomass. Converting temperate or boreal forests to farmland meant replacing ecosystems whose aboveground phytomass was mostly between 50 and 100 t C/ha with crop monocultures or low-diversity plant communities whose peak (preharvest) phytomass typically remained well below 10 t/ha, an order of magnitude loss, but the NPP of that old-growth forest was mostly between 5 and 8 t C/ha, compared to the 1–3 t C/ha of a low-yielding cereal crop. And replacing an old-growth forest with fast-growing tree species grown for fuelwood may have actually boosted such sites' NPP. But the overall global NPP had to decline, from perhaps as much as 70–75 Gt C/year about 8,000 years ago (at the beginning of the agricultural expansion) to less than 60 Gt C by the year 2000.

And although crop yields have risen, the only large cultivated regions with an NPP greater than 10 t C/ha are Western Europe, East Asia (Japan, South Korea, and eastern China), the central United States, and to a lesser extent parts of southern Brazil and northern Argentina (Monfreda, Ramankutty, and Foley 2008). The peak rates in excess of 20 t C/ha, made possible only by harvesting two or more crops a year, are recorded in the most intensively cultivated areas of Atlantic Europe (above all in the Netherlands) and in irrigated parts of the Middle East, the western United States, Java, and India. Modern well-managed tree plantations yield mostly between 10 and 15 t C/ha.

Many regional and countrywide summaries of crop and wood production have been available for decades in national statistics, but global assessments of phytomass harvests and their comparison to the biosphere's total primary productivity began only during the 1970s. This reflected ecology's beginnings as an examination of ponds and meadows and forest patches, small communities of plants, and heterotrophs. One of the great classics of the science, Lindeman's (1942) pioneering examination of energy transfers among trophic levels, focused on Wisconsin's Lake Mendota, and 15 years later Odum (1957) did a more thorough investigation of a another freshwater ecosystem in Florida. The advent of new environmental consciousness, satellite monitoring, and the emergence of global ecology shifted the focus all the way to the planet as a whole in general and to the biosphere's global production and carbon balance in particular (Lieth and Whittaker 1975; Bolin et al. 1979).

The best expression of this approach to biomass harvests has been the studies of human appropriation or co-option of NPP. There is no agreement on what should be included in such appropriated or co-opted totals, nor is there any uniform calculation procedure to follow. Moreover, NPP is a questionable choice for a common denominator because it is a theoretical construct that cannot be measured directly, whose annual rate fluctuates, and whose derivation, as much as it is now helped by satellite monitoring and complex dynamic plant productivity models, is not highly accurate. Global estimates used in different appropriation studies have differed by about 40%, ranging from about 57 Gt C/year, used by Imhoff et al. (2004a), to about 80 Gt C, used by Wright (1990): that is a substantial discrepancy for a value used as the key denominator. As with most studies of this kind, it is necessary to make entire chains of concatenated assumptions in order to aggregate the key subtotals of specific harvests. As expected, different authors have their own preferred numbers and simplifications.

The first study of this kind produced a range of appropriation shares from just 3% to about 39% (Vitousek et al. 1986). The second one (Wright 1990) produced a share of about 24%; the third one (Rojstaczer, Sterling, and Moore 2001), 32% (but its confidence interval indicated possible appropriations ranging from 10% to 55%); the fourth one (Imhoff et al. 2004a), a range of 14%–26%; and the last one (Haberl et al. 2007) added up to nearly 24% of potential NPP that is "appropriated" or "co-opted" by humans. There is no need to extend this small list of questionable rates: even the most careful and eminently defensible selections of analytical approaches and key data inputs could still result in shares differing by impractically wide margins (easily 50%), too uncertain to offer any practical policy guidance. And, as I have already noted in my detailed deconstruction of the concept in chapter 11, this poorly defined measure offers only rough quantitative impressions, as it completely ignores any qualitative consequences of phytomass harvests.

Accounting just for the biomass that is actually harvested by humans should be a simpler task, with more reliable results: after all, field crops, fish landings, and timber cuts are now overwhelming part of national and global markets, and their output and consumption are closely monitored. But as we have seen, some major uncertainties remain, and hence any claims of high accuracy must also be suspect. Historical records of crop harvests (often given as multiples of planted seed mass) are only good enough to trace centuries of very low and stagnating yields. Plant improvement proceeded very slowly until the first determined experiments of the seventeenth and eighteenth century, and it really took off only after Mendelian genetics opened up new opportunities (Kingsbury 2010).

The most important result of these efforts has been a steady rise in harvest indices (HI), and the most obvious outcome of that trend has been the shortening of cereal straws. Traditional cultivars were taller than an average person (see the figures at the beginning of chapter 4 and part II). Even in 1900 many wheat cultivars were still more than 1 m tall, while today the shortest varieties, carrying a dwarfing Norin-20 gene, are only about 50 cm tall. Higher HI values, denser planting, an optimum nutrient supply, and applications of herbicides and pesticides boosted the cereal yields (with national averages often more than doubling) during the twentieth century.

Better data allow fairly reliable global reconstructions of crop harvests for the entire twentieth century. In 1900 the worldwide harvests of food and feed crops amounted to about 400 Mt of dry matter. By 1950 that total had doubled, by 1975 it had doubled again, and at the beginning of the twenty-first century, when fields and permanent plantations claimed about 12% of the ice-free land, the global harvest of food, feed, and fiber crops was about 2.7 Gt. Their residues added about 3.7 Gt and forage crops totaled about 1.2 Gt, for a global total of about 7.6 Gt of herbaceous aboveground phytomass. Roughly half of this phytomass (4 Gt) was fed to animals and produced (all values as fresh weight) nearly 300 Mt of meat, almost 700 Mt of milk, and 65 Mt of eggs. And annual harvests of woody phytomass—including fuelwood, industrial roundwood, and pulpwood, as well as biomass directly destroyed, disturbed, or abandoned during harvesting—reached at least 13 Gt of dry matter by the year 2000.

During the first decade of the twenty-first century the annual harvest (and direct destruction) of terrestrial phytomass thus added up to roughly 20 Gt of dry matter, or about 10 Gt C; this translates to nearly 17% of the biosphere's annual terrestrial NPP when assuming the standard rate of 60 Gt C/year, and to a bit more than 13% when assuming a higher rate of 75 Gt C/year. And while the stocks of many aquatic species have been seriously depleted, the relative impact of fisheries and aquaculture on the ocean's NPP is lower than the impact of cropping and wood harvests on the terrestrial NPP. Catches (including discards) of wild invertebrates and fish and aquaculture of an increasing number of these species now surpass 130 Mt (fresh weight), or about 15 Mt C, and the production of this zoomass requires annual intakes of nearly 3 Gt C of aquatic phytomass, or roughly 6% of the total oceanic primary production.

Neither of these shares appears to be alarmingly high, but, as I have tried to demonstrate, the "appropriation" ratio is not the best way to measure the human impact on the biosphere because many qualitative implications and multifaceted ecosystem and social impacts of the phytomass harvests are utterly beyond its scope. In any case, the enormous latent demand for more phytomass associated with lifting the more than five billion of people living in low-income countries closer to the standard of living enjoyed by a minority of humans could keep raising the ratio for generations to come. There is plenty of other evidence of the enormous scope of the human transformation of the Earth, and future interventions may be further complicated by the unfolding climate change.

Anthropogenic Earth

Many complexities and uncertainties prevent us from developing accurate quantitative appraisals of the claim humans make on the biosphere's stocks and annual increments of biomass, but the essentials are clear. First, humans have transformed significant portions of all natural terrestrial biomes in destructive ways that have largely to completely eliminated natural vegetation and its attendant heterotrophic diversity. Second, humans have extensively altered—degraded, fragmented, and impoverished—natural vegetation in those biomes that still preserve, or more or less resemble, the ecosystems as they were before the worldwide diffusion of our species. Third, humans and their domesticated animals have become the dominant mammalian species on Earth, and the accompanying loss of wild zoomass has been relatively more pronounced than the transformation of the planet's plant cover.

Fourth, the two most easily observable consequences of this dominance, the expansion of agricultural and of built-up land, have progressed beyond local and regional scales, and since the 1970s it has been possible to monitor their global advances by satellite-based remote sensing. Agroecosystems (fields planted to annual crops and permanent plantations) have retained their soils (and some of them still harbor marginal vestiges of their original vegetation), but they are (often intensively) managed to maximize the yields of cultivated species. Such fields now cover 12% of all ice-free land, but that is a qualitatively misleading comparison because agricultural production has claimed virtually all the best soils in all large populated nations.

Urbanization and industrialization have either entirely eliminated primary production from the sites occupied by structures and paving or reduced it to a small percentage of the original extent. This process has reached its most expansive stage with the emergence of megacities and conurbations. These built-up areas—whose extent is perhaps most impressively illustrated by nighttime satellite images showing large patches and long corridors or amoeba-like formations of bright light—increased to cover about 5 million km^2 by the end of the twentieth century, an equivalent of half the United States or an area about 50% larger than India.

Harvesting the woody phytomass has been done in ways ranging from aggressive destruction (large-scale clear-cutting) to intensive, croplike cultivation of planted trees. Encouragingly, recent declines in the annual rate of tropical deforestation and the regrowth of temperate forests have slowed the overall retreat of natural forest biomes, but not before their precivilizational extent had been reduced by at least 20%. And aggressive, worldwide exploitation of marine fishes, invertebrates, and mammals has reduced the zoomass of what previously were some of the world's most abundant heterotrophs to levels that cause concerns about their very survival. These actions have brought an array of negative environmental changes whose impact has reached from deep aquifers and the ocean's abyss to the stratosphere (Foley et al. 2005).

A single-paragraph list of such notable transformations must include the following: changes of surface albedo and hence of soil and vegetation temperatures, evaporation, and evapotranspiration (Liang et al. 2010); altered CO_2 fluxes due to photosynthesis and respiration and increased emissions of CO_2 from biomass combustion (Houghton 2003; Bond-Lamberty and Thomson 2010; Dolman et al. 2010); reduced or increased emissions of volatile hydrocarbons and numerous trace gases, including CH_4, N_2O, NO_x, and SO_x (Houweling et al. 2008; Crutzen, Mosier, and Smith 2008; Penuelas and Staudt 2010); increased generation of terrigenic dust, lowered retention of soil moisture, enhanced soil erosion and sedimentation, and losses of organic soil nitrogen (Kellogg and Griffin 2006; Wilkinson and McElroy 2007; Lal 2007; Eglin et al. 2010); and declines in abundance and biodiversity affecting species ranging from soil microfauna to the top predators, with perhaps the outright extinction of some invertebrates and many vertebrate animals (Butchart et al. 2010; Rands et al. 2010; IUCN 2011).

Curiously, a recent assessment of a safe operating space for humanity—one that identifies quantifies the planetary boundaries that must not be transgressed if unacceptable environmental change is to be prevented (Rockström et al. 2009)—does not list either a minimum phytomass stock or primary productivities among its ten concerns. Rather, these concerns are climate change, disturbances of nitrogen and phosphorus cycles, stratospheric ozone depletion, ocean acidification, global freshwater use, atmospheric aerosol loading, and chemical pollution, and they address phytomass stocks and productivity only indirectly, via the categories of biodiversity loss and changes in land use (explicitly defined as the percentage of global land cover converted to cropland, with a proposed limit of no more than 15%, compared to the current 12%).

In my list the collective and cumulative impact of harvesting the biosphere would have been included explicitly and prominently because it has been—together with the increasing production, processing, and combustion of fossil fuels and the extraction and use of other minerals—the principal cause of the human transformation of the biosphere. The recent pace and extent of this process have led not only to questions about the degree of the human appropriation of the biosphere's primary production that I analyzed in some detail in chapter 11 but also to concerns about the "ecological overshoot" of the human economy (Wackernagel et al. 2002; Kitzes et al. 2008) and about the perilous human domination of Earth's ecosystems (Vitousek et al. 1997; Kareiva et al. 2007; McCarthy 2009).

These concerns have found their most extreme expressions in the works of Martin Rees and James Lovelock. Rees (2003) details in *Our Final Hour: A Scientist's Warning* how "terror, error and environmental disasters" threaten the very survival of humankind during the twenty-first century. And Lovelock (2009), in another "final warning," foresees the vanishing face of Gaia, a planetary goddess of his anthropomorphic constructs. Perhaps the two most radical expressions of these concerns in less catastrophically inclined academic publications have been the proposal to use a new label for the era we live in and to discard the traditional classification of biomes in favor of new human-dominated entities.

Crutzen (2002) has argued that the escalating effects of humans on the global environment—above all in the form of anthropogenic emissions of CO_2, whose atmospheric perseverance may change climate for millennia to come, but also in the form of tropical deforestation, ocean exploitation, and the mobilization of reactive nitrogen—are overwhelming the great forces of nature and that this process warrants labeling the period that began during the late eighteenth century the Anthropocene. His arguments were subsequently elaborated (Steffen, Crutzen, and McNeill 2007), and the first steps have been taken by the International Commission on Stratigraphy to formalize the new classification (Zalasiewicz, Williams, and Crutzen 2010); in 2016 the International Geological Congress will consider whether the changes have been distinctive and enduring enough to add a new epoch (Vince 2011).

Ruddiman (2005) supports the concept of an anthropogenic era but argues that it began 8,000 years ago, when CO_2 began its "anomalous increase," followed by the rise CH_4 concentrations that began some 5,000 years ago, driven first by forest clearance for farming and then by rice cultivation. Even that backdating is not enough for Smith, Elliott, and Lyons (2010), who assume that about 100 million North American herbivores were killed by hunting and claim that the resulting decline in CH_4 emissions could explain the observed decrease of atmospheric CH_4 concentrations dated to the beginning of the Younger Dryas, and that this presumed connection justifies a recalibration of the onset of the Anthropocene to 13,400 years before the present. Given the enormous uncertainties in estimating CH_4 emissions from modern cattle, it is not surprising that the range of their estimate for methane production by America's megafauna could explain as little as 10% or as much as 100% of the recorded decline, making their claim (already based on dubious assumptions of 100 million carcasses) a matter of arbitrary choice.

Approaching the problem in a different way, Ellis and Ramankutty (2008) believe that the conventional biome classification that sees ecosystem processes as a function of macroclimate is unrealistic because the biosphere is now dominated by anthromes (anthropogenic biomes), where the processes are *primarily* (their emphasis) a function of human populations and their ecosystem interactions (land use)—or, as they put it, "nature is now embedded within human systems." They identify 18 different categories, with more than 40% of all people living in the urban anthrome and another 40% living in six village anthromes. They claim not only that the anthromes now cover more than 75% of all ice-free land but that they incorporate 90% of terrestrial NPP.

Even more astonishingly, Ellis and Ramankutty (2008) state that the share of global NPP taking place in five anthromes in the croplands category is as high as in two forested anthromes (populated and remote forests, in their classification), namely, about 32%. This high share is explained by the fact that all natural growth within areas affected by humans falls into one of the five anthrome categories, as well as by the fact that the anthromes are the creations of computerized classification. Grid cells are first separated into "wild" and "anthropogenic" categories (based on land-use features), and the latter are then stratified into population density classes and assigned to one of the 18 anthromes that contain varying combination of land uses and land covers.

Obviously, different criteria and different assumptions could produce differently biased outcomes than these anthromic mosaics; one need only check the highest available resolution of this classification (available at http://ecotope.org/anthromes/ v2/maps/a2000) for an area with whose land cover one is very familiar, and one will be astounded by the degree of misattribution (forests labeled cropland, croplands labeled residential area) and by misleading classifications. The anthromic mapping classifies about 85% of the city I live in as "dense settlements," the rest as "residential rainfed croplands." But Winnipeg contains one of the largest urban "forests" in the temperate zone, with an estimated eight million trees (along the Red and Assiniboine Rivers, in extensive parks, and in street and residential plantings), including the largest number of surviving American elms (about 160,000), trees that have been destroyed elsewhere in North America by *Ophiostoma ulmi* (City of Winnipeg 2011).

My very conservative calculations put the total standing phytomass of Winnipeg's trees at 1.2 Mt C, prorated to about 40 t C/ha of actually settled area within the city's boundary. That phytomass density is only slightly lower than the expected aboveground storage in a boreal semiarid deciduous forest (Luyssaert et al. 2007), and Winnipeg could be thus be "anthromically" classified as "residential woodlands" rather than as "dense settlements." This urban forest is overwhelmingly a result of deliberate planting and continuous management, but its phytomass density is considerably higher than that of the original grassland (prairie) that covered the Red River floodplain before the westward push of European settlers. The result: an "anthrome" that harbors more phytomass and has a higher NPP than its natural predecessor.

This example (repeated either in similar or other guises worldwide) raises an obvious question: why should all ecosystems that have been strongly influenced by human actions be automatically considered inferior, or at least less desirable, plant assemblages than their natural predecessors? Not only will some disturbed, regenerated, or planted ecosystems have an NPP rivaling the NPP of their natural predecessors, they could be nearly as good from the functional heterotrophic perspective. For example, Edwards et al. (2010) surveyed 18 sites in Borneo using birds and dung beetles as indicators of biodiversity and found that 75% of all species that lived in untouched forests continued to live at twice-logged sites! And how is selective logging functionally different from damage that is periodically inflicted on forests by various natural disturbances and disasters? How fundamental is the difference between a clear-cut forest and an area whose plant cover was denuded by a strong hurricane, the latter being a periodic occurrence in eastern American forests as far north as New England (Foster and Aber 2004)? These realities make the anthromic/natural dichotomy a matter of choice, and often of a highly arguable one.

Arbitrary distinctions, definitions, and boundaries have been also involved in the global quantifications of the remaining "wilderness." The first such inventory defined it as contiguous blocks larger than 400,000 ha: during the late 1980s such blocks still covered more than a third of the global land surface (nearly 48 Gm^2), but 40% of that total was in the Arctic or Antarctic (McCloskey and Spalding 1989). National shares of the remaining wilderness ranged from 65% for Canada to less than 2% for Mexico and Nigeria. Sweden (with about 5%) was the only European country to qualify: no other nations had such large tracts of undisturbed land, nor did such tropical regions as the Guinean highlands, Madagascar, Java, and Sumatra, or the rich temperate broadleaf forests of eastern North America, China, or California.

The second study of "the last of the wild" used several criteria to calculate human influence index scores, assigning numbers to all areas with a population density greater than 1 person/km^2; agricultural land use; built-up areas and settlements; all land within 15 km of roads, railways, major rivers, and coastlines; and areas of nighttime light bright enough to be seen on satellite images. The authors admitted that they might have overestimated the spatial extent of human influence but added up their numbers anyway to conclude that 83% of the land surface is influenced by one or more of the above factors (Sanderson et al. 2002). Tundra and boreal forests had the lowest influence scores, and tropical and subtropical grasslands, mangroves, and temperate forests were the most affected biomes.

The most recent study of this kind mapped the world's intact forest landscapes, defined as area of at least 500 km² or 50,000 ha devoid of any signs of significant human activity (Potapov et al. 2008). The global extent of such landscapes was just over 13 Mkm², or slightly less than a quarter (23.5%) of all forests, with nearly 90% of the total roughly split between dense tropical and subtropical forests and the boreal biome. Large intact forest areas were found in 66 of 149 countries having forest biomes, but just three of them, Canada, Russia, and Brazil, accounted for nearly two-thirds. Unfortunately, less than 10% of the remaining intact forests were strictly protected, and less than a fifth had any form of protection.

And the real extent of intact forests may be substantially greater. Potapov et al. (2008) assumed that all fires near the vicinity of human settlements were set by people, and this rule led them to classify more than 400,000 km² of Canadian forests as "not intact." But Lee (2009) pointed out that such a rule, perhaps valid for Siberia, is inappropriate for Canadian forests, where fire, whether close to or far from any settlement, is an expected natural agent of change. If a simple assumption change can "reclaim" nearly half a million square kilometers of wilderness, what other rules in that study of forest intactness, and in other similar mappings, may result in less dramatic human impacts? All of these realities force us to conclude that while hybrid ecosystems are ubiquitous, the complexities of such landscapes raise challenging questions about the "naturalness" of our environment and about authenticity in nature (Dudley 2011).

As for those audacious claims that humans are now supreme and natural systems are somehow subserviently embedded within human systems, their authors would do well to remember that the fundamental geophysical variables that make the biosphere possible, that make the Earth habitable, that are the primary governors of climate, that subject it to Milanković cycles, and that are the pacemakers of the ice ages (ranging from the Sun's electromagnetic flux and the planet's tilt, orbital distance, path and its eccentricity to the Earth's shape and rotation speed) are *absolutely beyond* any human influences (Milankovitch 1941; Hays, Imbrie, and Shackleton 1976). Not surprisingly, Claussen et al. (2005) found no support for Ruddiman's (2005) claim that the early Holocene anthropogenic CO_2 and CH_4 emissions prevented a glaciation that should have taken place 4,000–5,000 years ago.

Nor can there be ever any human control of the fundamental planet-forming processes of plate tectonics that govern climate by redistributing land masses and the oceans and by elevating continents (Condie 1997)—or that generate volcanic eruptions and earthquakes. Earthquakes are the worst natural catastrophes in terms of casualties: between 1970 and 2005 they killed nearly 900,000 people (Swiss Reinsurance Company 2006), and the actual total was almost certainly over one million because the toll of China's Tangshan earthquake of July 28, 1976, was put as high as 655,000 people rather than the official admission of 242,219 casualties (Huixian et al. 2002). On a time scale of millennia there are also the nontrivial threats of collisions with large asteroids, events that surely (and catastrophically) do not fit the concept of nature embedded within human systems (Smil 2008).

As far as the naming of geological epochs goes, I would argue for a more detached appraisal. The average duration of the six elapsed epochs of the Cenozoic era (from the Paleocene to the beginning of the Holocene) was more than 10 million years, the last two, the Pliocene and Pleistocene, each lasted nearly three million years, and we are now just over 10,000 years into the Holocene. Does it make sense, considering these time scales, to rush into singling out an epoch that has lasted, so far, at best 8,000 but perhaps only 200 years? Would not it be prudent to wait at least another 10,000 years before we make more solidly founded judgments about the adaptability and hence longevity of human civilizations? Only an unreconstructed fan of anything-goes science fiction could feel certain that modern civilization, with its extraordinarily high energy and resource demands, will be around for 10,000 years. Or, to address both the anthrome and Anthropocene matters in a single simple judgment: in order to acknowledge the obvious reality, namely, a high degree of human interference in various biospheric processes, there is no need to make excessive and highly arguable claims.

What Next?

Global population growth, a key driver of future demand for phytomass, reached its relative peak (at just over 2%/year) during the 1960s, and the absolute annual addition peaked at nearly 90 million people a year during the late 1980s (Smil 1998). Most long-range forecasts have envisaged a comparatively modest increase before an eventual stabilization, but the most recent UN projections changed their fertility assumptions and assume continuous growth past the 10 billion mark during the twenty-first century (UN 2011). In any case, there is an enormous potential for increases in per capita consumption throughout the modernizing world, and that growth may lead to very large gains (even to a doubling) of currently consumed biomass in just two generations, well before any eventual leveling off of the global population count.

I will abstain from any time-specific forecasts, as they are bound to be wrong. But I will identify those major factors whose interaction will determine future developments, and hint at the probabilities of the most important long-term trends. Three major concerns stand out when probing the future of biomass productivity and storage from the perspective of natural responses and feedbacks. The first is the basic photosynthetic response, the future productivity of plants in a warmer world with higher CO_2 levels: will they do, on the whole, better than now, will there be little discernible difference on the global scale, or will the primary productivity suffer? The second one is a closely connected concern about carbon uptake by the biosphere, particularly by the forests: will they remain substantial sinks of carbon or will their storage capacities weaken or will they shift into being net sources of carbon to the atmosphere? The third one concerns the primary production in the ocean: its waters, too, will be warmer and also more acid, conditions that raise worries about the future phytoplanktonic productivity.

A great deal of recent research has explored the impacts of climate change on primary productivity, particularly as far as agriculture and forestry are concerned: convenient summaries are in the impact volume of the Intergovernmental Panel on Climate Change's Fourth Assessment Report (Parry et al. 2007). The earliest studies were confined to small-scale, short-term laboratory phytotron experiments; later came all-season field experiments with crops subjected to elevated CO_2 levels, and these arrangements were then replicated (with more difficulty and at a much greater expense) in enrichment trials with trees; at the same time, advances in computer modeling have provided somewhat more realistic global simulation of the biosphere's response to a warmer and CO_2-richer world. The existing atmospheric CO_2 concentration (now approaching 400 ppm, or 0.04% by volume) is much below the level needed to saturate the photosynthesis of C_3 plants(whose species dominate all forests) as well as to inhibit photorespiration, and we also know that higher concentrations of CO_2 will improve the water-use efficiency of C_3 species.

Warming will lengthen the growing seasons everywhere, and as it will intensify the global water cycle, many regions will receive higher precipitation. As a result, in a warmer world with more CO_2, the NPP could increase even in areas where precipitation would decrease, and entire biomes could continue to act as substantial net carbon sinks. And while autotrophic respiration (R_A) tends to increase with increasing temperature, this response may be only transient, and plants growing in warmer climates may acclimate and eventually respire at the same rate as those growing in cooler settings (King et al. 2006). Forest response remains especially unclear because their R_A:GPP ratio actually declines when the mean temperature is below about 11°C (Piao et al. 2010b). In any case, available data indicate that increases in forest GPP do not lead to preferential allocation to any particulate plant tissues (Litton, Raich, and Ryan 2007).

The combination of higher NPP, improved water-use efficiency, and unchanged (or only mildly elevated) R_A may thus bring noticeable improvements in primary productivity of annual crops and increase carbon storage on long-lived tissues. Some early studies confirmed such a response. For example, a model by Cao and Wood-ward (1998) indicated a much enhanced global NPP (by 25%) and substantially higher phytomass stocks (up by 20%) with doubled CO_2, and Nemani et al. (2003) concluded that between 1982 and 1999, changing climate had eased some key climatic constraints to plant growth (optimum temperature, precipitation), and as a result, global NPP increased by about 6% (3.4 Gt C) over those 18 years.

The effects of CO_2 "fertilization" have been found on scales raging from local to national. Higher annual productivity became evident throughout most of the United States during the latter half of the twentieth century (Nemani et al. 2002), and Tao et al. (2007) found that even in such a poorly forested country as China, the net ecosystem productivity (NEP) rose between 1981 and 2000, as did the overall carbon storage in plants. But greater carbon sinks may not be the norm, as photosynthesis may remain limited by shortages of key macronutrients (particularly nitrogen) or as the new photosynthate may produce mostly short-lived litter (foliage, fine roots) rather than long-lived stems and taproots. In particular, mature temperate trees may not respond to CO_2 enrichment in the same way smaller trees would. For example, a four-year experiment by Körner et al. (2005) found a sustained enhance-ment of carbon flux through the trees but no overall stimulation of stem growth or leaf litter production.

Northern forests are also particularly sensitive to changes in spring and fall tem-peratures, and the past increase of about 1°C may already have led to autumnal losses high enough to offset 90% of increased spring CO_2 uptake (Piao et al. 2008). Greater warming in the fall could thus substantially reduce the future capacity of northern ecosystems to store more carbon; additional losses may result from a higher frequency of regional wildfires and longer duration of droughts (Westerling et al. 2006). Further, a study by the International Union of Forest Research Orga-nizations maintains that the capacity of forests to store carbon could be lost entirely once the average global temperature rises by 2.5°C over the preindustrial level as the forests become a massive source of CO_2 (Seppälä, Buck, and Katila 2009). In any case, the net fluxes in the Amazon and Congo basins will be driven primarily by deforestation, and continued warming should eventually see a gradual expansion of forests into the latitudes and altitudes formerly occupied by tundra and mountain grasses, but the eventual extent of this process is unclear.

The outlook for crop response is also uncertain. There will be a direct effect on C_3 crops, as higher CO_2 levels increase the stomatal uptake of the gas, but the eventual boost in average yields could range from marginal to substantial. Early chamber studies with plants usually grown in pots indicated that a doubling of preindustrial CO_2 levels (concentrations of about 550 ppm) boosted rice and soybean yields by about 20% and wheat yields by as much as 30%, but numerous subsequent FACE (free-air concentration enrichment) studies (typically using 20-m-diameter plots within actual crop fields) indicated gains only about half those seen in enclosure experiments (Long et al. 2006). The rice yield could be thus less than 10% higher and wheat yields no more than 13% better. But higher gains are possible with concurrent nitrogen enrichment, and lower gains may be caused by excessive tropospheric ozone levels, now a major concern in many agricultural regions surrounding the megacities (Fuhrer 2009).

C_4 crops—corn, sorghum, and millets, amounting to about 40% of the world's grain harvest, and sugarcane—do not respond directly to higher CO_2 levels, and their yields will increase only in times and places of drought (Leakey 2009). And in food production, quality also matters. All hard spring wheat cultivars released between 1903 and 1996 responded to higher CO_2 levels with increased vegetative growth and increased seed yield, but the gains were lower with the most recent cultivars, and the quality of grain was lower in all cases (Ziska, Morris, and Goins 2004). Hogy et al. (2009) found not only significant decreases in total grain protein but, more important, a decline in gluten proteins, the all-important gliadins that make the best bread and pasta dough.

And plant carbon, whether in herbaceous or woody species, cannot be considered in isolation. All densely inhabited regions now receive unprecedented amounts of reactive nitrogen from atmospheric deposition, and this enrichment has already enhanced the growth and carbon storage of not only annual crops but also perennial herbaceous species and trees (Smil 2000; Elser et al. 2010; Thomas et al. 2010). At the same time, a warmer world could also produce a very worrisome feedback by releasing significant amounts of soil carbon. Soils store more than twice as much carbon as do biota or the atmosphere, and accelerated decomposition producing additional releases of CO_2 could reinforce the warming. This response was confirmed not only by models and small-scale experiments but also (as already noted) by some large-scale analyses (Bellamy et al. 2005). At the same time, an exceptional complexity of stores and fluxes of soil organic carbon precludes any unequivocal conclusions (Davidson and Janssens 2006).

All of this makes any large-scale, long-term appraisals of future net trends highly uncertain, and there is no consensus even as to the recent net global values. Between mid-1991 and mid-1997, the biosphere sequestered annually 1.4 ($\pm$0.8) and 2 ($\pm$0.6) Gt C, in contrast to the 1980s, when the land biota were basically neutral (Battle et al. 2000). Wigley and Schimel (2000) put the additional sequestration due to the CO_2 fertilizing effect at 1.2–2.6 Gt C/year, but Potter et al. (2003) concluded that between 1982 and 1998, the carbon flux for the terrestrial biosphere ranged widely between being a source of 0.9 Gt C/year to being a large sink of 2.1 Gt C, and Houghton (2003) thought that the terrestrial biomes were a source of 0.7 Gt C during the 1990s. But Ballantyne et al. (2012) concluded that between 1960 and 2010 the net global carbon uptake (by land and ocean) has increased significantly, by about 50 Mt C/year, and that the total uptake doubled to about 5 Gt C/year. Moreover, global terrestrial carbon fluxes appear to be about twice as variable (with precipitation and surface solar irradiance being the key drivers) as ocean fluxes, and at different time periods they can be dominated by either tropical or mid- and high-latitude ecosystems (Bousquet et al. 2000).

These are some of the basic qualitative conclusions that might withstand the test of time: the carbon sequestration potential of croplands has most likely been overestimated (Smith et al. 2005); the role of old forests (generally thought to be insignificant sinks of carbon) has been almost certainly (and significantly) underestimated (Carey et al. 2001); the availability of nitrogen will be a key factor in carbon sequestration in many forests (De Vries et al. 2006); and, given the pressures to reduce the nitrogen loading of European ecosystems, it is reasonable to expect that the continent's future net carbon intake will be limited (Janssens et al. 2005). Obviously, the actual outcomes will be critically dependent on the eventual level of CO_2 enrichment, on the degree of other atmospheric and biospheric changes associated with climate change, and on the supply of nitrogen.

Not surprisingly, when Schaphoff et al. (2006) tested a global vegetation model in conjunction with five general atmospheric circulation models, they found that the additional carbon storage by the year 2100 ranged from −106 to +201 Gt, making even the direction of the likely change uncertain. And when it is unclear whether the biosphere will respond as a source or a sink of carbon, then the only truly defensible conclusion regarding the long-term effects of global climate change on plant productivity, carbon sequestration, and associated changes in water supply, diseases, and pests is not a string of catastrophic predictions but an honest acknowledgment that our ignorance still trumps our knowledge. That is why I was glad to see (rather exceptionally, as most of these studies tend to have indefensibly alarming conclusions) the authors of a recent multidisciplinary appraisal of the impacts of climate change on China's water resources and agriculture admitting that "current understanding does not allow a clear assessment of the impact of anthropogenic climate change" (Piao et al. 2010a, 43).

The ocean's primary productivity will be affected by both the warming and the acidification. Bathythermograph measurements show a robust warming of the global upper ocean, a trend amounting to $0.64 \ W/m^2$ between 1993 and 2008 (Lyman et al. 2010). In 2006, Behrenfeld et al. (2006) analyzed satellite data from an ocean color sensor and found an initial brief increase in NPP (1.93 Gt C/year), followed by a prolonged decrease at an annual rate of 190 Mt C. Subsequently, Boyce, Lewis, and Worm (2010) combined ocean transparency measurements and in situ chlorophyll observations to estimate long-term trends of the global phytoplankton biomass since 1899.

In addition to the expected interannual and decadal fluctuations (observable thanks to the post-1978 satellite monitoring of phytoplankton), they concluded that phytoplankton mass has declined in eight out of ten ocean regions, and put the rate of decrease at about 1%/year. These temperature-driven declines in marine NPP should have already affected the productivity of marine zoomass, and if the trend were to continue it would amount to a fundamental shift in the biosphere productivity. Even if this trend were to slow down or reverse, other concerns would remain. Many fisheries have shown large-scale decadal and annual stock fluctuations (Laevastu 1993), and their future amplitudes may increase.

Marine fishes could be affected by a growing disparity between their demand for oxygen and the capacity of warmer waters to meet it, as well as by the freshening of Arctic waters, the changes in nutrient supply brought by upwelling, more persistent pathogens, and the loss of coral habitats as a result of the progressive acidification of the ocean (Parry et al. 2007; Pörtner and Knust 2007). Another concern has been the increase in the number of reported oxygen minimum zones whose existence, and persistence, limit the vertical habitat available for pelagic organisms (Bertrand, Ballo, and Chaigneau 2010). On the other hand, recent reports of changing NPP have included a positive effect of stronger monsoon winds blowing over the western Arabian Sea: because of the increased nutrient upwelling, the average summertime phytoplankton mass was boosted by more than 350% (Goes et al. 2005).

The future extent and pace of natural responses and feedbacks will be strongly influenced by human actions, and the future rate of population growth, pace of technical innovation, the progress of economic development, changes in per capita consumption, and the degree of readiness to forgo some material advances in order to preserve the biosphere's irreplaceable services will be among the most obvious factors on the list of technical, economic, and social factors that will determine the eventual balance between the biosphere and modern civilization. If current appropriations were already as high as 35%–45%, then future harvest gains would push them easily past the 50% mark, leaving less than half of the Earth's primary productivity outside human reach and imperiling the provision of many environmental services. Even if the actual claim is still no more than 20%, its qualitative impact has already been substantial (Millennium Ecosystem Assessment 2005), and it would be desirable not only to prevent its further expansion but to achieve its gradual reduction.

Assessing the likely trajectories of all major socioeconomic factors in some detail could be a topic for a number of books: in closing this one I will briefly concentrate only on the two categories of human action that will most directly affect the biosphere's productivity and storage, the future rate of vegetation losses and the level of agricultural output. More correctly, the first concern is with the net effect of two countervailing trends: on one hand, the continuing deforestation and conversions of other natural ecosystems to farmland and urban and industrial uses, harvests of fuelwood and roundwood, and new liquid and gaseous biofuels; on the other hand, continuing afforestation, more effective protection of existing forests, grasslands, and wetlands (in parks, reservations, or carefully managed areas), and natural reclamation of previously cultivated landscapes.

The only development that could slow down or even reverse the European, North American, or Japanese forest expansion would be the (unlikely) large-scale conversion of marginal farmland and currently forested land to biofuel plantations. In contrast, the best outcome in most tropical and some subtropical countries would be to have slower rates of deforestation and larger areas of remaining forests put under effective protection: that the overall net loss of forest phytomass will be most likely considerably lower during the next 50 years than it was during the past half century is obviously welcome, but in light of a still high absolute reduction (most likely in excess of 500 Mt C/year), it still remains a matter of serious concern because primary forests are irreplaceable repositories of tropical biodiversity (Gibson et al. 2011). On the other hand, it has now become clearer that productivity is a poor predictor of plant species richness (Adler et al. 2011), and hence the loss of some highly productive forests may not be associated with a similarly high loss of biodiversity.

In low-income countries the greatest savings of natural woody phytomass could come from the combination of four key strategies: the adoption of efficient rural stoves (such as those that are now widely used in China), a substantial reduction of illegal logging, a major expansion of suitable tree plantations, and the willingness of affluent countries to subsidize effective protection of the richest forest ecosystems in well-managed national parks. In high-income nations the most obvious routes toward reduced wood consumption are whole-tree utilization and the expanded production of engineered timber (Williamson 2001), higher rates of paper recycling (McKinney 1994), and a further shift from paper-based files to purely electronic records.

Because of the importance of agriculture in ecosystem transformations, the net outcome will depend largely on the countervailing effects of cropping intensification (higher yields requiring smaller planted areas) and a rising demand for food and feed crops (owing to the increasing global population and the worldwide dietary transition and higher demand for animal foods). Cropping intensification—not only higher yields and more frequent multicropping, but more recently also a greater extent of more energy- and labor-intensive cultivation in greenhouses or under plastic shields—has been the key reason limiting the demand for new cultivated land.

With average crop yields remaining at the 1900 level, the crop harvest in the year 2000 would have required nearly four times more land, and its total (nearly 60 Mkm^2) would have claimed nearly half of all ice-free continental area rather than less than 15% the agricultural lands claim today. A different perspective has been quantified by Burney, Davis, and Lobell (2010): they calculated that between 1961 and 2005, increased crop yields had a net effect of avoiding emissions of up to 161 Gt C. There is no doubt that this intensification will continue virtually everywhere, but its national and regional rates will continue to differ.

Most of the world, and Africa in particular, has average crop yields well below realistically achievable performance: Africa's recent average yields for corn, sorghum, and wheat have been 65%, 30%, and 25% below their respective global means (FAO 2011d). And even the most intensively farmed countries, whose best yields are much closer to the highest recorded rates, can do better (Sylvester-Bradley and Wiseman 2005). For example, even in the UK, where the average yield of wheat (main staple grain) is already high at 7.7 t/ha and that of rapeseed (main oil crop) is 3.2 t/ha, the application of management and genetic improvements based only on existing knowledge could raise these means to, respectively, 8.7 and 3.9 t/ha (Spink et al. 2009). Even then the long-term possibilities for genetic engineering would remain large: theoretical yield potentials for wheat and rapeseed are, respectively, 19.2 and 9.2 t/ha.

New ways to increase yields will be necessary because there are obvious biophysical limits to further increases in HI values: minimum needs for indispensable structural (roots and stems) and photosynthetic tissues (leaves) will make it impossible to increase them much beyond their current level. The most likely maxima for HI in cereals are between 0.60 and 0.65, as it would be impossible to support more than 65% of the total yield as grain on less than 35% of the overall phytomass. However, values up to 0.80 may be achieved with some root crops (Hay 1995). Future yield increases will thus have to come largely from increased production of total phytomass, which is limited by the amount of intercepted solar radiation and by the maximum rate of photosynthetic conversion, which is higher inherently higher in C_4 crops.

Maxima of new phytomass production in unstressed crops are about 1.7 g/MJ of intercepted solar radiation for corn, 1.4 g/MJ for rice, and 1.2 g/MJ for soybeans (Smil 2000). In terms of actual yields, the first rate would translate to harvests of up to 35 t/ha in the Corn Belt, the second one to yields of up to 25 t/ha of rice in East China. Actual maxima, even for crops grown under optimum conditions, will be considerably lower, but potential maxima can be judged by looking at the record harvests: 22 t/ha for corn (in Michigan in 1977), 14.1 t/ha for wheat (in Washington in 1965), and 13.2 t/ha for rice (in Japan in 1981). Of course, all of them required exceptionally high applications of nitrogenous fertilizers.

What difference future yield increases of staple grains will make will depend on the rate of population growth and on the progress, and eventual saturation levels, of the worldwide dietary transition, that is, primarily on the future rates of increased meat consumption. Although the UN has lowered every one of its regular long-term projections of the world's population, and although there is a fairly good probability that the global total will not surpass nine billion, the latest medium variant forecast is for 9.15 billion in the year 2050, nearly a third above the 2010 aggregate of 6.9 billion (UN 2011). Virtually all of this growth will take place in low- and medium-income countries: European and Japanese populations will shrink, North America might add about 100 million people, but Asia's total will go up by at least one billion, and so will Africa's population, nearly doubling to two billion.

Many worrisome scenarios can be constructed using these figures. For example, if Africa's average per capita meat consumption were to reach only today's Asian mean (going from about 16 kg/year in 2010 to 28 kg/year by 2050), the continent's total meat supply would have to more than triple in four decades. Even worse, if Asia in 2050 were to reach Europe's 2010 mean (that is, going from about 28 to 77 kg/capita), it would have to raise its meat supply roughly 3.5 times. Neither of these achievements is likely, and while the North American, European, and Japanese intakes of animal foods have reached their respective saturation points, even conservative expectations mean additional demand increases on the order of 30%–50% in both Asia and Africa. Such increases may not be fully met by higher yields and may require further expansion of cropland for feed production.

And rich but land-poor countries may add to this demand. Rich food importers, particularly the Middle Eastern hydrocarbons-exporting states (Kuwait, Saudi Arabia, Qatar), have been buying up large tracts of land abroad to grow food crops. They have been joined by China (which secured nearly 3 Mha abroad between 2006 and 2009) and South Korea (whose failed deal in Madagascar aimed at growing rice on 1.3 Mha), while British investors have shown some interest in acquiring foreign land for the production of biofuels (the UK to grow jatropha in Ethiopia and Mozambique). What is new about these deals is their size: 1.5 Mha in Sudan (including 690,000 ha for South Korea and 400,000 ha for the United Arab Emirates), 2 Mha in Zambia, and 2.8 Mha in the Republic of Congo (Knaup and Mittelstaedt 2009). Are these just a few exceptional events, or does this represent the beginning of a new megafarming trend on behalf of foreigners?

The production of liquid biofuels could further add to these claims. Irrational as it may be from environmental, economic, and technical points of view, a determination to expand the output of ethanol, biodiesel, and even biokerosene for jet flight could become a major contributor to further losses of natural ecosystems to cropland—and some fantasies even envision adding another 2.4 Gha of rain-fed arable land (on top of the already cultivated 1.5 Gha), most of it in the tropics, to avoid the continuing reliance on fossil fuels (Read 2008). Perhaps the most charitable assessment of such calls is what Marland and Obersteiner (2008, 335) wrote in an editorial: "But it is not now clear if his [Read's] vision is a dream or a nightmare." I do not hesitate to label it a vision beyond nightmare. Moreover, as Fargione et al. (2008) showed, such massive land conversions (mostly in Brazil, Southeast Asia, and the United States) could also release one to two orders of magnitude more CO_2 into the atmosphere than the mass of the gas displaced by using phytomass-based rather than refined oil liquids.

At the same time, there are enormous opportunities for reducing not only postharvest (storage, processing, and distribution) phytomass losses but also unacceptably high household food waste losses. Annual losses are about 100 kg/capita in

Europe and North America, with most food lost during the consumption stage: more than 10% for meat, about 25% for cereals, and 20%–30% for vegetables; absolute losses are much smaller in low-income countries, where they are dominated by waste in the postharvest processing and distribution stages, that can reduce foodstuffs reaching households by as much as 20% for cereals and 30% for tubers and seafood (Gustavsson et al. 2011). What this means in food energy terms can be appreciated by comparing supply (from food balance sheets) with actual intakes: in the EU and North America, the per capita rates for supply are in excess of 3,000 kcal/day, the rates for actual intake are 30%–40% lower.

A more accurate approach to quantifying this waste was used by Hall et al. (2009), who modeled metabolic and activity requirements in order to find the most likely food intake of the U.S. population between 1974 and 2003. That rate ranged from about 2,100 to 2,300 kcal/day, but during the same period the average food supply rose from about 3,000 to 3,700 kcal/day, and the country's food waste thus increased from 28% of the total retail supply in 1974 to about 40% three decades later. Similarly, a detailed survey in the UK found that British households waste about 31% of purchased food (WRAP 2009). And even Japan, the least wasteful of all affluent countries, now loses about 25% of its total daily food supply (Smil and Kobayashi 2011), and rising food losses are reported from China. The irrationality of wasting 20%–40% of all produced food is self-evident, particularly given the rising incidence of obesity.

Conditions needed to minimize human claims on the biosphere's productivity are not difficult to summarize:

- The global population should be stabilized at less than nine billion people.
- Food should be produced using the best agronomic practices, including optimized irrigation, fertilization, and the use of pesticides, reduced tillage, and crop rotations rather than monoculture.
- The average per capita food requirements should be limited to levels necessary to support healthy lives of decent longevity rather than to support excessive carnivory and obesity-inducing diets.
- Much more attention should be paid to postharvest food losses and to household food waste.
- We should pursue such difficult but highly rewarding challenges as developing cereal permacultures and staple grain crops able to fix nitrogen.
- Wood output should be limited by the long-term productive capacities of forest ecosystems and properly managed plantations.

- Wood demand should be set by maintaining sensible material comforts rather than by striving to perpetuate extravagant consumption delivered in wasteful throw-away packaging.
- As in the case of crops, we should be supporting research leading to new tree cultivars that will combine superior productivity with hardiness.
- And, unrealistic as it may seem, rational resource management on the global level should eventually incorporate a degree of sharing that would reduce the immensity of existing inter- and intranational inequalities.

Even a partial fulfillment of these rational desiderata would lead to an aggregate global demand for primary production in the year 2050 only marginally (25%–35%) higher than today's harvest of phytomass. In contrast, if the global means of resource claims were to rise to the level of per capita consumption that now prevails in the richest countries—U.S. meat intakes are now three times the global mean, Japanese seafood consumption is 3.5 times the world average, and the EU's consumption of industrial roundwood is three times the global mean—then the doubling or even tripling of global phytomass and zoomass harvests would be necessary.

But there could be no tripling of global phytomass harvests by 2050, and although doubling is not an impossible total to contemplate, it, too, is not very likely. Even the FAO's liberal scenarios foresee no global doubling of total food production, although they assume 9.1 billion people in 2050 and almost 9.4 billion in 2080 and anticipate a very high average food supply of almost 3,200 kcal/day (FAO 2012). They envisage a worldwide agricultural production increase of some 60% by 2050 and nearly 80% by 2080, although the latter achievement would require a 100% increase in overall agricultural output in today's low-income countries, as well as the doubling of 2010 global meat production.

Caution, foresight, and determination in pursuit of rational solutions could help us to keep the level of global harvests within acceptable limits. The continuation of many irrational trends and the pursuit of patently unrealistic expectations guarantee the opposite. The results are not preordained and the worst expectations are not inevitable, but to avoid many undesirable outcomes, the world's richest nations will have to modify their expectations and moderate their claims. If the billions of poor people in low-income countries were to claim even half the current per capita harvests prevailing in affluent economies, too little of the Earth's primary production would be left in its more or less natural state, and very little would remain for mammalian species other than ours.

Scientific Units and Prefixes

Units

˚C	degree of Celsius	temperature
g	gram	mass
J	joule	energy
m	meter	length
m^2	square meter	area
m^3	cubic meter	volume
t	tonne (metric ton)	mass
W	watt	power

Prefixes

m	milli	10^{-3}
c	centi	10^{-2}
h	hecta	10^2
k	kilo	10^3
M	mega	10^6
G	giga	10^9
T	tera	10^{12}
P	peta	10^{15}
E	exa	10^{18}
Z	zeta	10^{21}
Y	yota	10^{24}

References

Abboe, S., et al. 2009. Reconsidering domestication of legumes versus cereals in the ancient Near East. *Quarterly Review of Biology* 84:30–50.

Abel, W. 1962. *Geschichte der deutschen Landwirtschaft von frühen Mittelalter bis zum 19 Jahrhundert*. Stuttgart: Ulmer.

Adams, J. M., et al. 1990. Increases in terrestrial carbon storage from the Last Glacial Maximum to the present. *Nature* 348:711–714.

Adams, J. M., and H. Faure. 1998. A new estimate of changing carbon storage on land since the last glacial maximum, based on global land ecosystem reconstruction. *Global and Planetary Change* 16–17:3–24.

Adas, M. 2001. *Agricultural and Pastoral Societies in Ancient and Classical History*. Philadelphia: Temple University Press.

Adler, P. B. et al. 2011. Productivity is a poor predictor of plant species richness. *Science* 333:1750–1753.

Agnew, D. J., et al. 2009. Estimating the worldwide extent of illegal fishing. *PLoS ONE* 4 (2): e4570. doi:10.1371/journal.pone.0004570.

Aiello, L. C., and P. Wheeler. 1995. The expensive-tissue hypothesis. *Current Anthropology* 36:199–221.

Ajtay, G. L., et al. 1979. Terrestrial primary production and phytomass. In *The Global Carbon Cycle*, ed. B. Bolin et al., 129–181. New York: Wiley.

Allan, J. D. 1976. Life history patterns in zooplankton. *American Naturalist* 110:165–180.

Allan, W. 1965. *The African Husbandman*. Edinburgh: Oliver & Boyd.

Allen, J. R. M. 2010. Last glacial vegetation of northern Eurasia. *Quaternary Science Reviews* 29:2604–2618.

Allen, R. C. 2007. *How Prosperous Were the Romans? Evidence from Diocletian's Price Edict (301 AD)*. Oxford: Oxford University, Department of Economics.

Alroy, J. 2001. A multispecies overkill simulation of the end-Pleistocene megafaunal mass extinction. *Science* 292:1893–1896.

Alvard, M. S., and L. Kuznar. 2001. Deferred harvests: The transition from hunting to animal husbandry. *American Anthropologist* 103:295–311.

Alverson, D. L., et al. 1994. *A Global Assessment of Fisheries Bycatch and Discards.* Rome: FAO.

Alverson, D. L. 2005. Managing the catch of non-target species. In *Improving Fishery Management: Melding Science and Governance,* ed. W. S. Wooster and J. M. Quinn. Seattle: University of Washington, School of Marine Affairs.

Ambrose, S. H. 1998. Late Pleistocene human population bottlenecks: Volcanic winter, and the differentiation of modern humans. *Journal of Human Evolution* 34:623–651.

Amthor, J. S., et al. 1998. *Terrestrial Ecosystems.* Oak Ridge, TN: Oak Ridge National Laboratory.

Amthor, J. S., and D. D. Baldocchi. 2001. Terrestrial higher plant respiration and net primary production. In *Terrestrial Global Productivity,* ed. J. Roy, B. Saugier, and H. A. Mooney, 33–59. San Diego, CA: Academic Press.

Antoine, D., et al. 1996. Oceanic primary production. 2. Estimation at global scale from satellite (coastal zone color scanner) chlorophyll. *Global Biogeochemical Cycles* 10:57–69.

Antongiovanni, M., and C. Sargentini. 1991. Variability in chemical composition of straws. *Options Méditerranéennes* 16:49–53.

Armelagos, G. J., and K. N. Harper. 2005. Genomics at the origins of agriculture: Part one. *Evolutionary Anthropology* 14:68–77.

Arpi, G. 1953. The supply with charcoal of the Swedish iron industry from 1830 to 1950. *Geografiska Annaler* 35:11–27.

Assibey, E. O. A. 1974. Wildlife as a source of protein in Africa south of the Sahara. *Biological Conservation* 6:32–39.

Atalay, S., and C. A. Hastorf. 2006. Food, meals, and daily activities: Food *habitus* at Neolithic Çatalhöyük. *American Antiquity* 71:283–319.

Atkinson, A., et al. 2009. A re-appraisal of the total biomass and annual production of Antarctic krill. *Deep-sea Research. Part I, Oceanographic Research Papers* 56:727–740.

Augustine, D. J., and S. K. McNaughton. 2006. Interactive effects of ungulate herbivores, soil fertility, and variable rainfall on ecosystem processes in a semiarid savanna. *Ecosystems (New York.)* 9:1242–1256.

Baars, C. 1973. *De geschiedenis van de landbouw in de Beijerlanden.* Wageningen: Centrum voor landbouwpublikaties en landbouwdocumentatie.

Baes, C. F., et al. 1976. *The Global Carbon Dioxide Problem.* Oak Ridge, TN: Oak Ridge National Laboratory.

Bailey, R. C., et al. 1989. Hunting and gathering in tropical rain forest: Is it possible? *American Anthropologist* 91:59–82.

Bailey, R. C., and T. N. Headland. 1991. The tropical rain forest: Is it a productive environment for human foragers? *Human Ecology* 19:261285.

Ballantyne, A. P., et al. 2012. Increase in observed net carbon dioxide uptake by land and oceans during the past 50 years. *Nature* 488:70–72.

Barbosa, P. M., D. Stroppiana, and J.-M. Grégoire. 1999. An assessment of vegetation fire in Africa (1981–1991): Burned areas, burned biomass, and atmospheric emissions. *Global Biogeochemical Cycles* 13:933–949.

Barnosky, A. D. 2008. Megafauna biomass tradeoff as a driver of Quaternary and future extinctions. *Proceedings of the National Academy of Sciences of the United States of America* 105:11543–11548.

Bar-Yosef, O. 2002. The Upper Paleolithic revolution. *Annual Review of Anthropology* 31:363–393.

Barreveld, W. H. 1989. *Rural Use of Lignocellulosic Residues*. Rome: FAO.

Barros, E., et al. 2002. Effects of land-use system on the soil macrofauna in western Brazilian Amazonia. *Biology and Fertility of Soils* 35:338–347.

Bassino, J. 2006. The growth of agricultural output, and food supply in Meiji Japan: Economic miracle or statistical artifact? *Economic Development and Cultural Change* 54:503–521.

Bath, D., et al. 1997. Byproducts and unusual feedstuffs. *Feedstuffs* 69 (30): 32–38.

Battle, M., et al. 2000. Global carbon sinks and their variability from atmospheric O_2 and $\delta^{13}C$. *Science* 287:2467–2470.

Bavington, D. L. Y. 2010. *Managed Annihilation: An Unnatural History of the Newfoundland Cod Collapse*. Vancouver: University of British Columbia Press.

Bazilevich, N. I., L. E. Rodin, and N. N. Rozov. 1971. Geographical aspects of biological productivity. *Soviet Geography* 12:293–317.

BBC News. 2002. Rare species campaign targets chefs. *BBC News*, February 1. http://news.bbc.co.uk/2/hi/asia-pacific/1795388.stm.

BBC News. 2012. Japan tuna sale smashes record. *BBC News*, January 5. http://www.bbc.co.uk/news/world-asia-pacific-16421231.

Beer, C., et al. 2006. Small net carbon dioxide uptake by Russian forests during 1981–1999. *Geophysical Research Letters* 33 (15): L15403

Beer, C., et al. 2010. Terrestrial gross carbon dioxide uptake: Global distribution and covariation with climate. *Science* 329:834–838.

Beerling, D. J. 1999. New estimates of carbon transfer to terrestrial ecosystems between the last glacial maximum and the Holocene. Terra Nova 11:162–167.

Behrenfeld, M. J., and P. G. Falkowski. 1997. Photosynthetic rates derived from satellite-based chlorophyll concentration. *Limnology and Oceanography* 42:1–20.

Behrenfeld, M. J., et al. 2006. Climate-driven trends in contemporary ocean productivity. *Nature* 444:752–755.

Bellamy, P. H., et al. 2005. Carbon losses from all soils across England and Wales 1978–2003. *Nature* 437:245–248.

Bennett, M. K. 1935. British wheat yield per acre for seven centuries. *Economic History* 3 (10):12–29.

Bermejo, M. 1999. Status and conservation of primates in Odzala National Park, Republic of the Congo. *Oryx* 33:323–331.

Bernhardt, A. 1966. *Geschichte des Waldeigentums, der Waldwirtschaft und Forstwissenschaft in Deutschland*. Aalen: Scientia Verlag.

Bernstein, I. S., and E. O. Smith, eds. 1979. *Primate Ecology and Human Origins: Ecological Influences on Social Organization*. New York: Garland STPM Press.

Bertrand, A., M. Ballo, and A. Chaigneau. 2010. Acoustic observation of living organisms reveals the upper limit of the oxygen minimum zone. *PLoS ONE* 5 (e10330): 1–9.

Best, P. B. 1983. Sperm whale stock assessments and the relevance of historical whaling records. *Report of the International Whaling Commission* 5 (Special Issue): 41–55.

Bhardwaj, P. 2006. Threatened tiger, sickly dragon. *Asia Times*, March 21.

Bifarin, J. O., M. E. Ajibola, and A. A. Fadiyimu. 2008. Analysis of marketing bush meat in Idanre Local Government area of Ondo state, Nigeria. *African Journal of Agricultural Research* 3:667–671.

Birdsey, R. A. 1996. Carbon storage in United States forests. In *Forests and Global Change*, vol. 2, ed. N. Sampson and D. Hair, 1–25. Washington, DC: American Forests.

Biringuccio, V. 1990. *1540: Pirotechnia*, trans. C. S. Smith and M. T. Gnudi. New York: Dover Publications.

Blanc, J. J., et al. 2007. *African Elephant Status Report 2007: An Update from the African Elephant Database*. Gland: IUCN.

Blanco-Canqui, H., et al. 2007. Soil hydraulic properties influenced by corn stover removal from no-till corn in Ohio. *Soil and Tillage Research* 92:144–155.

Bloch, M. 1931. *Les caractères originaux de l'histoire rurale française*. Paris: Librairie Aramnd Colin.

Boesch, C. 1994. Chimpanzees—red colobus: A predator-prey system. *Animal Behaviour* 47:1135–1148.

Boesch, C., and H. Boesch-Achermann. 2000. *The Chimpanzees of the Taï Forest*. Oxford: Oxford University Press.

Bolin, B. 1970. The carbon cycle. *Scientific American* 223:125–132.

Bolin, B., et al., eds. 1979. *The Global Carbon Cycle*. New York: Wiley.

Bolin, B., et al. 2001. Global perspective, in land use, land-use change, and forestry. In *A Special Report of the IPCC*, ed. R. T. Watson et al., 23–52. Cambridge: Cambridge University Press.

Bond-Lamberty, B., et al. 2007. Fire as the dominant driver of central Canadian boreal forest carbon balance. *Nature* 450:89–92.

Bond-Lamberty, B., and A. Thomson. 2010. Temperature-associated increases in the global soil respiration record. *Nature* 464:579–582.

Bondeau, A., et al. 2007. Modelling the role of agriculture for the 20th century global terrestrial carbon balance. *Global Change Biology* 13:679–706.

Bousquet, P., et al. 2000. Regional changes in carbon dioxide fluxes of land and oceans since 1980. *Science* 290:1342–1346.

Bowen, H. J. M. 1966. *Trace Elements in Biochemistry*. London: Academic Press.

Bowman, D. M. J. S., et al. 2009. Fire in the Earth system. *Science* 324:481–484.

Boyce, D. G., M. R. Lewis, and B. Worm. 2010. Global phytoplankton decline over the past century. *Nature* 466:591–596.

BP (British Petroleum). 2011. *BP Statistical Review of World Energy*. http://www.bp.com/assets/bp_internet/globalbp/globalbp_uk_english/reports_and_publications/statistical_energy _review_2011/STAGING/local_assets/pdf/statistical_review_of_world_energy_full_report _2011.pdf.

Bradford, J. B., W. K. Lauenroth, and I. C. Burke. 2005. The impact of cropping on primary production in the U.S. Great Plains. *Ecology* 86:1863–1872.

Branch, E. D. 1929. *The Hunting of the Buffalo*. Lincoln: University of Nebraska Press.

Branch, E. D. 1997. *The Hunting of the Buffalo*. Lincoln: University of Nebraska Press.

Branch, T. A., et al. 2010. The trophic fingerprint of marine fisheries. *Nature* 468:431–435.

Brasier, C., and J. Webber. 2010. Sudden larch death. *Nature* 466:824–825.

Bratbak, G., and I. Dundas. 1984. Bacterial dry matter content and biomass estimations. *Applied and Environmental Microbiology* 48:755–757.

Bray, W. 1977. From foraging to farming in early Mexico. In *Hunters, Gatherers and First Farmers Beyond Europe*, ed. J. V. S. Megaw, 225–250. Leicester: Leicester University Press.

Bressan, A., and E. Contini. 2007. Brazil: A pioneer in biofuels. Paper presented at the workshop, "Global Biofuel Developments: Modelling the Effects on Agriculture." Washington, DC, February 27. http://www.farmfoundation.org/projects/documents/Brazilpanelpres.pdf.

Brian, M. V., ed. 1978. *Production Ecology of Ants and Termites*. Cambridge: Cambridge University Press.

Brook, B. W., and D. M. J. S. Bowman. 2004. The uncertain blitzkrieg of Pleistocene megafauna. *Journal of Biogeography* 31:517–523.

Brown, K. S., et al. 2009. Fire as an engineering tool of early modern humans. *Science* 325:859–862.

Brown S., and A. E. Lugo. 1982. The storage and production of organic matter in tropical forests and their role in the global carbon cycle. *Biotropica* 14:161–187.

Brown, S., P. Schroeder, and R. Birdsey. 1997. Aboveground biomass distribution of US eastern hardwood forests and the use of large trees as an indicator of forest development. *Forest Ecology and Management* 96:31–47.

Brum, J. R. 2005. Concentration, production and turnover of viruses and dissolved DNA pools at Stn ALOHA, North Pacific Subtropical Gyre. *Aquatic Microbial Ecology* 42: 103–113.

Buck, E. H. 1995. *Atlantic Bluefin Tuna: International Management of a Shared Resource. CRS Report for Congress*. Washington, DC: U.S. Congress.

Buck, J. L. 1930. *Chinese Farm Economy*. Nanking: University of Nanking.

Buck, J. L. 1937. *Land Utilization in China*. Nanjing: University of Nanking.

Buerk, R. 2010. Tuna hits highest price in nine years at Tokyo auction. *BBC News*. http://news.bbc.co.uk/2/hi/8440758.stm.

Burney, J., S. J. Davis, and D. B. Lobell. 2010. Greenhouse gas mitigation by agricultural intensification. *Proceedings of the National Academy of Sciences of the United States of America* 107:12052–12057.

Butchart, S. H. M., et al. 2010. Global biodiversity: Indicators of recent declines. *Science* 328:1164–1168.

Butzer, K. W. 1976. *Early Hydraulic Civilization in Egypt*. Chicago: University of Chicago Press.

Cachel, S. 1997. Dietary shifts and the European Upper Paleolithic transition. *Current Anthropology* 38:579–603.

Caddy, J. F., et al. 1998. How pervasive is "fishing down marine food webs"? *Science* 282:1383.

Cairns, M. A., et al. 1997. Root biomass allocation in the world's upland forests. *Oecologia* 111:1–11.

Campbell, M. S., and M. Overton. 1993. A new perspective on medieval and early modern agriculture: Six centuries of Norfolk farming c.1250–c.1850. *Past & Present* 141: 38–105.

Campbell-Platt, G. 1980. African locust bean (*Parkia* species) and its West African fermented food product dawadawa. *Ecology of Food and Nutrition* 9:123.

Campos, P. F., et al. 2010. Ancient DNA analyses exclude humans as the driving force behind the late Pleistocene musk ox (*Ovibos moschatus*) population dynamics. *Proceedings of the National Academy of Sciences of the United States of America* 107:5675–5680.

Cao, M., and F. I. Woodward. 1998. Net primary and ecosystem production and carbon stocks of terrestrial ecosystems and their response to climate change. *Global Change Biology* 4:185–198.

Carder, A. 1995. *Forest Giants of the World: Past and Present*. Markham, ON: Fitzhenry & Whiteside.

Carey, E. V., et al. 2001. Are old forests underestimated as global carbon sinks? *Global Change Biology* 7:339–344.

Carson, W. P., and S. A. Schnitzer, eds. 2008. *Tropical Forest Community Ecology*. Oxford: Blackwell.

Carter, W. E. 1969. *New lands and Old Traditions: Kekchi Cultivators in the Guatemalan Lowlands*. Gainesville: University of Florida Press.

Chadare, F. J., et al. 2009. Baobab food products: A review of their composition and nutritional value. *Critical Reviews in Food Science and Nutrition* 49:254–274.

Chakrabarti, D. K. 1992. *The Early Use of Iron in India*. New Delhi: Oxford University Press.

Chang, K. C., ed. 1977. *Food in China's Culture*. New Haven, CT: Yale University Press.

Chapin, F. S., P. Matson, and H. A. Mooney. 2002. *Principles of Terrestrial Ecosystem Ecology*. New York: Springer-Verlag.

Chapin, F. S., et al. 2006. Reconciling carbon-cycle concepts, terminology, and methods. *Ecosystems (New York)* 9:1041–1050.

Chessa, B., et al. 2009. Revealing the history of sheep domestication using retrovirus integrations. *Science* 324:531–536.

Childe, V. G. 1951. The Neolithic revolution. In *Man Makes Himself*, ed. V. G. Childe, 67–72. London: C. A. Watts.

Chorley, G. P. H. 1981. The agricultural revolution in Northern Europe, 1750–1880: n = Nitrogen, legumes, and crop productivity. *Economy and History* 34:71–93.

Christanty, L., D. Mailly, and J. P. Kimmins. 1996. "Without bamboo, the land dies": Biomass, litterfall, and soil organic matter dynamics of a Javanese bamboo talun-kebun system. *Forest Ecology and Management* 87:75–88.

Christensen, L. B. 2006. *Marine Mammal Populations: Reconstructing Historical Abundances at the Global Scale.* Vancouver, BC: University of British Columbia, Fisheries Centre.

Christiansen, P. 2004. Body size in proboscideans, with notes on elephant metabolism. *Zoological Journal of the Linnean Society* 140:523–549.

Churkina, G., D. G. Brown, and G. Keolian. 2010. Carbon stored in human settlements: The coterminous United States. *Global Change Biology* 16:135–143.

Ciais, P., et al. 2005. Europe-wide reduction in primary productivity caused by the heat and drought in 2003. *Nature* 437:529–533.

Ciais, P., et al. 2010. The European carbon balance. Part 2. Croplands. *Global Change Biology* 16:1409–1428.

City of Winnipeg. 2011. *Winnipeg Tree Facts.* Winnipeg: Public Works Department. http://winnipeg.ca/publicworks/Forestry/PAMPHLETS/wpgtreefctspamph%20(2).pdf.

Clark, D. A. 2002. Are tropical forests an important carbon sink? Reanalysis of the long-term plot data. *Ecological Applications* 12:3–7.

Clark, G., M. Huberman, and P. Lindert. 1995. A British food puzzle, 1770–1850. *Economic History Review* 48:215–237.

Claussen, M., et al. 2005. Did humans prevent a Holocene glaciation? *Climatic Change* 69:409–417.

Clout, H. 1983. *The Land of France.* London: George Allen & Unwin.

Cochrane, M. A., and M. D. Schulze. 1999. Fire as a recurrent event in tropical forests of the eastern Amazon: Effects on forest structure, biomass, and species composition. *Biotropica* 31:2–16.

Coe, M. J., D. H. Cumming, and J. Phillipson. 1976. Biomass and production of large African herbivores in relation to rainfall and primary production. *Oecologia* 22:341–354.

Coleman, D. C., and D. A. Crossley. 1996. *Fundamentals of Soil Ecology.* San Diego, CA: Academic Press.

Condie, K. C. 1997. *Plate Tectonics and Crustal Evolution.* Oxford: Butterworth Heinemann.

Conklin, H. C. 1957. *Hanunoo Agriculture.* Rome: FAO.

Connell, J. H., and M. D. Lowman. 1989. Low-diversity tropical rain forests: Some possible mechanisms for their existence. *American Naturalist* 134:88–119.

Consigny, T. 2010. *Yasei Kinoko—Wild Mushrooms of Japan.* http://www.kamimoku.com/index_fungi.html.

Constantinou, G. 1982. Geological features and ancient exploitation of cupriferous sulphide ore bodies in Cyprus. In *Early Metallurgy in Cyprus, 4000–500 BC,* ed. J. D. Muhly, R. Maddin, and V. Karageorghis, 12–34. Nicosia: Pierides Foundation.

Cook, R. M., A. Sinclair, and G. Stefansson. 1997. Potential collapse of North Sea cod stocks. *Nature* 385:521–522.

Cornwell, W. K., et al. 2009. Plant traits and wood fates across the globe: Rotted, burned, or consumed? *Global Change Biology* 15:2431–2449.

Costa-Leonardo, A. M., F. E. Casarin, and J. Ferreira. 2003. Estimates of foraging population of *Heterotermes tenuis* colonies using mark-release-recapture. *Sociobiology* 42:807–814.

Coughenour, M. B., et al. 1985. Energy extraction and use in a nomadic pastoral ecosystem. *Science* 230:619–624.

Cramer, W., et al. 1999. Comparing global models of terrestrial net primary productivity (NPP): Overview and key results. *Global Change Biology* 5:1–15.

Crawley, M. J. 1983. *Herbivory: The Dynamics of Animal-Plant Interactions*. Berkeley: University of California Press.

Cristoffer, C., and C. A. Peres. 2003. Elephants versus butterflies: The ecological role of large herbivores in the evolutionary history of two tropical worlds. *Journal of Biogeography* 30:1357–1380.

Crutzen, P. J. 2002. Geology of mankind. *Nature* 415:23.

Crutzen, P. J., A. R. Mosier, and K. A. Smith. 2008. N$_2$O release from agro-biofuel production negates global warming reduction by replacing fossil fuels. *Atmospheric Chemistry and Physics* 8:389–395.

Cuntz, M. 2011. A dent in carbon's gold standard. *Science* 477:547–548.

Currie, D. J., and J. T. Fritz. 1993. Global patterns of animal abundance and species energy use. *Oikos* 67:56–68.

Curtis, W. H. 1919. *Wood Ship Construction*. New York: McGraw-Hill.

Cushing, D. H. 1969. *Upwelling and Fish Production*. Rome: FAO.

Dalrymple, D. G. 1986. *Development and Spread of High-yielding Wheat Varieties in Developing Countries*. Washington, DC: Bureau for Science and Technology.

Damuth, J. 1981. Population density and body size in mammals. *Nature* 290:699–700.

Darby, H. C. 1956. The clearing of the woodland of Europe. In *Man's Role in Changing the Face of the Earth*, ed. W. L. Thomas, 183–216. Chicago: University of Chicago Press.

Datta, A., M. O. Anand, and R. Naniwadekar. 2008. Empty forests: Large carnivore and prey abundance in Namdapha National Park, north-east India. *Biological Conservation* 141:1429–1435.

Davidson, E. A., and I. A. Janssens. 2006. Temperature sensitivity of soil carbon decomposition and feedbacks to climate change. *Nature* 440:165–173.

de Carvalho, C. S. M., C. Sales-Campos, and M. C. N. de Andrade. 2010. Mushrooms of the *Pleurotus* genus: A review of cultivation techniques. *Interciencia* 35:177–182.

De Leo, F. C., et al. 2010. Submarine canyons: Hotspots of benthic biomass and productivity in the deep sea. *Proceedings, Biological Sciences*. doi:10.1098/rspb.2010.0462. http://rspb.royalsocietypublishing.org/content/early/2010/05/04/rspb.2010.0462.full.pdf+html.

De Oliveira, M. E. D., B. E. Vaughan, and E. J. Rykiel. 2005. Ethanol as fuel: Energy, carbon dioxide balances, and ecological footprint. *BioScience* 55:593–602.

De Vooys, C. G. N. 1979. Primary production in aquatic systems. In *The Global Carbon Cycle*, ed. B. Bolin, E. T. Degens, S. Kempe, and P. Ketner, 259–292. New York: Wiley.

de Vos, A. 1977. Game as food. *Unasylva* 29 (111): 2–12.

De Vries, W., et al. 2006. The impact of nitrogen deposition on carbon sequestration in European forests and forest soils. *Global Change Biology* 12:1151–1173.

Deevey, E. S. 1960. The human population. *Scientific American* 203 (3): 194–204.

Demirbaş, A. 2003. Sustainable cofiring of biomass with coal. *Energy Conversion and Management* 44:1465–1479.

Demirbaş, A. 2009. Biofuels from agricultural biomass. *Energy Sources Part A* 31:1573–1582.

Demographia. 2010. International data. http://www.demographia.com.

Denevan, W. M. 1982. Hydraulic agriculture in the American tropics: Forms measures and recent research. In *Maya Subsistence*, ed. K. V. Flannery, 181–203. New York: Academic Press.

Denevan, W. M. 1992. The pristine myth: The landscape of the Americas in 1492. *Annals of the Association of American Geographers, Association of American Geographers* 82:369–385.

Detling, J. K. 1988. Grasslands and savannas: Regulation of energy flow and nutrient cycling by herbivores. In *Concepts of Ecosystem Ecology*, ed. L. R. Pomeroy and J. J. Alberts, 131–148. New York: Springer-Verlag.

Devos, C., et al. 2008. Comparing ape densities and habitats in Northern Congo: Surveys of sympatric gorillas and chimpanzees in the Odzala and Ndoki regions. *American Journal of Primatology* 70:439–451.

Di Blasi, C., C. Branca, and A. Galgano. 2010. Biomass screening for the production of furfural via thermal decomposition. *Industrial & Engineering Chemistry Research* 49: 2658–2671.

Do, T. V., A. Osawa, and N. T. Thang. 2010. Recovery process of a mountain forest after shifting cultivation in Northwestern Vietnam. *Forest Ecology and Management* 259: 1650–1659.

Dolman, A. J., et al. 2010. A carbon cycle science update since IPCC AR-4. *Ambio* 39: 402–412.

Domínguez-Rodrigo, M. 2002. Hunting and scavenging by early humans: The state of the debate. *Journal of World Prehistory* 16:1–54.

Donald, C. M., and J. Hamblin. 1984. The convergent evolution of annual seed crops in agriculture. *Advances in Agronomy* 36:97–143.

Dong, H., et al. 2006. Emissions from livestock and manure management. In *2006 IPCC Guidelines for National Greenhouse Gas Inventories*. Vol. 4, *Agriculture, Forestry and Other Land Use*, 10.1–10.87. http://www.ipcc-nggip.iges.or.jp/public/2006gl/pdf/4_Volume4/V4_10_Ch10_Livestock.pdf.

Doughty, C. E., and C. B. Field. 2010. Agricultural net primary production in relation to that liberated by the extinction of Pleistocene mega-herbivores: An estimate of agricultural carrying capacity? *Environmental Research Letters*. doi:10.1088/1748-9326/5/4/044001.

Dove, M. R. 1985. *Swidden Agriculture in Indonesia: The Subsistence Strategies of the Kalimantan Kantu*. Berlin: Mouton.

Duby, G. 1968. *Rural Economy and Country Life in the Medieval West*. Philadelphia: University of Pennsylvania Press.

Dudley, N. 2011. *Authenticity in Nature: Making Choices about the Naturalness of Ecosystems*. London: Routledge.

Dupin, H., et al. 1984. Evolution of the French diet: Nutritional aspects. *World Review of Nutrition and Dietetics* 44:57–84.

Durgin, F. A. 1962. The virgin lands programme, 1954–1960. *Soviet Studies* 13:255–280.

Duvigneaud, P. 1962. Écosystémes et biosphère. In *L'ècologie, science moderne de synthèse*. Brussels: Ministère de l'éducation nationale et de la culture.

Ebermayer, E. W. F. 1882. *Die Bestandteile der Pflanzen*. Vol. 1, *Naturgesetzliche Grundlagen des Wald- und Ackerbaues*. Berlin: Springer-Verlag.

Eckardt, E. E., ed. 1968. *Functioning of Terrestrial Ecosystems at the Primary Production Level*. Paris: UNESCO.

Eckert, A. W. 1965. *The Silent Sky: The Incredible Extinction of the Passenger Pigeon*. Boston: Little, Brown.

Edmonds, R. L., ed. 1974. An initial synthesis of results in the coniferous biome, 1970–1973. *Bulletin No. 7 Coniferous Forest Biome Ecosystem Analysis Studies*.

Edmonds, R. L., ed. 1982. *Analysis of Coniferous Forest Ecosystems in the Western United States*. New York: Van Nostrand Reinhold.

Edwards, C. A., and J. R. Lofty. 1972. *Biology of Earthworms*. London: Chapman and Hall.

Edwards, D. P. et al. 2010. Degraded lands worth protecting: The biological importance of Southeast Asia's repeatedly logged forests. *Proceedings of the Royal Society, Biology*. doi:10.1098/rspb.2010.1062.

Eglin, T., et al. 2010. Historical and future perspectives of global soil carbon response to climate and land-use changes. *Tellus. Series B, Chemical and Physical Meteorology* 62 (Special Issue):700–718.

Ellis, E. C., and N. Ramankutty. 2008. Putting people in the map: Anthropogenic biomes of the world. *Frontiers in Ecology and the Environment* 6:439–447.

Ellis, K. J. 2000. Human body composition: In vivo methods. *Physiological Reviews* 80: 649–680.

Elser, J. J., et al. 2010. Biological stoichiometry of plant production: Metabolism, scaling and ecological response to global change. *New Phytologist* 186:593–608.

Elvidge, C. D., et al. 2004. U.S. constructed area approaches the size of Ohio. *Eos* 85:233–240.

Elvidge, C. D., et al. 2007. Global distribution and density of constructed impervious surfaces. *Sensors* 7:1962–1979.

Enghoff, I. B., et al. 2007. The Danish fish fauna during the warm Atlantic period (ca. 7000–3900 BC): Forerunner of future change? *Fisheries Research* 87:167–180.

Erwin, T. L. 1982. Tropical forests: Their richness in Coleoptera and other arthropod species. *Coleopterists Bulletin* 36:74–82.

Esser, G. 1987. Sensitivity of global carbon pools and fluxes to human and potential climatic impacts. *Tellus* 39B:245–260.

Evangelou, P. 1984. *Livestock Development in Kenya's Maasailand*. Boulder, CO: Westview Press.

Fa, J. E., and D. Brown. 2009. Impacts of hunting on mammals in African tropical moist forests: A review and synthesis. *Mammal Review* 39:231–264.

Fahey, T. J., and A. K. Knapp. 2007. *Principles and Standards for Measuring Primary Production*. Oxford: Oxford University Press.

Fang, J., et al. 2005. Biomass carbon accumulation by Japan's forests from 1947 to 1995. *Global Biogeochemical Cycles* 19:1–10.

FAO. 2001. *Global Forest Resources Assessment 2000*. Rome: FAO. http://www.fao.org/docrep/004/y1997e/y1997e00.htm.

FAO. 2005. *Global Forest Resources Assessment 2005*. Rome: FAO. http://www.fao.org/docrep/008/a0400e/a0400e00.htm.

FAO. 2010. *Euphausia superba*. FAO Species Fact Sheets. Rome: FAO.

FAO. 2011a. FAOSTAT: FisheriesSTAT. http://www.fao.org/fishery/statistics/en.

FAO. 2011b. FAOSTAT: Food balance sheets. http://faostat.fao.org/site/368/default.aspx#ancor.

FAO. 2011c. FAOSTAT: Live animals. http://faostat.fao.org/site/573/default.aspx#ancor.

FAO. 2011d. FAOSTAT: Crops. http://faostat.fao.org/site/567/default.aspx#ancor.

FAO. 2011e. FAOSTAT: ForesSTAT. http://faostat.fao.org/site/377/default.aspx#ancor.

FAO. 2011f. FAOSTAT: Land. http://faostat.fao.org/site/377/default.aspx#ancor.

FAO. 2011g. *Global Forest Resources Assessment 2010*. Rome: FAO. http://www.fao.org/docrep/013/i1757e/i1757e.pdf.

FAO. 2012. *World Agriculture in the 21st Century*. Rome: FAO.

Fargione, J., et al. 2008. Land clearing and the biofuel carbon debt. *Science* 319:1235–1238.

Fernandes, S. D., et al. 2007. Global biofuel use, 1850–2000. *Global Biogeochemical Cycles* 21:GB2019. doi:10.1029/2006GB002836.

Fernandez-Cornejo, J., and M. Caswell. 2006. *The First Decade of Genetically Engineered Crops in the United States*. Washington, DC: USDA.

Fernández-González, F. 2006. *Ship Structures under Sail and under Gunfire*. Madrid: Universidad Politécnica de Madrid. http://oa.upm.es/1520/1/PONEN_FRANCISCO_FERNANDEZ_GONZALEZ_01.pdf.

Ferreira, O. C. 2000. The future of charcoal in metallurgy. *Energy Economy* 21:1–5.

Ferrett, G. 2007. Biofuels "crime against humanity." *BBC Online*, October 27. http://news.bbc.co.uk/2/hi/7065061.stm.

Fiedel, S. J. 2005. Man's best friend—mammoth's worst enemy? A speculative essay on the role of dogs in Paleoindian colonization and megafaunal extinction. *World Archaeology* 37:11–25.

Fiedel, S., and G. Haynes. 2004. A premature burial: Comments on Grayson and Meltzer's "Requiem for overkill." *Journal of Archaeological Science* 31:121–131.

Field, C. B., et al. 1998. Primary production of the biosphere: Integrating terrestrial and oceanic components. *Science* 281:237–240.

Fleuret, A. 1979. The role of wild foliage plants in the diet: A case study from Lushoto, Tanzania. *Ecology of Food and Nutrition* 8:87–93.

Foley, J., et al. 2005. Global consequences of land use. *Science* 309:570–574.

Foley, J. A. 1994. Net primary productivity in the terrestrial biosphere: the application of a global model. *Journal of Geophysical Research* 99:20773–20783.

Forbes, R. J. 1966. Heat and heating. In *Studies in Ancient Technology*, Vol. 6, 1–103. Leiden: E. J. Brill.

Forest Watch, EU. 2007. Estimates of the percentage of "illegal timber" in the imports of wood-based products from selected countries. http://www.globaltimber.org.uk/IllegalTimber Percentages.doc.

Foster, D. R., and J. D. Aber. 2004. *Forests in Time: The Environmental Consequences of 1,000 Years of Change in New England.* New Haven, CT: Yale University Press.

Francescato, V. et al. 2008. *Wood Fuels Handbook.* Legnaro: Italian Agriforestry Energy Association.

François, L. M., et al. 1999. Carbon stocks and isotopic budgets of the terrestrial biosphere at mid-Holocene and last glacial maximum times. *Chemical Geology* 159:163–189.

Freeman, D. 1980. *Iban Agriculture: A Report on the Shifting Cultivation of Hill Rice by the Iban of Sarawak.* New York: AMS Press.

Freinkel, S. 2007. *American Chestnut: The Life, Death, and Rebirth of a Perfect Tree.* Berkeley: University of California Press.

French, N. R., et al. 1976. Small mammal energetic in grassland ecosystems. *Ecological Monographs* 46:201–220.

Freudenberg, K., and A. C. Nash, eds. 1968. *Constitution and Biosynthesis of Lignin.* Berlin: Springer-Verlag.

Friedlingstein, P., et al. 1995. Carbon-biosphere-climate interactions in the last glacial maximum climate. *Journal of Geophysical Research—Atmospheres* 100:7203–7221.

Fuentes, J. D., et al. 2000. Biogenic hydrocarbons in the atmospheric boundary layer: A review. *Bulletin of the American Meteorological Society* 81:1537–1575.

Fuhrer, J. 2009. Ozone risk for crops and pastures in present and future climates. *Naturwissenschaften* 96:173–194.

Fuhrman, J. A. 1999. Marine viruses and their biogeochemical and ecological effects. *Nature* 399:541–548.

Gales, B., et al. 2005. Energy consumption in Europe over the long run: A comparative approach. Paper presented at the First Workshop of the Economic History RTN, "Europe's Growth and Development Experience." Warwick, October 28–30. http://www.cepr.org/meets/wkcn/1/1635/papers/Kander.pdf.

Gales, R., N. Brothers, and T. Reid. 1998. Seabird mortality in the Japanese tuna longline fishery around Australia. *Biological Conservation* 86:37–56.

Galloway, J. A., D. Keene, and M. Murphy. 1996. Fuelling the city: Production and distribution of firewood and fuel in London's region, 1290–1400. *Economic History Review* 49:447–472.

Galloway, J. N., et al. 2007. International trade in meat: The tip of the pork chop. *Ambio* 36:622–629.

Garrett, W. N., and N. Hinman. 1969. Re-evaluation of the relationship between carcass density and body composition of beef steers. *Journal of Animal Science* 28:1–5.

Gates, C. C., et al. 2005. *The Ecology of Bison Movements and Distribution in and beyond Yellowstone National Park*. Calgary, AB: University of Calgary.

Gay, K. 2001. *Rainforests of the World: A Reference Handbook*. Santa Barbara, CA: ABC-CLIO.

Geider, R. J., et al. 2001. Primary productivity of planet earth: Biological determinants and physical constraints in terrestrial and aquatic habitats. *Global Change Biology* 7:849–882.

Ghiglieri, M. P. 1984. *The Chimpanzees of Kibale Forest*. New York: Columbia University Press.

Gibbs, H. K., et al. 2007. Monitoring and estimating tropical forest carbon stocks: Making REDD a reality. *Environmental Research Letters* 2:1–13.

Gibson, L., et al. 2011. Primary forests are irreplaceable for sustaining tropical biodiversity. *Nature* 478:378–381.

Gifford, R. M., and L. T. Evans. 1981. Photosynthesis, carbon partitioning, and yield. *Annual Review of Plant Physiology* 32:485–509.

Gillespie, R. 2008. Updating Martin's global extinction model. *Quaternary Science* 27: 2522–2529.

Ginouvès, R. 1962. *Balaneutikè: Recherches sur le bain dans l'antiquité grecque*. Paris: de Boccard.

Glaser, B. 2007. Prehistorically modified soils of central Amazonia: A model for sustainable agriculture in the twenty-first century. *Philosophical Transactions of the Royal Society B* 362:187–196.

Glass, S. V., and S. L. Zelinka. 2010. Moisture relations and physical properties of wood. In USDA, *Wood Handbook*, part 4, 1–19.

Global Timber. 2011. The global timber trade. http://www.globaltimber.org.uk.

Goes, J. I., et al. 2005. Warming of the Eurasian landmass is making the Arabian Sea more productive. *Science* 308:545–547.

Gold, S. 2003. *The development of European Forest Resources, 1950 to 2000: A Better Information Base*. Geneva: UNECE.

Golley, F. B. 1968. Secondary productivity in terrestrial communities. *American Zoologist* 8:53–59.

Golley, F. B., et al., eds. 1975. *Small Mammals: Their Productivity and Population Dynamics*. Cambridge: Cambridge University Press.

Gomes, C. M., and C. Boesch. 2009. Wild chimpanzees exchange meat for sex on a long-term basis. *PLoS ONE* 4 (4): e5116.

Goren-Inbar, N., et al. 2004. Evidence of hominin control of fire at Gesher Benot Ya'aqov, Israel. *Science* 304:725–727.

Gorham, E. 1991. Northern peatlands: Role in the carbon cycle and probable responses to climatic warming. *Ecological Applications* 1:182–195.

Goudriaan, J., and P. Ketner. 1984. A simulation study for the global carbon cycle, including man's impact on the biosphere. *Climatic Change* 6:167–192.

Goudsblom, J. 1992. *Fire and Civilization*. London: Allen Lane.

Grayson, D. K., and D. J. Meltzer. 2003. A requiem for North American overkill. *Journal of Archaeological Science* 30:585–593.

Greenwood, W. H. 1907. *Iron*. London: Cassell.

Grün, R., et al. 2009. ESR and U-series analyses of faunal material from Cuddie Springs, NSW, Australia: Implications for the timing of the extinction of the Australian megafauna. *Quaternary Science Reviews* 29:596–610.

Gu, L., et al. 2003. Response of a deciduous forest to the Mount Pinatubo eruption: Enhanced photosynthesis. *Science* 299:2035–2038.

Guldemond, R., and R. Van Aarde. 2008. A meta-analysis of the impact of African elephants on savanna vegetation. *Journal of Wildlife Management* 72:892–899.

Gupta, M. C., B. M. Gandhi, and B. N. Tan. 1974. An unconventional legume—Prosopis cineraria. *American Journal of Clinical Nutrition* 27:1035–1036.

Gustafson, T. 1979. Khrushchev and the development of Soviet agriculture: Virgin lands programme. *Problems of Communism* 28:45–50.

Gustavsson, J., et al. 2011. *Global Food Losses and Food Waste*. Rome: FAO.

Guthrie, R. D. 2004. Radiocarbon evidence of mid-Holocene mammoths stranded on an Alaskan Bering Sea island. *Nature* 429:746–749.

Guthrie, R. D. 2006. New carbon dates link climatic change with human colonization and Pleistocene extinctions. *Nature* 441:207–209.

Gutman, G., and A. Ignatov. 1995. Global land monitoring from AVHRR: potential and limitations. *International Journal of Remote Sensing* 16:2301–2309.

Gutman, G., et al. 1995. The enhanced NOAA global data set from Advanced Very High Resolution Radiometer. *Bulletin of the American Meteorological Society* 76:1141–1156.

Gymnosperm Database. 2011. Sequoia sempervirens. http://www.conifers.org/cu/Sequoia.php.

Haberl, H. 1997. Human appropriation of net primary production as an environmental indicator: Implications for sustainable development. *Ambio* 26:143–146.

Haberl, H., et al. 2006. The energetic metabolism of the European Union and the United States. *Journal of Industrial Ecology* 10:151–171.

Haberl, H., et al. 2007. Quantifying and mapping the human appropriation of net primary production in earth's terrestrial ecosystems. *Proceedings of the National Academy of Sciences of the United States of America* 104:12942–12947.

Hairston, N., F. E. Smith, and L. B. Slobodkin. 1960. Community structure, population control and competition. *American Naturalist* 94:421–425.

Hall, D. O., and J. M. O. Scurlock. 1993. Biomass production and data. In *Photosynthesis and Production in a Changing Environment: A Field and Laboratory Manual*, ed. D. O. Hall, et al., 425–444. London: Chapman and Hall.

Hall, K. D., et al. 2009. The progressive increase of food waste in America and its environmental impact. *PLoS ONE* 4 (11): e7940. doi:10.1371/journal.pone.0007940.

Hamilton, A. J., et al. 2010. Quantifying uncertainty in estimation of tropical arthropod species richness. *American Naturalist* 176:90–95.

Hanby, J. P., et al. 1995. Ecology, demography and behavior of lions in two contrasting habitats: Ngorongoro Crater and the Serengeti Plains. In *Serengeti II*, ed. A. R. E. Sinclair and P. Arcese, 315–331. Chicago: University of Chicago Press.

Harpending, H. C., et al. 1998. Genetic traces of ancient demography. *Proceedings of the National Academy of Sciences* 95:1961–1967.

Harris, D., ed. 1996. *Origins and Spread in Agriculture and Pastoralism in Eurasia*. London: UCL Press.

Harris, N. L., et al. 2012. Baseline map of carbon emissions from deforestation in tropical regions. *Science* 336:1573–1576.

Hartenstein, R. 1986. Earthworm biotechnology and global biogeochemistry. *Advances in Ecological Research* 15:379–409.

Hartmann, F. 1923. *L'agriculture dans l'ancienne Egypte*. Paris: Libraire-Imprimerie Réunies.

Hassan, F. A. 1984. Environment and subsistence in Predynastic Egypt. In *From Hunters to Farmers: The Causes and Consequences of Food Production in Africa*, ed. J. D. Clark and S. A. Brandt, 57–64. Berkeley: University of California Press.

Hawks, J., et al. 2000. Population bottlenecks and Pleistocene human evolution. *Molecular Biology and Evolution* 17:2–22.

Hay, R. K. M. 1995. Harvest index: A review of its use in plant crop physiology. *Annals of Applied Biology* 126:197–216.

Hayden, B. 1981. Subsistence and ecological adaptations of modern hunter/gatherers. In *Omnivorous Primates*, ed. R. S. O. Harding and G. Teleki, 344–421. New York: Columbia University Press.

Hays, J. D., J. Imbrie, and N. J. Shackleton. 1976. Variations in the Earth's orbit: Pacemaker of the Ice Ages. *Science* 194:1121–1132.

Hayward, M. W., J. O'Brien, and G. I. H. Kerley. 2007. Carrying capacity of large African predators: Predictions and tests. *Biological Conservation* 139:219–229.

He, F., et al. 2007. Quantitative of analysis on forest dynamics of China in recent 300 years. *Acta Geographica Sinica* 62:30–62.

Heckenberger, M. J., et al. 2003. Amazonia1492: Pristine forest or cultural parkland? *Science* 301:1710–1714.

Heidenreich, C. 1971. *Huronia: A History and Geography of the Huron Indians*. Toronto: McClelland and Stewart.

Heilig, G. K. 1997. Anthropogenic factors in land-use change in China. *Population and Development Review* 23:139–168.

Heinsch, F. A., et al. 2003. *User's Guide: GPP and NPP (MOD17A2/A3) Products NASA MODIS Land Algorithm*. Washington, DC: NASA.

Helland, J. 1980. *Five Essays on the Study of Pastoralists and the Development of Pastoralism*. Bergen: Universitet i Bergen.

Hietz, P., et al. 2011. Long-term change in the nitrogen cycle of tropical forests. *Science* 334:664–666.

Hilborn, R., et al. 2003. State of the world's fisheries. *Annual Review of Environment and Resources* 28:359–399.

Hodgson, J., and A. W. Illius. 1996. *The Ecology and Management of Grazing Systems.* Wallingford, UK: CABI.

Hogy, P., et al. 2009. Does elevated atmospheric CO_2 allow for sufficient wheat grain quality in the future? *Journal of Applied Botany and Food Quality* 82:114–121.

Hohmann, G., and B. Fruth. 2008. Capture and meat eating by bonobos at Lui Kotale, Salonga National Park, Democratic Republic of Congo. *Folia Primatologica* 79:103–110.

Holdo, R. M., et al. 2007. Plant productivity and soil nitrogen as a function of grazing, migration and fire in an African savanna. *Journal of Ecology* 95:115–128.

Holdo, R. M., R. D. Holt, and J. M. Fryxell. 2009. Grazers, browsers, and fire influence the extent and spatial pattern of tree cover in the Serengeti. *Ecological Applications* 19:95–109.

Holmén, K. 1992. The global carbon cycle. In *Global Biogeochemical Cycles*, ed. S. S. Butcher, 239–262. London: Academic Press.

Holman, T., J. Muya, and E. Røskaft. 2007. Local law enforcement and illegal bushmeat hunting outside the Serengeti. *Environmental Conservation* 34:55–63.

Holt, J. A., and J. F. Easy. 1993. Numbers and biomass of mound-building termites (Isoptera) in a semiarid tropical woodland near Charters Towers, North Queensland, Australia. *Sociobiology* 21:281–286.

Houghton, R. A. 2003. Revised estimates of the annual net flux of carbon to the atmosphere from changes in land use and land management 1850–2000. *Tellus* 55B:378–390.

Houghton, R. A., et al. 2001. The spatial distribution of forest biomass in the Brazilian Amazon: A comparison of estimates. *Global Change Biology* 7:731–746.

Houghton, R. A., and S. J. Goetz. 2008. New satellites help quantify carbon sources and sinks. *Eos* 89:417–418.

Houghton, R.A., F. Hall, and S. J. Goetz. 2009. Importance of biomass in the global carbon cycle. *Journal of Geophysical Research* 114 (G00E03): 1–13.

House, J. I., I. C. Prentice, and C. Le Quéré. 2002. Maximum impacts of future reforestation or deforestation on atmospheric CO_2. *Global Change Biology* 8:1047–1052.

Houweling, S., et al. 2008. Early anthropogenic CH_4 emissions and the variation of CH_4 and (CH_4)-C-13 over the last millennium. *Global Biogeochemical Cycles* 22:GB1002.

Huang, Y., et al. 2010. Marshland conversion to cropland in northeast China from 1950 to 2000 reduced the greenhouse effect. *Global Change Biology* 16:680–695.

Huixian, L., et al. 2002. The Great Tangshan Earthquake 1976. Pasadena, CA: California Institute of Technology.

Hubbes, T. 1999. The American elm and Dutch elm disease. *Forestry Chronicle* 75: 265–273.

Humphreys, W. F. 1979. Production and respiration in animal populations. *Journal of Animal Ecology* 48:427–453.

Hyde, C. K. 1977. *Technological Change and the British Iron Industry 1700–1870*. Princeton, NJ: Princeton University Press.

HYDE. 2011. *History Database of the Global Environment*. http://themasites.pbl.nl/en/themasites/hyde/index.html.

ICR (Institute of Cetacean Research). 2010. *Japan's Whale Research Programs*. http://www.icrwhale.org/generalinfo.htm.

IEA (International Energy Agency). 2010. Statistics by country/region. http://www.iea.org/stats/index.asp.

Imafidon, G. I., and F. W. Sosulski. 1990. Nucleic acid nitrogen of animal and plant foods. *Journal of Agricultural and Food Chemistry* 38:118–120.

IMF (International Monetary Fund). 2010. Principal global indicators. http://www.principalglobalindicators.org/default.aspx.

Imhoff, M. L., et al. 2004a. The consequences of urban land transformations on net primary productivity in the United States. *Remote Sensing of Environment* 89:434–443.

Imhoff, M. L., et al. 2004b. Global patterns in human consumption of net primary production. *Nature* 429:870–873.

Imhoff, M. L., and L. Bounoua. 2006. Exploring global patterns of net primary production carbon supply and demand using satellite observations and statistical data. *Journal of Geophysical Research* 111:D22S12. doi:10.1029/2006JD007377.

Inoue, T., et al. 2001. The abundance and biomass of subterranean termites (Isoptera) in a dry evergreen forest of northeast Thailand. *Sociobiology* 37:41–52.

INPE (Instituto Nacional de Pesquisas Espaciais). 2011. DETER shows 593 km2 of deforestation on Amazon region in March and April. http://150.163.105.7/ingles/news/news_dest164.php.

IPCC (Intergovernmental Panel on Climate Change). 1990. *IPCC First Assessment Report*. Geneva: IPCC.

IPCC. 1995. *IPCC Guidelines for National Greenhouse Gas Inventories*. Paris: IPCC/OECD/IEA Inventory Programme.

Isenberg, A. C. 2000. *The Destruction of the Buffalo: An Environmental History, 1750–1920*. Cambridge: Cambridge University Press.

Ishige, N. 2001. *The History and Culture of Japanese Food*. London: Kegan Paul.

Ishihara, A., and J. Yoshii. 2003. A survey of the commercial trade in whale meat products in Japan. Tokyo: TRAFFIC East Asia–Japan. http://www.traffic.org.

Ito, A. 2011. A historical meta-analysis of global terrestrial net primary productivity: Are estimates converging? *Global Change Biology*. doi:10.1111/j.1365-2486.2011.02450.x.

IUCN (International Union for Conservation of Nature). 2011. *IUCN Red List*. http://www.iucnredlist.org.

IWC (International Whaling Commission). 2010. Whale population estimates. http://iwcoffice.org/conservation/estimate.htm

Jackson, A. 2009. Fish in—fish out (FIFO) ratios explained. *Aquaculture Europe* 34:5–10.

Jackson, J. A., and B. J. S. Jackson. 2007. Once upon a time in American ornithology. *Wilson Journal of Ornithology* 119:767–772.

Jackson, R. B., et al. 1996. A global analysis of root distributions for terrestrial biomes. *Oecologia* 108:389–411.

Jackson, R. B., H. A. Mooney, and E.-D. Schulze. 1997. A global budget for fine root biomass, surface area, and nutrient contents. *Proceedings of the National Academy of Sciences of the United States of America* 94:7362–7366.

James, H. 2001. Use and production of solid sawn timber in the United States. *Forest Products Journal* 51 (7/8): 23–28.

Janssens, I. A., et al. 2005. The carbon balance of terrestrial ecosystems as country scale: A European case study. *Biogeosciences* 2:15–26.

Jenkins, B. M., S. Q. Turn, and R. B. Williams. 1992. Atmospheric emissions from agricultural burning in California: Determination of burn fractions, distribution factors, and crop specific contributions. *Agriculture, Ecosystems, and Environment* 38:313–330.

Jennings, S. D., et al. 2008. Global-scale predictions of community and ecosystem properties from simple ecological theory. *Proceedings, Biological Sciences* 275:1375–1383.

Johannsen, O. 1953. *Geschichte des Eisens*. Düsseldorf: Stahleisen.

Johnson, C. N. 2009a. Ecological consequences of Late Quaternary extinctions of megafauna. *Proceedings, Biological Sciences* 276:2509–2519.

Johnson, C. 2009b. Megafaunal decline and fall. *Science* 326:1072–1073.

Junqueira, A. B, G. H. Shepard, and C. R. Clement. 2010. Secondary forests on anthropogenic soils in Brazilian Amazonia conserve agrobiodiversity. *Biodiversity and Conservation* 19:1933–1961.

Kadam, K. L., and J. D. McMillan. 2003. Availability of corn stover as a sustainable feedstock for bioethanol production. *Bioresource Technology* 88:17–25.

Kahuranga, J. 1981. Population estimates, densities and biomass of large herbivores in Simanjiro Plains, Northern Tanzania. *African Journal of Ecology* 19:225–238.

Kalland, A., and B. Moeran. 1992. *Japanese Whaling: End of an Era?* London: Curzon Press.

Kammen, D. M. 1995. Cookstoves for the developing world. *Scientific American* 273 (1): 72–75.

Kaplan, J. O., et al. 2002. Modeling the dynamics of terrestrial carbon storage since the Last Glacial Maximum. *Geophysical Research Letters* 29:2074.

Kaplan, J. O., K. M. Krumhardt, and N. Zimmermann. 2009. The prehistoric and preindustrial deforestation in Europe. *Quaternary Science Reviews* 28:3016–3034.

Karanth, K. U., and M. E. Sunquist. 1992. Population structure, density and biomass of large herbivores in the tropical forests of Nagarahole, India. *Journal of Tropical Ecology* 8: 21–35.

Kareiva, P., et al. 2007. Domesticated nature: Shaping landscapes and ecosystems for human welfare. *Science* 316:1866–1869.

Karkanas, P., et al. 2007. Evidence for habitual use of fire at the end of the Lower Paleolithic: Site-formation processes at Qesem Cave, Israel. *Journal of Human Evolution* 53:197–212.

Kastner, T. 2009. Trajectories in human domination of ecosystems: Human appropriation of net primary production in the Philippines during the 20th century. *Ecological Economics* 69:260–269.

Kato, M., et al. 2005. World at work: Charcoal producing industries in northeastern Brazil. *Occupational and Environmental Medicine* 62:128–132.

Kawanishi, K., and M. E. Sunquist. 2004. Conservation status of tigers in a primary rainforest of Peninsular Malaysia. *Biological Conservation* 120:329–344.

Kawashima, C. 1986. *Minka: Traditional Houses of Rural Japan.* Tokyo: Kōdansha.

Kay, C. E. 2007. Are lightning fires unnatural? A comparison of aboriginal and lightning ignition rates in the United States. In *Proceedings of the 23rd Tall Timbers Fire Ecology Conference: Fire in Grassland and Shrubland Ecosystems,* ed. R. E. Masters and K. E. M. Galley, 16–28. Tallahassee, FL: Tall Timbers Research Station.

Keith, H., B. G. Mackey, and D. B. Lindenmayer. 2009. Re-evaluation of forest biomass carbon stocks and lessons from the world's most carbon-dense forests. *Proceedings of the National Academy of Sciences of the United States of America* 106:11635–11640.

Kelleher, K. 2005. *Discards in the World's Marine Fisheries: An Update.* Rome: FAO. ftp://ftp.fao.org/docrep/fao/008/y5936e/y5936e00.pdf.

Kellogg, C. A., and D. W. Griffin. 2006. Aerobiology and the global transport of desert dust. *Trends in Ecology and Evolution* 21:638–644.

Kim, S., and B. E. Dale. 2002. Allocation procedure in ethanol production system from corn grain. *International Journal of Life Cycle Assessment* 7:237–243.

King, A. W., et al. 2006. Plant respiration in a warmer world. *Science* 312:536–537.

Kingsbury, N. 2010. *Hybrid: The History and Science of Plant Breeding.* Chicago: University of Chicago Press.

Kiple, K. F. 2000. The question of Paleolithic nutrition and modern diet. In *The Cambridge World History of Food,* ed. K. F. Kiple and K. C. Ornelas, 1704–1709. Cambridge: Cambridge University Press.

Kirch, P. V. 2005. Archaeology and global change: The Holocene record. *Annual Review of Environment and Resources* 30:409–440.

Kitzes, J., et al. 2008. Shrink and share: Humanity's present and future ecological footprint. *Philosophical Transactions of the Royal Society B: Biological Sciences* 363:464–475.

Klein Goldewijk, K. 2001. Estimating global land use change over the past 300 years: The HYDE Database. *Global Biogeochemical Cycles* 15:417–433.

Klein Goldewijk, K., A. Beusen, and P. Janssen. 2010. Long-term dynamic modelling of global population and built-up area in a spatially explicit way: HYDE 3.1. *Holocene* 20: 565–573.

Klein Goldewijk, K., and G. Van Drecht. 2006. HYDE 3: Current and historical population land cover. In *Integrated Modelling of Global Environmental Change,* ed. A. F. Bouwman, T. Kram, and K. Klein Goldewijk, 93–111. Bilthoven: Netherlands Environmental Assessment Agency.

Knaup, H., and J. von Mittelstaedt. 2009. Foreign investors buy up African farmland. *Bloomberg Businessweek,* August 3. http://www.businessweek.com/globalbiz/content/aug2009/gb2009083_487801.htm.

Koblents-Mishke, O. I., et al. 1968. New data on the magnitude of primary production in the oceans. *Doklady Akademii Nauk SSSR, Seria Biologicheskaya* 183:1189–1192.

Koch, P. L., and A. D. Barnosky. 2006. Late Quaternary extinctions: State of the debate. *Annual Review of Ecology Evolution and Systematics* 37:215–250.

Köhler, P., and H. Fischer. 2004. Simulating changes in the terrestrial biosphere during the last glacial/interglacial transition. *Global and Planetary Change* 43:33–55.

Körner, C., et al. 2005. Carbon flux and growth in mature deciduous forest trees exposed to elevated CO_2. *Science* 309:1360–1362.

Koval'chenko, I. D. 2004. *Agrarnyi stroi Rossii vtoroii poloviny XIX—nachala XX veka.* Moscow: ROSSPEN.

Kovda, V. A. 1971. The problem of biological and economic productivity of the earth's land areas. *Soviet Geography* 12:6–23.

Krausmann, F. 2001. Land use and industrial modernization: An empirical analysis of human influence on the functioning of ecosystems in Austria 1830–1995. *Land Use Policy* 18: 17–26.

Krausmann, F., and H. Haberl. 2002. The process of industrialization from an energetic metabolism point of view: Socio-economic energy flows in Austria 1830–1995. *Ecological Economics* 41:177–201.

Kristiansen, P., A. Taji, and J. Reganold, eds. 2006. *Organic Agriculture: A Global Perspective.* Ithaca, NY: Comstock Publishing Associates.

Kümpel, N. F., et al. 2010. Incentives for hunting: The role of bushmeat in the household economy in rural Equatorial Guinea. *Human Ecology* 38:251–264.

Kurz, W. A., and M. J. Apps. 1994. The carbon budget of Canadian forests: A sensitivity analysis of changes in disturbance regimes, growth rates, and decomposition rates. *Environmental Pollution* 83:55–61.

Kuzmin, Y. V. 2009. Extinction of the woolly mammoth (*Mammuthus primigenius*) and woolly rhinoceros (*Coelodonta antiquitatis*) in Eurasia: Review of chronological and environmental issues. *Boreas* 39:247–261.

Laevastu, T. 1993. *Marine Climate, Weather, and Fisheries.* New York: Halsted Press.

Lal, R. 2007. Anthropogenic influences on world soils and implications to global food security. *Advances in Agronomy* 93:69–93.

Lam, P. S., et al. 2008. Bulk density of wet and dry wheat straw and switchgrass particles. *Applied Engineering in Agriculture* 24:351–358.

Lamlom, S. H., and R. A. Savidge. 2003. A reassessment of carbon content in wood: Variation within and between 41 North American species. *Biomass and Bioenergy* 25:381–388.

Lauk, C., and K.-H. Erb. 2009. Biomass consumed in anthropogenic vegetation fires: Global patterns and processes. *Ecological Economics* 69:301–309.

Laybourn-Parry, J. 1992. *Protozoan Plankton Ecology.* London: Chapman & Hall.

Leakey, A. D. B. 2009. Rising atmospheric carbon dioxide concentration and the future of C4 crops for food and fuel. *Proceedings, Biological Sciences* 276:2333–2343.

Lee, P. G. 2009. Caution against using intact forest-landscapes data at regional scales. *Ecology and Society* 14:1.

Legge, A. J., and P. A. Rowley-Conwy. 1987. Gazelle killing in Stone Age Syria. *Scientific American* 257 (2): 88–95.

Lehmann, J., and C. Boesch. 2003. Social influences on ranging patterns among chimpanzees (*Pan troglodytes verus*) in the Taï National Park, Côte d'Ivoire. *Behavioral Ecology* 14:642–649.

Lehmann, N., et al. 2003. Classification of Amazonian dark earths and other ancient anthropic soils. In *Amazonian Dark Earths: Origin, Properties, and Management*, ed. N. Lehmann et al., 77–102. Berlin: Springer-Verlag.

Lehsten, V., et al. 2009. Estimating carbon emissions from African wildfires. *Biogeosciences* 6:349–360.

Lemaire, G., et al., eds. 2000. *Grassland Ecophysiology and Grazing Ecology*. Oxford: CABI.

Levine, J. S., ed. 1991. *Global Biomass Burning*. Cambridge, MA: MIT Press.

Lewis, S. L., et al. 2009. Increasing carbon storage in intact African tropical forests. *Nature* 457:1003–1006.

Li, Y., et al. 2000. Illegal wildlife trade in the Himalayan region of China. *Biodiversity and Conservation* 9:901–918.

Li, Y., and D. S. Wilcove. 2005. Threats to vertebrate species in China and the United States. *Bioscience* 55:147–153.

Liang, S. L., et al. 2010. Impacts of climate change and land use changes on land Surface radiation and energy budgets. *IEEE Journal of Selected Topics in Applied Earth Observations and remote Sensing* 3:219–224.

Liebenberg, L. 2006. Persistence hunting by modern hunter-gatherers. *Current Anthropology* 47:1017–1025.

Liebig, J. von 1862. *Die Naturgesetze des Feldbaues*. Braunschweig: Vieweg.

Lieth, H. 1973. Primary production: terrestrial ecosystems. *Human Ecology* 1:303–332.

Lieth, H. 2005. A reinterpretation of the earliest quantification of global plant productivity by von Liebig (1862). *New Phytologist* 167:641–644.

Lieth, H., and R. H. Whittaker, eds. 1975. *Primary Productivity of the Biosphere*. New York: Springer-Verlag.

Lindeman, R. 1942. The trophic-dynamic aspect of ecology. *Ecology* 23:399–418.

Linebaugh, P. 1993. *The London Hanged: Crime and Civil Society in the Eighteenth Century*. Cambridge: Cambridge University Press.

Litton, C. M., J. W. Raich, and M. G. Ryan. 2007. Carbon allocation in forest ecosystems. *Global Change Biology* 13:2089–2109.

Liu, J., et al. 2005. China's changing landscape during the 1990s: Large-scale land transformation estimated with satellite data. *Geophysical Research Letters* 32:L02405. doi: 10.1029/2004GL021649.

Liu, M., and H. Tian. 2010. China's land cover and land use change from 1700 to 2005: Estimations from high-resolution satellite data and historical archives. *Global Biogeochemical Cycles* 24. doi:10.1029/2009GB003687.

Lobell, D. B., et al. 2002. Satellite estimates of productivity and light use efficiency in United States agriculture, 1982–98. *Global Change Biology* 8:722–735.

Long, S. P., et al. 2006. Food for thought: Lower-than-expected crop yield stimulation with rising CO_2 concentrations. *Science* 312:1918–1921.

Lorenzen, E. D., et al. 2011. Species-specific response of Late Quaternary megafauna to climate and humans. *Nature* 479:359–364.

Lovelock, J. 2009. *The Vanishing Face of Gaia: A Final Warning*. New York: Basic Books.

Ludwick, A. E. 2000. High yield alfalfa: 24 tons irrigated . . . 12 tons non-irrigated. *Better Crops* 84 (1): 18–19.

Luo, Z. 1998. Biomass energy consumption in China. *Wood Energy News* 13 (3): 3–4.

Luxmoore, R., et al. 1989. The volume of raw ivory entering international trade from African producing countries from 1979 to 1988. In *The Ivory Trade and the Future of the African Elephant*, ed. S. Cobb. Oxford: Ivory Trade Review Group.

Luyssaert, S., et al. 2007. CO_2 balance of boreal, temperate, and tropical forests derived from a global database. *Global Change Biology* 13:2509–2537.

Luyssaert, S., et al. 2008. Old-growth forests as global carbon sinks. *Nature* 455: 2123–2215.

Luyssaert, S., et al. 2009. The European carbon balance: Part 3. Forests. *Global Change Biology*. doi:10.1111/j.1365-2486.2009.02056.x.

Lykke, A. M., O. Mertz, and S. Ganaba. 2002. Food consumption in rural Burkina Faso. *Ecology of Food and Nutrition* 41:119–153.

Lyman, J. M., et al. 2010. Robust warming of the global upper ocean. *Nature* 465:334–337.

Lyons, S. K., et al. 2004. Was a "hyperdisease" responsible for the late Pleistocene megafaunal extinction? *Ecology Letters* 7:859–868.

Macedo, I. C., M. R. L. V. Leal, and J. E. A. R. da Silva. 2004. *Assessment of Greenhouse Gas Emissions in the Production and Use of Fuel Ethanol in Brazil*. São Paulo: Government of the State of São Paulo.

MacPhee, R. D. E., and P. A. Marx. 1997. Humans, hyperdisease, and first-contact extinctions. In *Natural Change and Human Impact in Madagascar*, ed. S. M. Goodman and B. D. Patterson, 169–217. Washington, DC: Smithsonian Institution Press.

Maddison, A. 2007. *Contours of the World Economy, 1–2030 AD*. Oxford: Oxford University Press.

Madsen, T., et al. 2006. Size matters: Extraordinary rodent abundance on an Australian tropical flood plain. *Austral Ecology* 31:361–365.

Mao Yushi, et al. 1996. *Economic Costs of China's Environmental Change. Report prepared for the Project on Environmental Change and Security*. Cambridge, MA: American Association of Arts and Sciences.

Marland, G., and M. Obersteiner. 2008. Large-scale biomass for energy, with considerations and cautions: An editorial comment. *Climatic Change* 87:335–342.

Marlon, J. R., et al. 2009a. Climate and human influences on global biomass burning over the past two millennia. *Nature Geoscience* 1:697–702.

Marlon, J. R., et al. 2009b. Wildfire responses to abrupt climate change in North America. *Proceedings of the National Academy of Sciences of the United States of America* 106: 2519–2524.

Marlowe, F. W. 2005. Hunter-gatherers and human evolution. *Evolutionary Anthropology* 14:54–67.

Martin, P. S. 1958. Pleistocene ecology and biogeography of North America. *Zoogeography* 151:375–420.

Martin, P. S. 1967. Overkill at Olduvai Gorge. *Nature* 215:212–213.

Martin, P. S. 1990. 40,000 years of extinctions on the "planet of doom." *Palaeogeography, Palaeoclimatology, Palaeoecology* 82:187–201.

Martin, P. S. 2005. *Twilight of the Mammoths*. Berkeley: University of California Press.

Martius, C. 1994. Diversity and ecology of termites in Amazonia forests. *Pedobiologia* 38:407–428.

Maslin, M. A., and E. Thomas. 2003. Balancing the deglacial global carbon budget: The hydrate factor. *Quaternary Science Reviews* 22:1729–1736.

Mather, A. S. 2005. Assessing the world's forests. *Global Environmental Change* 15:267–280.

Matthews, E. 1983. Global vegetation and land use: New high-resolution data bases for climatic studies. *Journal of Climate and Applied Meteorology* 22:474–487.

Matthews, E. 1984. Global inventory of the pre-agricultural and present biomass. In *Interactions between Climate and Biosphere*, ed. H. Lieth, R. Fantechi and H. Schnitzler, 237–246. Lisse: Swets and Zeitlinger.

Matthews, E., et al. 2000. *Pilot Analysis of Global Ecosystems*. Washington, DC: World Resources Institute.

McCarthy, J. J. 2009. Reflections on: Our planet and its life, origins, and futures. *Science* 326:1646–1655.

McClain, C. 2010. An empire lacking food. *American Scientist* 98:470–477.

McCloskey, J. M., and H. Spalding. 1989. A reconnaissance-level inventory of the amount of wilderness remaining in the world. *Ambio* 18:221–227.

McEvedy, C., and R. Jones. 1978. *Atlas of World Population History*. London: Allen Lane.

McGranahan, G., and F. Murray. 2003. *Air Pollution and Health in Rapidly Developing Countries*. London: Earthscan.

McGrath, D. G. 1987. The role of biomass in shifting cultivation. *Human Ecology* 15:221–242.

McHenry, H. M., and K. Coffing. 2000. *Australopithecus* to *Homo*: transformations in body and mind. *Annual Review of Anthropology* 29:125–146.

McHugh, T. 1972. *The Time of the Buffalo*. Lincoln: University of Nebraska Press.

McKinney, R., ed. 1994. *Technology of Paper Recycling*. New York: Springer-Verlag.

McLauchlan, K. 2003. Plant cultivation and forest clearance by prehistoric North Americans: Pollen evidence from Fort Ancient, Ohio, USA. *Holocene* 13:557–566.

MEDEA. 1997. *China Agriculture: Cultivated Land Area, Grain Projections, and Implications*. Washington, DC: National Intelligence Council.

Meentemeyer, V., et al. 1982. World patterns and amounts of terrestrial plant litter production. *Bioscience* 32:125–128.

MEF (Ministry of Environment and Forests). 2009. *State of Forest Report 2009*. Dehradun: Government of India.

Mellars, P. 2006. Why did modern human populations disperse from Africa ca. 60000 years ago? A new model. *Proceedings of the National Academy of Sciences of the United States of America* 103:9381–9386.

Melville, H. 1851. *Moby-Dick*. New York: Harper & Brothers.

Menzel, P., and F. D'Aluisio. 1998. *Man Eating Bugs: The Art and Science of Eating Insects*. Berkeley, CA: Ten Speed Press.

Mertz, O., et al. 2008. A fresh look at shifting cultivation: Fallow length an uncertain indicator of productivity. *Agricultural Systems* 96:75–84.

Milankovitch, M. 1941. *Kanon der Erdbestrahlung und seine Andwendung auf das Eiszeiten-Problem*. Beograd: Royal Serbian Academy.

Millennium Ecosystem Assessment. 2005. *Ecosystems and Human Well-being: General Synthesis*. Washington, DC: Island Press.

Miller, G. H., et al. 2005. Ecosystem collapse in Pleistocene Australia and a human role in megafaunal extinction. *Science* 309:287–290.

Milligan, K., S. S. Ajayi, and J. B. Hall. 1982. Density and biomass of the large herbivore community in Kainji Lake National Park, Nigeria. *African Journal of Ecology* 20:1–12.

Milner-Gulland, E. J., and J. R. Beddington. 1993. The exploitation of elephants for the ivory trade; An historical perspective. *Proceedings, Biological Sciences* 252:29–37.

Milner-Gulland, E. J., and R. H. Mace. 1991. The impact of the ivory trade on the elephant population of the trade, as assessed by data from the trade. *Biological Conservation* 55:215–229.

Misra, R. V., and P. R. Hesse. 1983. *Comparative Analyses of Organic Manures*. Rome: FAO.

Mitchell, A. D., A. M. Scholz, and J. M. Conway. 1998. Body composition analysis of pigs from 5 to 97 kg by dual-energy X-ray absorptiometry. *Applied Radiation and Isotopes* 49:5210523.

Mitscherlich, G. 1963. *Zustand, Wachstum und Nutzung des Waldes im Wandel der Zeit*. Freiburg: Freiburger Universitätsreden.

Miyamoto, M. 2004. Quantitative aspects of Tokugawa ecoomy. In *The Economic History of Japan: 1600–1990* Vol. 1, *Emergence of Economic Society in Japan, 1600–1859*, ed. A. Hayami, O. Satō, and R .P. Toby, 36–84. Oxford: Oxford University Press.

Moiseev, P. A. 1969. *The Living Resources of the World Ocean*. Moscow: Pishchevaia promyshlennost.

Mollicone, D., H. D. Eva, and F. Achard. 2006. Human role in Russian wild fires. *Nature* 440:436–437.

Monfreda, C., N. Ramankutty, and J. A. Foley. 2008. Farming the planet: 2. Geographic distribution of crop areas, yields, physiological types, and net primary production in the year 2000. *Global Biogeochemical Cycles* 22:1–19.

Mora, C., et al. 2011. How many species are there on Earth and in the ocean? *PLoS Biology* 9 (8): 1–8.

Morgan, B. J. 2007. Group size, density and biomass of large mammals in the Reserve de Faune du Petit Loango, Gabon. *African Journal of Ecology* 45:508–518.

Morse, E. S. 1886. *Japanese Homes and Their Surroundings*. Boston: Ticknor and Co..

Mosimann, J. E., and P. S. Martin, P.S. 1975. Simulating overkill by Paleoindians. *American Scientist* 63:304–313.

Mukhopadhyay, D., and B. Nandi. 1979. Biodegradation of rice stumps by soil mycolfora. *Plant and Soil* 53:215–218.

Musel, A. 2009. Human appropriation of net primary production in the United Kingdom, 1800–2000. *Ecological Economics* 2009:270–281.

Myneni, R. B., et al. 1997. Estimation of global leaf area index and absorbed PAR using radiative transfer models. *IEEE Transactions on Geoscience and Remote Sensing* 4:1380–1393.

Nabuurs, G.-J., et al. 2003. Temporal evolution of the European forest carbon sink from 1950 to 1999. *Global Change Biology* 9:152–160.

Nagy, K. A. 2005. Field metabolic rate and body size. *Journal of Experimental Biology* 208:1621–1625.

Naylor, R. L., S. L. Williams, and D. R. Strong. 2000. Aquaculture: A gateway for exotic species. *Science* 294:1655–1656.

NBSC (National Bureau of Statistics of China). 2010. *China Statistical Yearbook*. Beijing: China Statistics Press.

Ndibalema, V. G., and A. N. Songorwa. 2008. Illegal meat hunting in Serengeti: Dynamics in consumption and preferences. *African Journal of Ecology* 46:311–319.

Needham, J. 1964. *The Development of Iron and Steel Technology in China*. London: Newcomen Society.

Nemani, R. R. et al. 2002. Recent trends in hydrologic balance have enhanced the terrestrial carbon sink in the United States. *Geophysical Research Letters* 2:106–1-106–4.

Nemani, R. R., et al. 2003. Climate-driven increases in global terrestrial net primary production from 1982 to 1999. *Science* 300:1560–1563.

Nepstad, D., et al. 2009. The end of deforestation in the Brazilian Amazon. *Science* 326:1350–1351.

Njiforti, H. L. 1996. Preferences and present demand for bushmeat in north Cameroon: Some implications for wildlife conservation. *Environmental Conservation* 23:149–155.

Noble, I. R., and R. Dirzo. 1997. Forests as human-dominated ecosystems. *Science* 277: 522–525.

Noddack, W. 1937. Der Kohlenstoff im Haushalt der Natur. *Angewandte Chemie* 50:505–510.

NOO (National Obesity Observatory). 2009. International comparisons of adult obesity prevalence. http://www.noo.org.uk/NOO_about_obesity/international.

NRC (National Research Council). 1996. *Nutrient Requirements of Beef Cattle*. Washington, DC: National Academy Press.

Ntiamoa-Baidu, Y. 1998. *Sustainable Harvesting, Production and Use of Bushmeat*. Accra: Ministry of Lands and Forestry.

NTSG (Numerical Terradynamic Simulation Group). 2004. Four years of MOD17 annual NPP. Missoula, MT: NTSG. http://images.ntsg.umt.edu.

Nuckolls, A. E., et al. 2009. Hemlock declines rapidly with hemlock woolly adelgid infestation: Impacts on the carbon cycle of southern Appalachian forests. *Ecosystems* 12:179–190.

Nyahongo, J. W., et al. 2005. Benefits and costs of illegal grazing and hunting in the Serengeti ecosystem. *Environmental Conservation* 32:326–332.

Odum, H. T. 1957. Trophic structure and productivity of Silver Springs, Florida. *Ecological Monographs* 27:55–112.

Ogden, C. L., et al. 2004. Mean body weight, height, and body mass index, United States1960–2002. *Advanced Data from Vital and Health Statistics* 347:1–20.

Ogoye-Ndegwa, C., and J. Aagaard-Hansen. 2003. Traditional gathering of wild vegetables among the Luo of western Kenya: A nutritional anthropology project. *Ecology of Food and Nutrition* 42:69–89.

Ogutu, J. O., and H. T. Dublin. 2002. Demography of lions in relation to prey and habitat in the Maasai Mara National Reserve, Kenya. *African Journal of Ecology* 40:120–129.

Okigbo, B. N. 1984. Improved production systems as an alternative to shifting intermittent cultivation. In *Improved Production Systems as an Alternative to Shifting Cultivation*, 1–100. Rome: FAO.

Olson, J. S., J. A. Watts, and L. J. Allison. 1983. *Carbon in Live Vegetation of Major World Ecosystems*. Oak Ridge, TN: Oak Ridge National Laboratory.

Ometto, J., et al. 2005. Amazonia and the modern carbon cycle: Lessons learned. *Oecologia* 143:483–500.

O'Neill, D. W., and D. J. Abson. 2009. To settle or protect? A global analysis of net primary production in parks and urban areas. *Ecological Economics* 69:319–327.

Orme, B. 1977. The advantages of agriculture. In *Hunters, Gatherers and First Farmers beyond Europe*, ed. J. V. S. Megaw, 41–49. Leicester: Leicester University Press.

ORNL (Oak Ridge National Laboratory). 2010. Net primary productivity (NPP) data sets. http://daac.ornl.gov/cgi-bin/dataset_lister.pl?p=13.

Otto, D., et al. 2002. Biospheric carbon stocks reconstructed at the last glacial maximum: Comparison between general circulation models using prescribed and computed sea surface temperature. *Global and Planetary Change* 33:117–138.

Owen, R. 1861. *Palaeontology, or, A Systematic Study of Extinct Animals and Their Geological Relations*. Edinburgh: Adam and Charles Black.

Padoch, C., et al. 2008. The demise of swidden in Southeast Asia? Local realities and regional ambiguities. *Geografisk Tidskrift* 107:29–41.

Palm, C., et al. 2004. Mitigating GHG emissions in the humid tropics: Case studies from the alternatives to slash-and-burn program (ASB). *Environment, Development and Sustainability* 6:145–162.

Pan, Y., et al. 2004. New estimates of carbon storage and sequestration in China's forests: Effects of age class and method on inventory-based carbon estimation. *Climatic Change* 67:211–236.

Pang, J. F., et al. 2009. mtDNA data indicate a single origin for dogs south of Yangtze River, less than 16,300 years ago, from numerous wolves. *Molecular Biology and Evolution* 26:2849–2864.

Parker, I. S. C. 1979. *Ivory Trade.* Washington, DC: Department of Fisheries and Wildlife. Parkes. R. J., et al. 1994. Deep bacterial biosphere in Pacific Ocean sediments. *Nature* 371:410–413.

Parry, M. L., et al., eds. 2007. *Contribution of Working Group II to the Fourth Assessment Report of the Intergovernmental Panel on Climate Change.* Cambridge: Cambridge University Press.

Pauly, D. 2009. Beyond duplicity and ignorance in global fisheries. *Scientia Marina* 73:215–224.

Pauly, D., and V. Christensen. 1995. Primary production required to sustain global fisheries. *Nature* 374:255–257.

Pauly, D., et al. 1998. Fishing down marine food webs. *Science* 279:860–863.

Pauly, D., et al. 2000. Fishing down aquatic food webs. *American Scientist* 88:46–51.

Peláez-Samaniegoa, M. R. 2008. Improvements of Brazilian carbonization industry as part of the creation of a global biomass economy. *Renewable & Sustainable Energy Reviews* 12:1063–1086.

Peng, C. H., Guiot, J., and E. Van Campo. 1998. Estimating changes in terrestrial vegetation and carbon storage: Using palaeoecological data and models. *Quaternary Science Reviews* 17:719–735.

Penner, M., et al. 1997. *Canada's Forest Biomass Resources: Deriving Estimates from Canada's Forestry Inventory.* Victoria, BC: Pacific Forestry Centre.

Penuelas, J., and M. Staudt. 2010. BVOCs and global change. *Trends in Plant Science* 15:133–134.

Perkins, D. H. 1969. *Agricultural Development in China, 1368–1968.* Chicago: Aldine.

Perren, R. 1985. The retail and wholesale meat trade 1880–1939. In *Diet and Health in Modern Britain*, ed. D. J. Oddy and D. S. Miller, 45–65. London: Croom Helm.

Pershing, A. J. et al. 2010. The impact of whaling on the ocean carbon cycle: Why bigger was better. *PLoS ONE* 5 (8): 1–9 e12444

Piao, S., et al. 2008. Net carbon dioxide losses of northern ecosystems in response to autumn warming. *Nature* 451:49–52.

Piao, S., et al. 2009. The carbon balance of terrestrial ecosystems in China. *Nature* 458: 1009–1013.

Piao, S. et al. 2010a. Forest annual carbon cost: A global-scale analysis of autotrophic respiration. *Ecology* 91:652–661.

Piao, S., et al. 2010b. The impacts of climate change on water resources and agriculture in China. *Nature* 467:43–51.

PIK (Potsdam Institute for Climate Impact Research). 2012. LPJmL: Lund-Potsdam-Jena managed Land Dynamic Global Vegetation and Water Balance Model. http://www.pik-potsdam.de/research/climate-impacts-and-vulnerabilities/models/lpjml.

Pimentel, D. 2003. Ethanol fuels: Energy balance, economics, and environmental impacts are negative. *Natural Resources Research* 12:127–134.

Piperno, D. R., et al. 2004. Processing of wild cereal grains in the Upper Palaeolithic revealed by starch grain analysis. *Nature* 430:670–673.

Plumptre, A. J., and D. Cox. 2006. Counting primates for conservation: primate surveys in Uganda. *Primates* 47:65–73.

Plumptre, A. J., and S. Harris. 1995. Estimating the biomass of large mammalian herbivores in a tropical montane forest. *Journal of Applied Ecology* 32:111–120.

Pongratz, J., et al. 2008. A reconstruction of global agricultural areas and land cover for the last millennium. *Global Biogeochemical Cycles* 22. doi:10.1029/2007GB003153.

Pongratz, J. et al. 2009. Effects of anthropogenic land cover change on the carbon cycle of the last millennium. *Global Biogeochemical Cycles* 23 doi: 10.1029/2009GB003488.

Pordesimo, L. O., W. C. Edens, and S. Sokhansanj. 2004. Distribution of aboveground biomass in corn stover. *Biomass and Bioenergy* 26:337–343.

Pörtner, H. O., and R. Knust. 2007. Climate change affects marine fishes through the oxygen limitation of thermal tolerance. *Science* 315:95–97.

Post, W. M., et al. 1982. Soil carbon pools and world life zones. *Nature* 298:156–159.

Post, W. M., A. W. King, and S. D. Wullschleger. 1997. *Historical Variations in Terrestrial Biospheric Carbon Storage.* Oak Ridge, TN: Oak Ridge National Laboratory.

Potapov, P. A., et al. 2008. Mapping the world's intact forest landscapes by remote sensing. *Ecology and Society* 13 (2): 51. http://www.ecologyandsociety.org/vol13/iss2/art51.

Potere, D., et al. 2009. Mapping urban areas on a global scale: Which of the eight maps now available is more accurate? *International Journal of Remote Sensing* 24:6531–6558.

Potter, C. S. 1999. Terrestrial biomass and the effects of deforestation on the global carbon cycle. *BioScience* 49:769–778.

Potter, C. S., et al. 2003. Continental scale comparisons of terrestrial carbon sinks estimated from satellite data and ecosystem modeling 1982–1998, *Global Planetary Change* 39: 201–213.

Potter, C., et al. 2005. Variability in terrestrial carbon sinks over two decades: Part 2. Eurasia. *Global and Planetary Change* 49:177–186.

Potter, C., et al. 2007. Satellite-derived estimates of potential carbon sequestration through afforestation of agricultural lands in the United States. *Climatic Change* 80:232–236.

Potter, C., et al. 2008. Storage of carbon in U.S. forests predicted from satellite data, ecosystem modeling, and inventory summaries. *Climatic Change* 90:269–282.

Prasad, V. K., and K. V. S. Badarinth. 2004. Land use changes and trends in human appropriation of above ground net primary production (HANPP) in India (1961–98). *Geographical Journal* 170:51–63.

Prentice, K. C., and I. Fung. 1990. The sensitivity of terrestrial carbon storage to climate change. *Nature* 346:48–50.

Prentice, I. C., et al. 1993. Modelling global vegetation patterns and terrestrial carbon storage at the last glacial maximum. *Global Ecology and Biogeography Letters* 3:67–76.

Prescott, G. W., et al. 2012. Quantitative global analysis of the role of climate and people in explaining late Quaternary megafaunal extinctions. *Proceedings of the National Academy of Sciences* 109:4527–4531.

Price, L. L. 1997. Wild plant food in agricultural environments. *Human Organization* 56: 209–221.

Prins, H. H. T., and J. M. Reitsma. 1989. Mammalian biomass in an African equatorial rain forest. *Journal of Animal Ecology* 58:851–861.

Pusey, A. E., et al. 2005. Influence of ecological and social factors on body mass of wild chimpanzees. *International Journal of Primatology* 26:3–31.

Pushkina, D., and P. Raia. 2007. Human influence on distribution and extinctions of the late Pleistocene Eurasian megafauna. *Journal of Human Evolution* 54:769–782.

Puyravaud, J.-P., P. Davidar, and W. F. Laurance. 2010. Cryptic loss of India's native forests. *Science* 329:32.

PVE (Product Boards for Livestock, Meat and Eggs). 2010. *Livestock, Meat and Eggs in the Netherlands*. Zoetermeer: PVE.

Qiu, J. 2010. It's a microbial world: Worldwide census ups diversity estimates for marine microbes one-hundred-fold. *Nature*. doi:10.1038/news.2010.190.

Ramankutty, N., and J. A. Foley. 1999. Estimating historical changes in global land cover: Croplands from 1700 to 1992. *Global Biogeochemical Cycles* 13:997–1027.

Ramankutty, N., et al. 2008. Farming the planet: 1. Geographic distribution of global agricultural lands in the year 2000. *Global Biogeochemical Cycles* 22:1–19.

Ramírez-Moreno, L., and R. Olvera-Ramírez. 2006. Traditional and present use of *Spirulina* sp. (*Arthrospira* sp.). *Interciencia* 31:657–663.

Randerson, J. T., et al. 2002. Net ecosystem production: A comprehensive measure of net carbon accumulation by ecosystems. *Ecological Applications* 12:937–947.

Rands, M. R. W., et al. 2010. Biodiversity conservation: Challenges beyond 2010. *Science* 329:1298–1303.

Rangarajan, M. 1998. The Raj and the natural world: The campaign against "dangerous beasts" in colonial India. *Studies in History* 14:266–299.

Rappaport, R. A. 1968. *Pigs for the Ancestors: Ritual in the Ecology of New Guinea People*. New Haven, CT: Yale University Press.

Read, P. 2008. Biosphere carbon stock management: Addressing the threat of abrupt climate change in the next few decades. An editorial essay. *Climatic Change* 87:305–320.

Reagan, D. P., and R. B. Waide, eds. 1996. *The Food Web of a Tropical Rain Forest*. Chicago: University of Chicago Press.

Reaman, G. E. 1970. *A History of Agriculture in Ontario*. Don Mills, ON: Saunders.

Rees, M. 2003. *Our Final Hour. A Scientist's Warning. How Terror, Error, and Environmental Disaster Threaten Humankind's Future in This Century—On Earth and Beyond*. New York: Basic Books.

Reichle, D. E., ed. 1981. *Dynamic Properties of Forest Ecosystems*. Cambridge: Cambridge University Press.

Reumer, J. W. F. 2007. Habitat fragmentation and the extinction of mammoths (*Mammuthus primigenius*, Proboscidea, Mammalia): Arguments for a causal relationship. In *Late Neogene and Quaternary Biodiversity and Evolution: Regional Developments and Interregional Correlations*, vol. 2, ed. R. D. Kahlke, L. C. Maul, and P. P. A. Mazza, 279–286. Frankfurt: Courier Forschungsinstitut.

Richards, J. F. 1990. Land transformation. In *The Earth as Transformed by Human Action*, ed. B. L. Turner et al., 163–178. New York: Cambridge University Press.

Richerson, P. J., R. Boyd, and R. L. Bettinger. 2001. Was agriculture impossible during the Pleistocene but mandatory during the Holocene? A climate change hypothesis. *American Antiquity* 66:387–411.

Rick, T. C., M. Erlandson, and R. L. Vellanoweth. 2001. Paleocoastal Marine fishing on the Pacific coast of the Americas: Perspectives from Daisy Cave, California. *American Antiquity* 66: 595–613.

Riggs, T. J., et al. 1981. Comparison of spring barley varieties grown in England and Wales between 1880 and 1980. *Journal of Agricultural Science* 97:599–610.

Riginos, C., and J. B. Grace. 2008. Savanna tree density, herbivores, and the herbaceous community: Bottom-up vs. top-down effects. *Ecology* 89:2228–2238.

Riley, G. A. 1944. The carbon metabolism and photosynthetic efficiency of the Earth as a whole. *American Scientist* 32:129–134.

Roberts, C. 2007. *The Unnatural History of the Sea*. Washington, DC: Island Press.

Roberts, G., M. J. Wooster, and E. Lagoudakis. 2009. Annual and diurnal African biomass burning temporal dynamics. *Biogeosciences* 6:849–866.

Roberts, R. G., et al. 2001. New ages for the last Australian megafauna: Continentwide extinction about 46,000 years ago. *Science* 292:1888–1892.

Rockström, J., et al. 2009. A safe operating space for humanity. *Nature* 461:472–475.

Rodriguez, L. 2008. *A Global Perspective on the Total Economic Values of Pastoralism*. Nairobi: WISP, GEF, UNDP and IUCN.

Rogan, E., and M. Mackey. 2007. Megafauna bycatch in drift nets for albacore tuna (*Thunnus alalunga*) in the NE Atlantic. *Fisheries Research* 86:6–14.

Rojstaczer, S., S. M. Sterling, and N. Moore. 2001. Human appropriation of photosynthesis products. *Science* 294:2549–2552.

Roman, J., and S. R. Palumbi. 2003. Whales before whaling in the North Atlantic. *Science* 301:508–510.

Roxburgh, S. H., et al. 2005. What is NPP? Inconsistent accounting of respiratory fluxes in the definition of net primary production. *Functional Ecology* 19:378–382.

Roy, J., B. Saugier, and H. A. Mooney, eds. 2001. *Terrestrial Global Productivity*. San Diego, CA: Academic Press.

Ruddiman, W. F. 2005. *Plows, Plagues and Petroleum*. Princeton, NJ: Princeton University Press.

Ruddle, K. 1974. *The Yukpa Cultivation System*. Berkeley: University of California Press.

Running, S. W., et al. 2004. A continuous satellite-derived measure of global terrestrial primary productivity. *Bioscience* 54:547–560.

RWEDP (Regional Wood Energy Development Programme in Asia). 1997. *Regional Study of Wood Energy Today and Tomorrow*. Rome: FAO-RWEDP.

Ryder, M. L. 1969. Remains of fishes and other aquatic animals. In *Science in Archaeology*, ed. D. Brothwell and E. Higgs, 376–394. London: Thames and Hudson.

Ryder, M. L. 1983. *Sheep & Man*. London: Duckworth.

Saatchi, S., et al. 2011. Benchmark map of forest carbon stocks in tropical regions across three continents. *Proceedings of the National Academy of Sciences of the United States of America* 108:9899–9904.

Sachdev, H. S., et al. 2005. Anthropometric indicators of body composition in young adults: Relation to size at birth and serial measurements of body mass index in childhood in the New Delhi birth cohort. *American Journal of Clinical Nutrition* 82:456–466.

Sagan, C., et al. 1993. A search for life on Earth from the Galileo spacecraft. *Nature* 365:715–721.

Sanders, W. T., J. R. Parsons, and R. S. Santley. 1979. *The Basin of Mexico: Ecological Processes in the Evolution of a Civilization*. New York: Academic Press.

Sanderson, E. W., et al. 2002. The human footprint and the last of the wild. *Bioscience* 52:891–904.

Sandford, S. 1983. *Management of Pastoral Development in the Third World*. Chichester: Wiley.

Sarhage, D., and J. Lundbeck. 1992. *A History of Fishing*. Berlin: Springer-Verlag.

Sassaman, K. E. 2004. Complex hunter-gatherers in evolution and history: A North American perspective. *Journal of Archaeological Research* 12:227–280.

Saugier, B., J. Roy, and H. A. Mooney. 2001. Estimations of global terrestrial productivity: Converging toward a single number? In *Terrestrial Global Productivity*, ed. J. Roy, B. Saugier, and H. A. Mooney, 543–557. San Diego, CA: Academic Press.

Sawyer, J., and A. Mallarino. 2007. Nutrient removal when harvesting corn stover. In *Integrated Crop Management*, 251–253. Ames: Iowa State University Press.

SB (Statistics Bureau). 2010. *Historical Statistics of Japan*. Tōkyō: SB. http://www.stat.go.jp/english/data/chouki/index.htm.

Schaller, G. B. 1972. *The Serengeti Lion*. Chicago: University of Chicago Press.

Schaphof, S., et al. 2006. Terrestrial biosphere carbon storage under alternative climate projections. *Climatic Change* 74:97–122.

Scheidel, W., and S. J. Friesen. 2009. The size of the economy and the distribution of income in the Roman Empire. *Journal of Roman Studies* 99:61–91.

Schiere, J. B., and J. deWit. 1995. Feeding urea ammonia treated rice straw in the tropics. II. Assumptions on nutritive value and their validity for least cost ration formulation. *Animal Feed Science and Technology* 51:45–63.

Schlebecker, J. T. 1975. *Whereby We Thrive: A History of American Farming, 1607–1972*. Ames: Iowa State University Press.

Schlüter, O. 1952. *Die Siedlungsräume Mitteleuropas in frühgeschichtlicher Zeit: Part 1. Forschungen zur Deutschen Landeskunde*. Hamburg: Atlantik Verlag.

Schmidt-Vogt, D., et al. 2009. An assessment of trends in the extent of swidden in Southeast Asia. *Human Ecology* 37:269–280.

Schmitt, C. B., et al. 2009. Global analysis of the protection status of the world's forests. *Biological Conservation* 142:2122–2130.

Schneider, A., M. A. Friedl, and D. Potere. 2009. Monitoring urban areas globally using MODIS 500m data: New methods based on "urban ecoregions." *Remote Sensing of Environment* 114:1733–1746.

Schneider, V., and D. Pearce. 2004. What saved the whales? An economic analysis of 20th century whaling. *Biodiversity and Conservation* 13:543–562.

Schorger, A. W. 1955. *The Passenger Pigeon: Its Natural History and Extinction*. Madison: University of Wisconsin Press.

Schroeder, H. 1919. Die jährliche Gesamtproduktion der grünen Pflanzdecke der Erde. *Naturwissenschaften* 7:8–12.

Schumacher, E. F. 1973. *Small Is Beautiful: Economics as if People Mattered*. New York: Harper & Row.

Schurr, S. H., and B. C. Netschert. 1960. *Energy in the American Economy 1850–1975*. Baltimore, MD: Johns Hopkins University Press.

Schurgers, G., et al. 2009. European emissions of isoprene and monoterpenes from the Last Glacial Maximum to present. *Biogeosciences* 6:2779–2797.

Schwarzmüler, E. 2009. Human appropriation of aboveground net primary production in Spain, 1955–2003: An empirical analysis of the industrialization of land use. *Ecological Economics* 69:282–291.

Schwidetzky, I., B. Chiarelli, and O. Necrasov, eds. 1980. *Physical Anthropology of European Populations*. The Hague: Mouton.

Scudder, T. 1976. Social anthropology and the reconstruction of prehistoric land use in tropical Africa: A cautionary case study from Zambia. In *Origins of African Plant Domestication*, ed. J. R. Harlan et al., 357–381. The Hague: Mouton.

Scurlock, J. M. O., G. P. Asner, and S. T. Gower. 2001. *Global Leaf Area Index Data from Field Measurements, 1932–2000*. Oak Ridge, TN, Oak Ridge National Laboratory. http://www.daac.ornl.gov/VEGETATION/lai_des.html.

Scurlock, J. M. O., K. Johnson, and R. J. Olson. 2002. Estimating net primary productivity from grassland biomass dynamics measurements. *Global Change Biology* 8:736–753.

Seppälä, R., A. Buck, and P. Katila, eds. 2009. *Adaptation of Forests and People to Climate Change. A Global Assessment Report*. Helsinki: IUFRO.

SFA (State Forestry Administration). 2009. *China Forestry Statistical Yearbook 2009*. Beijing: China Forestry Publishing House.

Shackleton, C. M. 2003. The prevalence of use and value of wild herbs in South Africa. *South African Journal of Science* 99:23–25.

Shah, M. B. 2005. *Manipur, Jhum and Eco-degradation*. New Delhi: B. R. Publishing Corp.

Shearman, P. L., et al. 2009. Forest conversion and degradation in Papua New Guinea 1972–2002. *Biotropica* 41:379–390.

Sheehan, G. W. 1985. Whaling as an organizing focus in Northwestern Eskimo society. In *Prehistoric Hunter-Gatherers*, ed. T. D. Price and J. A. Brown, 123–154. Orlando, FL: Academic Press.

Shi, H. T., et al. 2008. Evidence for the massive scale of turtle farming in China. *Oryx* 42:147–150.

Shiba, T. 1986. *Kujira to Nihonjin*. [Whales and the Japanese.] Tōkyō: Yōsensha.

Shigo, A. L. 1986. *A New Tree Biology*. Durham, NH: Shigo and Trees.

Sieferle, R. P. 2001. *The Subterranean Forest*. Cambridge: White Horse Press.

Sillitoe, P. 2002. Always been farmer-foragers? Hunting and gathering in the Papua New Guinea Highlands. *Anthropological Forum* 12:45–76.

Silva, M., and J. A. Downing. 1995. The allometric scaling of density and body mass: A nonlinear relationship for terrestrial mammals. *American Naturalist* 145:704–727.

Singh, I. D., and N. C. Stoskopf. 1971. Harvest index in cereals. *Agronomy Journal* 63:224–226.

Sinsin, B., et al. 2002. Abundance and species richness of larger mammals in Pendjari National Park in Benin. *Mammalia* 66:369–380.

Sitch, S., et al. 2003. Evaluation of ecosystem dynamics, plant geography and terrestrial carbon in the LPJ dynamic global vegetation model. *Global Change Biology* 9:161–185.

Skole, D., and C. Tucker. 1993. Tropical deforestation and habitat fragmentation in the Amazon: Satellite data from 1978 to 1988. *Science* 260:1905–1909.

Slicher van Bath, B. H. 1963. *The Agrarian History of Western Europe, A.D. 500–1850*. London: Arnold.

Smil, V. 1987. *Energy Food Environment*. Oxford: Oxford University Press.

Smil, V. 1993. *China's Environmental Crisis*. Armonk, NY: M. E. Sharpe.

Smil, V. 1994. *Energy in World History*. Boulder, CO: Westview.

Smil, V. 1998. Global population: Milestones, hopes, and concerns. *Medicine and Global Survival* 5:105–108.

Smil, V. 1999a. China's agricultural land. *China Quarterly* 158:414–429.

Smil, V. 1999b. Crop residues: Agriculture's largest harvest. *Bioscience* 49:299–308.

Smil, V. 2000. *Feeding the World*. Cambridge, MA: MIT Press.

Smil, V. 2004. *China's Past, China's Future*. London: RutledgeCurzon.

Smil, V. 2005. *Transforming the Twentieth Century: Technical Innovations and Their Consequences*. Cambridge, MA: MIT Press.

Smil, V. 2008. *Energy in Nature and Society: Energetics of Complex Systems*. Cambridge, MA: MIT Press.

Smil, V. 2010a. *Energy Myths and Realities*. Washington, DC: American Enterprise Institute.

Smil, V. 2010b. *Energy Transitions*. Santa Barbara, CA: Praeger.

Smil, V. 2010c. *Why America Is Not a New Rome*. Cambridge, MA: MIT Press.

Smil, V., and K. Kobayashi. 2012. *Japan's Dietary Transition and Its Impacts*. Cambridge, MA: MIT Press.

Smil, V., P. Nachman, and T. V. Long. 1983. *Energy Analysis in Agriculture*. Boulder, CO: Westview.

Smith, A. D. M., et al. 2011. Impacts of fishing low-trophic level species on marine ecosystems. *Science* 333:1147–1150.

Smith, C. S., and P. J. Urness. 1984. Small mammal abundance on native and improved foothill ranges, Utah. *Journal of Range Management* 37:353–357.

Smith, F. A., S. M. Elliott, and S. K. Lyons. 2010. Methane emissions from extinct megafauna. *Nature Geoscience* 3:374–375.

Smith, M. D., et al. 2010. Sustainability and global seafood. *Science* 327:784–786.

Smith, P., et al. 2005. Carbon sequestration potential in European croplands has been over-estimated. *Global Change Biology* 11:2153–2163.

Smith, T. D., and R. R. Reeves. 2002. Estimating historical humpback whale removals from the North Atlantic. *Journal of Cetacean Research and Management* 4 (Supplement): 242–255.

Smith, T. D., and R. R. Reeves. 2004. Estimating whaling catch history. IWC Scientific Committee, Document SC56/O 22, Cambridge, England. *Contributions to World Whaling History* 3. http://www.WorldWhalingHistory.org.

Solomon, B. D., J. R. Barnes, and K. E. Halvorsen. 2007. Grain and cellulosic ethanol: History, economics, and energy policy. *Biomass and Bioenergy* 31:416–425.

Sommath, S. S., et al. 1995. A review of wood railroad ties performance. *Forest Products Journal* 45 (9): 55.

Sørensen, B. 2009. Energy use by Eem Neanderthals. *Journal of Archaeological Science* 36:2201–2205.

Spector, D. 1983. *Agriculture on the Prairies, 1870–1940*. Ottawa: Environment Canada.

Speth, J. D. and K. A. Spielmann. 1983. Energy source, protein metabolism, and hunter-gatherer subsistence strategies. *Journal of Anthropological Archaeology* 2:1–31.

Spink, J., et al. 2009. The potential to increase crop productivity of wheat and oilseed rape in the UK. http://www.bis.gov.uk/go-science/science-in-government/global-issues/food/~/media/5C4E476342334B608B748767805B1115.ashx.

SSB (State Statistical Bureau). 2000. *Zhongguo tongji nianjian.* [*China Statistical Yearbook.*] Beijing: Zhongguo tongji chubanshe.

Stanford, C. B. 1999. *The Hunting Apes: Meat Eating and the Origins of Human Behavior*. Princeton, NJ: Princeton University Press.

Stanhill, G. 1976. Trends and deviations in the yield of the English wheat crop during the last 750 years. *Agro-ecosystems* 3:1–10.

Staniforth, A. R. 1979. *Cereal Straw*. Oxford: Clarendon Press.

Stanton, N. L. 1988. The underground in grasslands. *Annual Review of Ecology and Systematics* 19:573–589.

Starbuck, A. 1989. *History of the American Whale Fishery*. Seacaucus, NJ: Castle Books.

Steen, A. S., et al. 1994. *The Straw Bale House Book*. White River Junction, VT: Chelsea Green Publishing.

Steenblik, R. 2007. *Biofuels—At What Cost? Government Support for Ethanol and Biodiesel in Selected OECD Countries*. Geneva: IISD.

Steffen, W., P. J. Crutzen, and J. R. McNeill. 2007. The Anthropocene: Are humans now overwhelming the great forces of nature? *Ambio* 36:614–621.

Steinfeld, H., et al. 2006. *Livestock's Long Shadow*. Rome: FAO.

Stenkjaer, N. 2009. Straw ovens. http://www.folkecenter.net/gb/rd/biogas/technologies/straw_ovens.

Stephens, B. B., et al. 2007. Weak northern and strong tropical land carbon uptake from vertical profiles of atmospheric CO_2. *Science* 316:1732–1735.

Stephens, S. L., R. E. Martin, and N. E. Clinton. 2007. Prehistoric fire area and emissions from California's forests, woodlands, shrublands, and grasslands. *Forest Ecology and Management* 251:205–216.

Stergiou, K. I., A. C. Tsikliras, and D. Pauly. 2008. Farming up the Mediterranean food webs. *Conservation Biology* 23:230–232.

Stiller, M., et al. 2010. Withering away: 25,000 years of genetic decline preceded cave bear extinction. *Molecular Biology and Evolution* 27:975–978.

Stiner, M. C., and R. Barkai, and A. Gopher. 2009. Cooperative hunting and meat sharing 400–200 kya at Qesem cave, Israel. *Proceedings of the National Academy of Sciences of the United States of America* 106:13207–13212.

Straker, E. 1969. *Wealden Iron*. New York: Augustus M. Kelley.

Strawbale. 2010. StrawBale. http://www.strawbale.com/.

Stuart, A. J. 2005. The extinction of woolly mammoth (*Mammuthus primigenius*) and straight-tusked elephant (*Palaeoloxodon antiquus*) in Europe. *Quaternary International* 126–128:171–177.

Stuart, A. J., and A. M. Lister. 2007. Patterns of Late Quaternary megafaunal extinctions in Europe and northern Asia. In *Late Neogene and Quaternary Biodiversity and Evolution: Regional Developments and Interregional Correlations*, vol. 2, ed. R. D. Kahlke, L. C. Maul, and P. P. A. Mazza, 287–297. Frankfurt: Courier Forschungsinstitut.

Stuart, A. J., et al. 2002. The latest woolly mammoths (*Mammuthus primigenius* Blumenbach) in Europe and Asia: A review of the current evidence. *Quaternary Science Reviews* 21:1559–1569.

Stuart, A. J., et al. 2004. Pleistocene to Holocene extinction dynamics in giant deer and woolly mammoth. *Nature* 431:684–689.

Suttie, J. M., S. G. Reynolds, and C. Batello. 2005. *Grasslands of the World*. Rome: FAO.

Swartz, W. K. 2000. *Global Maps of the Growth of Japanese Marine Fisheries and Fish Consumption*. Vancouver: University of British Columbia Press.

Swiss Reinsurance Company. 2006. Natural catastrophes and man-made disasters 2005: High earthquake casualties, new dimension in windstorm losses. *Sigma* 2:1–40.

Sylvester-Bradley, R., and J. Wiseman, eds. 2005. *Yields of Farmed Species: Constraints and Opportunities in the 21st Century*. Nottingham: Nottingham University Press.

Tabuti, J. R. S., S. S. Dhillion, and K. A. Lye. 2004. The status of wild food plants in Bulamogu County, Uganda. *International Journal of Food Sciences and Nutrition* 55:485–498.

Tacon, G. 2008. Compound aqua feeds in a more conservative market: Alternative protein sources for a more sustainable future. In *Avances en nutrition acuicola IX*, ed. E. Cruz Suarez et al., 1–5. Monterrey: Universidad Autonoma de Nuevo Leon.

Tahir, S. N. A., and M. Rafique. 2009. Emission of greenhouse gases (GHGs) from burning of biomass in brick kilns. *Environmental Forensics* 10:265–267.

Tao, B., et al. 2007. Spatial pattern of terrestrial net ecosystem productivity in China during 1981–2000. *Science in China Series D: Earth Sciences* 50:745–753.

Teleki, G. 1973. *The Predatory Behavior of Wild Chimpanzees*. Lewisburg, PA: Bucknell University Press.

Tennie, C., I. C. Gilby, and R. Mundry. 2009. The meat-scrap hypothesis: Small quantities of meat may promote cooperative hunting in wild chimpanzees (*Pan troglodytes*). *Behavioral Ecology and Sociobiology* 63:421–431.

Thieme, H. 1997. Lower Paleolithic hunting spears from Germany. *Nature* 385:807–810.

Thomas, M., et al. 2008. DNA from pre-Clovis human coprolites in Oregon, North America. *Science* 230:786–789.

Thomas, R. Q., et al. 2010. Increased tree carbon storage in response to nitrogen deposition in the US. *Nature Geoscience* 3:13–17.

Thomas, S. C. 1991. Population densities and patterns of habitat use among anthropoid primates of the Ituri Forest, Zaire. *Biotropica* 23:68–83.

Tollefson, J. 2010. The global farm. *Nature* 466:554–556.

Tollefson, J. 2012. Brazil set to cut forest protection. *Nature* 485:19.

Toutain, J.-C. 1971. La consommation alimentaire en France de 1789 a 1964. *Économies et Sociéteés* 5:1909–2049.

Trinkhaus, E. 2005. Early modern humans. *Annual Review of Anthropology* 34:207–230.

Trueman, C. N., et al. 2005. Prolonged coexistence of humans and megafauna in Pleistocene Australia. *Proceedings of the National Academy of Sciences of the United States of America* 102:8381–8385.

Trumbore, S. 2006. Carbon respired by terrestrial ecosystems: Recent progress and challenges. *Global Change Biology* 12:141–153.

Tsagarakis, K., et al. 2010. Food-web traits of the North Aegean Sea ecosystem (Eastern Mediterranean) and comparison with other Mediterranean ecosystems. *Estuarine, Coastal and Shelf Science* 88:233–248.

Tsikliras, A., D. Moutopoulos, and K. Stergiou. 2007. Reconstruction of Greek marine fisheries landings: National versus FAO statistics. In *The Marine Fisheries of China: Development and Reported Catches*, ed. R. Watson, L. Pang, and D. Pauly, 121–137. Vancouver, BC: Fisheries Centre, University of British Columbia.

Tunved, P., et al. 2006. High natural aerosol loading over boreal forests. *Science* 312:261–263.

Turner, D. P., et al. 2006. Evaluation of MODIS NPP and GPP products across multiple biomes. *Remote Sensing of Environment* 102:282–292.

Turkenburg, W. C., et al. 2000. Renewable energy technologies. In *World Energy Assessment: Energy and the Challenge of Sustainability*, ed. J. Goldemberg et al., 219–227. New York: United Nations Developmental Programme.

Twain, M. 2010. *Autobiography of Mark Twain*. Berkeley: University of California Press.

Twede, D. 2002. The packaging technology and science of ancient transport amphoras. *Packaging Technology and Science* 15:181–195.

Twede, D. 2005. The cask age: The technology and history of wooden barrels. *Packaging Technology and Science* 18:253–264.

UN (United Nations). 2011. *World Population Prospects: The 2010 Revision.* http://esa.un.org/wpp/unpp/panel_population.htm.

UNECE (United Nations Economic Commission for Europe). 2010. *Forest Product Conversion Factors for the UNECE Region.* Geneva: UN.

USBC (U.S. Bureau of the Census). 1975. *Historical Statistics of the United States.* Washington, DC: USBC.

USDA (U.S. Department of Agriculture). 2000. *2000 Agricultural Statistics Annual.* http://www.nass.usda.gov/Publications/Ag_Statistics/2000/index.asp.

USDA. 2001. *U.S. Forest Facts and Historical Trends.* Washington, DC: Forest Service.

USDA. 2003. *U.S. Timber Production, Trade, Consumption, and Price Statistics 1965–2002.* Washington, DC: Forest Service.

USDA. 2009. *2009 Agricultural Statistics Annual.* http://www.nass.usda.gov/Publications/Ag_Statistics/2009/index.asp.

USDA. 2010. *Wood Handbook.* Madison, WI: USDA. http://www.fpl.fs.fed.us/documnts/fplgtr/fpl_gtr190.pdf.

USGS (United State Geological Survey). 2000. National land cover dataset. U.S. Geological Survey fact sheet 108–00. http://erg.usgs.gov/isb/pubs/factsheets/fs10800.html.

USGS. 2010. Gross primary productivity 8-day global 1 km MOD17A2. https://lpdaac.usgs.gov/lpdaac/products/modis_products_table/gross_primary_productivity/8_day_l4_global_1km/mod17a2.

Valiela, I. 1984. *Marine Ecological Processes.* New York: Springer-Verlag.

van Gemerden, B. S., et al. 2003. The pristine rain forest? Remnants of historical human impacts on current tree species composition and diversity. *Journal of Biogeography* 30:1381–1390.

Vartanyan, S. L., et al. 1995. Radiocarbon dating evidence for mammoths on Wrangel Island, Arctic Ocean, until 2000 BC. *Radiocarbon* 37:1–6.

Vasconcellos, A. 2010. Biomass and abundance of termites in three remnant areas of Atlantic forest in northeastern Brazil. *Revista Brasileira de Entomologia* 54:455–461.

Verhoeven, J. D., A. Pendray, and W. E. Dauksch. 1998. The key role of impurities in ancient Damascus steel blades. *Journal of Metallurgy* 50 (9): 58–64.

Vernadskii, V. I. 1926. *Biosfera.* Leningrad: Nauchnoe khimiko-tekhnicheskoye izdatel'stvo.

Vernadskii, V. I. 1940. *Biogeokhimicheskie ocherki, 1922–1932.* Moscow: Izdatel'stvo Akademii Nauk SSSR.

Vince, G. 2011. An epoch debate. *Science* 334:32–37.

Vitousek, P. M., et al. 1986. Human appropriation of the products of photosynthesis. *Bioscience* 36:368–373.

Vitousek, P., et al. 1997. Human domination of Earth's ecosystems. *Science* 277:494–499.

Voeten, M. M., and H. H. T. Prins. 1999. Resource partitioning between sympatric wild and domestic herbivores in the Tarangire region of Tanzania. *Oecologia* 120:287–294.

Vogel, G. 2010. Europe tries to save its eels. *Science* 329:505–507.

Vuichard, N., et al. 2008. Carbon sequestration due to the abandonment of agriculture in the former USSR since 1990. *Global Biogeochemical Cycles* 22:GB4018. doi:10.1029/2008GB003212.

Wackernagel, M., et al. 2002. Tracking the ecological overshoot of the human economy. *Proceedings of the National Academy of Sciences of the United States of America* 99: 9266–9271.

Wald, W. J., P. F. Ritchie, and C. B. Purves. 1947. The elementary composition of lignin in Northern pine and black spruce woods, and of the isolated klason and periodate lignins. *Journal of the American Chemical Society* 69:1371–1377.

Wang, H. 1998. Deforestation and desiccation of China. In *The Economic Costs of China's Environmental Degradation*, ed. V. Smil and Y. Mao, 14–40. Boston: American Academy of Arts and Sciences.

Wang, M., and Y. Ding. 1998. Fuel-saving stoves in China. *World Energy News* 13 (3): 9–10.

Warde, P. 2007. *Energy Consumption in England and Wales, 1560–2004*. Napoli: Consiglio Nazionale della Ricerche.

Wasser, S. K., et al. 2008. Combating the illegal trade in African elephant ivory with DNA forensics. *Conservation Biology* 22:1065–1071.

Wasser, S. K., et al. 2010. Elephants, ivory, and trade. *Science* 327:1331–1332.

Waters, M. R., and T. W. Stafford. 2007. Redefining the Age of Clovis: Implications for the peopling of the Americas. *Science* 315:1122–1124.

Waters, M. R., et al. 2011. Pre-Clovis mastodon hunting 13,800 years ago at the Manis Site, Washington. *Science* 334:351–353.

Watson, R. and D. Pauly, 2001. Systematic distortions in world fisheries catch trends. *Nature* 414:534–436.

Watson, R., L. Pang, and D. Pauly, eds. 2001. *The Marine Fisheries of China: Development and Reported Catches*. Vancouver, BC: Fisheries Centre, University of British Columbia.

Watt, B. K., and A. L. Merrill. 1963. *Handbook of the Nutritional Contents of Foods*. Washington, DC: U.S. Department of Agriculture.

Watters, R. F. 1971. *Shifting Cultivation in Latin America*. FAO: Rome.

Watts, P. 1905. *The Ships of the Royal Navy as They Existed at the Time of Trafalgar*. London: Institution of Naval Architects.

WBGU (Wissenschaftlicher Beirat der Bundesregierung Globale Umwelveränerungen). 1998. *WBGU Special Report: The Accounting of Biological Sinks and Sources in the Kyoto Protocoal*. Bremerhaven: WBGU.

Webb, D. A. 2011. *The Tie Guide*. Fayetteville, GA: Railway Tie Association. http://www.rta.org/Portals/0/Documents/Tie%20Basics/TieGuide%20Revised%209%2005.pdf.

Webb, S. 2008. Megafauna demography and late Quaternary climatic change in Australia: A predisposition to extinction. *Boreas* 37:329–345.

Webster, C. R., M. A. Jenkins, and S. Jose. 2006. Woody invaders and the challenges they pose to forest ecosystems in the eastern United States. *Journal of Forestry* 104:366–374.

Weilemann de Tau, M. E., and J. Luquez. 2000. Variation for biomass, economic yield and harvest index among soybean cultivars of maturity groups III and IV in Argentina. *Soybean Genetics Newsletter 27.*

Welp, L., et al. 2011. Interannual variability in the oxygen isotopes of atmospheric CO_2 driven by El Niño. *Nature* 477:579–582.

West, P. W. 2009. *Tree and Forest Measurement.* Berlin: Springer-Verlag.

Westerling, A. L., et al. 2006. Warming and earlier spring increase Western U.S. forest wildfire activity. *Science* 313:940–943.

White, L. J. T. 1994. Biomass of rain-forest mammals in the Lope Reserve, Gabon. *Journal of Animal Ecology* 63:499–512.

Whitehead, H. 2002. Estimates of the current global population size and historical trajectory for sperm whales. *Marine Ecology Progress Series* 242:295–304.

Whitman, W. B., et al. 1998. Prokaryotes: the unseen majority. *Proceedings of the National Academy of Sciences* 95:6578–6583.

Whittaker, R. H., and G. E. Likens. 1973. Carbon in the biota. In *Carbon and the Biosphere*, ed. G. M. Woodwell and E. V. Pecan, 281–300. Washington, DC: U.S. Atomic Energy Commission.

Whittaker, R. H., and G. E. Likens. 1975. The biosphere and man. In *Primary Productivity of the Biosphere*, ed. H. Lieth and R. H. Whittaker, 305–328. New York: Springer-Verlag.

Wigley, T. M. L., and D. S. Schimel, eds. 2000. *The Carbon Cycle.* Cambridge: Cambridge University Press.

Wilkinson, B. H., and B. J. McElroy. 2007. The impact of humans on continental erosion and sedimentation. *Geological Society of America Bulletin* 119:140–156.

Williams, J. M., et al. 2002. Female competition and male territorial behaviour influence female chimpanzees' ranging patterns. *Animal Behaviour* 63:347–360.

Williams, M. 1990. Forests. In *The Earth as Transformed by Human Action*, ed. B. L. Turner II et al., 179–201. Cambridge: Cambridge University Press.

Williams, M. 2006. *Deforesting the Earth: From Prehistory to Global Crisis.* Chicago: University of Chicago Press.

Williamson, T. 2001. *APA Engineered Wood Handbook.* New York: McGraw-Hill.

Wilson, R. W., et al. 2009. Contribution of fish to the marine inorganic carbon cycle. *Science* 323:359–362.

Wirsenius, S. 2000. *Human Use of Land and Organic Materials.* Göteborg: Chalmers University of Technology.

Wood, S. M., and D. B. Layzell. 2003. *A Canadian Biomass Inventory: Feedstocks for a Bio-Based Economy.* Kingston, ON: Biocap Canada.

Woodbury, P. B., J. E. Smith, and L. S. Heath. 2007. Carbon sequestration in the US forest sector from 1990 to 2010. *Forest Ecology and Management* 241:14–27.

Woodman, N., and N. B. Athfield. 2009. Post-Clovis survival of American mastodon in the southern Great Lake region of North America. *Quaternary Research* 72:359–363.

Worm, B., et al. 2005. Global patterns of predator diversity in the open oceans. *Science* 309:1365–1369.

Worm, B., et al. 2006. Impacts of biodiversity loss on ocean ecosystem services. *Science* 314:787–790.

Worm, B., et al. 2009. Rebuilding global fisheries. *Science* 325:578–585.

Wrangham, R. 2009. *Catching Fire: How Cooking Made Us Human.* New York: Basic Books.

WRAP. 2009. *Household Food and Drink Waste in the UK.* http://www.e-alliance.ch/fileadmin/user_upload/docs/Household_food_and_drink_waste_in_the_UK_-_report.485524e8.8048.pdf.

Wright, D. H. 1990. Human impacts on energy flow through natural ecosystems, and implications for species endangerment. *Ambio* 19:189–194.

Wu, H. X. 1996. Wining and dining at public expense in post-Mao China from the perspective of sayings. *East Asia Forum* (Fall): 1–37.

Wu, H., et al. 2009. New coupled model used inversely for reconstructing past terrestrial carbon storage from pollen data: Validation of model using modern data. *Global Change Biology* 15:82–96.

Yang, Y., et al. 2010. Large-scale pattern of biomass partitioning across China's grasslands. *Global Ecology and Biogeography* 19:268–277.

Yesner, D. R. 2004. Prehistoric maritime adaptations of the subarctic and subantarctic zone: The Aleutian/Fuegian connection reconsidered. *Arctic Anthropology* 41:76–97.

Yevich, R., and J. A. Logan. 2003. An assessment of biofuel use and burning of agricultural waste in the developing world. *Global Biogeochemical Cycles* 17:1095. doi:10.1029/2002GB001952.

Yin, S. 2001. *People and Forests: Yunnan Swidden Agriculture in Human-ecological Perspective.* Kunming: Yunnan Education Publishing House.

Youngsteadt, E. 2012. How a fungus boosts a beetle's invasion. *American Scientist* 100:24–25.

Yule, J. V. 2009. North American late Pleistocene megafaunal extinctions: Overkill, climate change, or both? *Evolutionary Anthropology* 18:159–160.

Zalasiewicz, J., M. Williams, and P. Crutzen. 2010. The new world of the Anthropocene. *Environmental Science & Technology* 44:2228–2231.

Zeder, M. 2008. Domestication and early agriculture in the Mediterranean Basin: Origins, diffusion, and impact. *Proceedings of the National Academy of Sciences of the United States of America* 105:11597–11604.

Zeller, D., and D. Pauly. 2005. Good news, bad news: Global fisheries discards are declining, but so are total catches. *Fish and Fisheries* 6:156–159.

Zeller, D., and D. Pauly eds. 2007. *Reconstruction of Marine Fisheries Catches for Key Countries and Regions (1950–2005).* Vancouver, BC: Fisheries Centre Research Report 15(2).

Zeller, D., et al. 2009. Trends in global marine fisheries: A critical view. In *Fisheries, Trade and Development*, ed. P. Wrammer, H. Ackefors, and M. Cullberg, 55–77. Stockholm: Royal Swedish Academy of Agriculture and Forestry.

Zhang, Y., and S. Wang. 2010. Differences in development among children and adolescents in eastern and western China. *Annals of Human Biology* 37:658–667.

Zhang, Y., et al. 2009. Global pattern of NPP to GPP ratio derived from MODIS data: Effects of ecosystem type, geographical location and climate. *Global Ecology and Biogeography* 18:280–290.

Zhao, M., et al. 2005. Improvements of the MODIS terrestrial gross and net primary production global data set. *Remote Sensing of Environment* 95:164–176.

Zhao, M., and S. W. Running. 2010. Drought-induced reduction in global terrestrial net primary production from 2000 through 2009. *Science* 329:940–943.

Zimmerman, P. R., et al. 1982. Termites: A potentially large source of atmospheric methane, carbon dioxide, and molecular hydrogen. *Science* 218:563–565.

Zimov, S. A., et al. 1995. Steppe-tundra transition: A herbivore-driven biome shift at the end of the Pleistocene. *American Naturalist* 146:765–794.

Ziska, L. H., C. F. Morris, and E. W. Goins. 2004. Quantitative and qualitative evaluation of selected wheat varieties released since 1903 to increasing atmospheric carbon dioxide: Can yield sensitivity to carbon dioxide be a factor in wheat performance? *Global Change Biology* 10:1810–1819.

Zohary, D., and M. Hopf. 2000. *Domestication of Plants in the Old World: The Origin and Spread of Cultivated Plants in West Asia, Europe, and the Nile Valley*. Oxford: Oxford University Press.

Zon, R., and W. Sparhawk. 1923. *Forest Resources of the World*. New York: McGraw-Hill.

Subject Index

Species Index